GENETICS OF DYSLIPIDEMIA

BASIC SCIENCE FOR THE CARDIOLOGIST

1. B. Swynghedauw (ed.): *Molecular Cardiology for the Cardiologist*. Second Edition. 1998   ISBN: 0-7923-8323-0

2. B. Levy, A. Tedgui (eds.): *Biology of the Arterial Wall*. 1999
                                                        ISBN 0-7923-8458-X

3. M.R. Sanders, J.B. Kostis (eds): *Molecular Cardiology in Clinical Practice*. 1999.   ISBN 0-7923-8602-7

4. B. Ostadal, F. Kolar (eds.): *Cardiac Ischemia: From Injury to Protection*. 1999
                                                        ISBN 0-7923-8642-6

5. H. Schunkert, G.A.J. Riegger (eds.): *Apoptosis in Cardiac Biology*. 1999
                                                        ISBN 0-7923-8648-5

6. A. Malliani, (ed.): *Principles of Cardiovascular Neural Regulation in Health and Disease*. 2000   ISBN 0-7923-7775-3

7. P. Benlian : G*enetics of Dyslipidemia*. 2001

                                                        ISBN 0-7923-7362-6

8. D. Young : *Role of Potassium in Preventive Cardiovascular Medicine*. 2001
                                                        ISBN 0-7923-7376-6

KLUWER ACADEMIC PUBLISHERS - DORDRECHT/BOSTON/LONDON

GENETICS OF DYSLIPIDEMIA

by

Pascale Benlian
St. Antoine Hospital, France

KLUWER ACADEMIC PUBLISHERS
Boston / Dordrecht / London

Distributors for North, Central and South America:
Kluwer Academic Publishers
101 Philip Drive
Assinippi Park
Norwell, Massachusetts 02061 USA
Telephone (781) 871-6600
Fax (781) 681-9045
E-Mail <kluwer@wkap.com>

Distributors for all other countries:
Kluwer Academic Publishers Group
Distribution Centre
Post Office Box 322
3300 AH Dordrecht, THE NETHERLANDS
Telephone 31 78 6392 392
Fax 31 78 6546 474
E-Mail <services@wkap.nl>

 Electronic Services <http://www.wkap.nl>

Library of Congress Cataloging-in-Publication Data

Benlian, Pascale.
  Genetics of dyslipidemia / by Pascale Benlian.
    p. ; cm. – (Basic science for the cardiologist ; 7)
  Includes bibliographical references and index.
  ISBN 0-7923-7362-6 (alk. Paper)
    1. Lipids—Metabolism—Disorders. I. Title II. Series.
  [DNLM: 1. Hyperlipidemia—genetics. 2. Lipoproteins—metabolism. WD 200.5.H8
B468g2001]
  RC632.L54 B46 2001
  616.3'997042—dc21

                                                        2001037593

Printed on acid-free paper.   Printed in the United States of America

GENETICS OF DYSLIPIDEMIA

PREFACE

Cardiovascular diseases represent a heavy morbidity and mortality burden worldwide. They generally result form the late complications of a silent and chronic arterial disease: atherosclerosis. The patient often experiences the disease by a sudden event as brutal as unexpected (myocardial infarction, stroke etc.). In other cases, the level of individual awareness on cardiovascular risk factors brings one to adopt a somewhat constraining preventive lifestyle towards a disease, which remains unfelt. For the expert or the practitioner, it is thus a question of adapted diagnostic and therapeutic responses to prevent a disease as much common as it is complex. Dyslipidemia or lipid disorders are major cardiovascular risk factors and are themselves multifactorial diseases, resulting from interactions between genetic and environmental factors. If environmental factors may be uneasy to handle in humans, the study of genetic factors has recently made a decisive turn as the consequence of dramatic technological advances in the study and manipulation of DNA, as an experimental object. The cloning of the first genes of lipoprotein metabolism led to explore their variations and their contribution to the pathogenesis of human dyslipidemia. More than fifty genes have been identified not only in sequence but also most importantly in their physiological or pathological actions. The richness and the diversity of observations, which rose from these studies, have greatly refined our knowledge on the control of lipid metabolism in living organisms. They gave place to a true upheaval of former phenotypic classifications (based on plasma lipid measurements), distinguishing new clinical entities and defining new therapeutic strategies, targeted on authentic molecular causes or genetic risk factors.

At the turn of the millennium, with the recent release of the human genome sequence, an analysis in retrospect of the past progress, of novel areas opened by the study of the genetic component of these complex and common disorders might be welcome. This work aims at exploring the genetic basis of dyslipidemia by illustrative examples taken from the extraordinary advances of the past two decades. The objectives were to guide the non-familiar or non-expert reader through this extremely rich domain of science and medicine. The abundance of the related literature could not hold in this volume, due to a rapid and permanent expansion that largely exceeds the limits of cardiovascular disease. Therefore, we hope that the expert reader will forgive several omissions or schematic descriptions, and that one might find some information useful for his education, his scientific research or his medical practice.

I-   THE METABOLISM OF LIPOPROTEINS

Since the discovery of cholesterol at the end of the 18th century, the multifactorial and heterogeneous nature of dyslipidemia has been a recurrent source of debate about their pathogenesis, implying alternately causal genetic and/or environmental factors. Indeed dyslipoproteinemia, commonly named dyslipidemia or lipid disorders, are disorders of the metabolism of lipoproteins circulating in human plasma. Their clinical manifestations may vary from highly specific and rare symptoms like cutaneous xanthoma, to the various forms of expression of atherosclerosis, whose cardiovascular complications are most prevalent in a growing number of populations worldwide.

This diversity of expression results from the complexity of lipoprotein metabolism and of its regulations. This open metabolic system ensures the transport of lipids (which are hydrophobic by essence) in plasma (an aqueous milieu) along pathways adapted to respond to various physiological and environmental conditions. Lipids being essential components of cellular life, pathways set up for their transport in extracellular spaces have gradually diversified along with evolution of living species, saving the energetic cost of intracellular lipid biosynthesis, facilitating and modulating their cellular accessibility. Lipoproteins are complex edifices, with the unique capacity of loading or unloading lipids exchanged with cells under the control of their apolipoprotein content. In addition, apolipoproteins are structural components of lipoprotein particles, most of them being exchangeable between various lipoproteins. Lipoproteins are thus in permanent remodeling, representing a source of difficulties in medical practice to connect a measurable quantitative or qualitative anomaly of lipid transport, with a precise pathogenic mechanism.

The diversity of mechanisms, which control plasma levels of lipoproteins thus, determines the phenotypic heterogeneity of dyslipoproteinemia, but also their genetic heterogeneity, as much by the potential number of genes (candidate genes) as by the ways they are involved (permanent interactions with the environment). This chapter will underline the most revealing elements of lipoprotein metabolism with a relevance to the pathogenesis of dyslipoproteinemia.

1.1- HISTORICAL LANDMARKS

Dyslipidemia are diseases of the 20th century, their denomination being of common practice only since the end of the "sixties" [1]. The beginning of their history could be dated back to the 18th century with the first suspicions of cholesterol existence.

1.1.1-   Cholesterol: A "*Janus-faced*" Molecule

Since its discovery cholesterol has been the subject of intense controversies sometimes given a beneficial role, sometimes given harmful role in human health. At the end of the 18th century, chemical analyses became of more common use for the study of the living matter. In 1733, VALISNIERI in Padova, had observed that gallstones could dissolve in a mixture "*of spirit of wine and spirits of turpentine*". POULLETIER DE LA SALLE, a doctor in Paris, interested by these observations, came to study the composition of "*stones from the gallbladder, or bile stones*". FOURCROY, one of his disciples, described the circumstances of this discovery which he dates about 1758 "*... Returned to his place he hastens to put some of gallstone powders into alcohol; he helps its action by the soft heat of a sand bath, and confirms the assertion of VALISNIERI. Then letting the dissolution to cool down, he sees a very great quantity of small white blades, crystalline, bright, that he collects carefully. He collects as much as he can of this matter, by testing all the human gallstones he could gather, and finds that all, indistinctly, provide this crystalline substance... After several years, having collected some large amount of this matter, he tries to unravel its nature; he notes its volatility, extreme lightness, the reduction to white vapor by ignited coals, but it was not possible for him to determine the exact composition of this novel matter.*" [2,3].

In 1786, after FOURCROY had become a member of the French Society of Medicine, he was given a strange occasion to discover the presence of this substance in the composition of human fat: " *the cemetery of the Innocents (infant's cemetery) was located at the center of Paris, a most attended district surrounded by high houses where the air was allowed to circulate only with difficulty, and of which the ground was overloaded, crammed with corpses, which one had never ceased depositing there for three centuries; this enclosure, closed since a few years, was given a disastrous reputation for its influence on life and health. The goal was to convert the place into a ventilated and useful place: a market. It was thus necessary to stir up the ground, to cleanse it, and strengthen it before paving it with stones. The administration invited masters of the art to cure for their help, to prevent any accident that such an excavation would threaten to produce. The Society of Medicine was charged to take care of this great operation, and to guarantee by all precautions and all possible means of disinfection, all harms foreseeable by the Society."* [4]. The *"waxy substances"* produced after putrid degradation presented similarities of solubility in alcohol with substances from the gallstones. However, there were differences in their melting point, reasons for which FOURCROY proposed to name them *"adipocires"*. Thus cholesterol was marked from the beginnings of its history by a morbid connotation, having at best a status of waste.

CHEVREUL a chemist applying the new analytical methods developed by the French chemist LAVOISIER, identified this substance in 1814 by its unsaponificable nature, which distinguishes it from other bile components and animal "fat" [2,5]. He quoted *"to designate chemical substances which had been confused so far, alike several others which I was first to reveal the existence, I found useful for myself so far to use of periphrases, while waiting for the real nature of these substances to be better determined; Now, these observations have multiplied enough so that I may substitute special names for these periphrases, thereby allowing faster speed to the speech, together with having a better feeling of their relationships the ones with the others; I shall name cholesterin after "χολη " bile and "στερεοξ " solid, the crystallized substance of human gallstones... ".* He will determine later that it's composition was made out of carbon, oxygen, and hydrogen, and its physicochemical constants (nonacid character, melting point etc.). The precision of his descriptions offered to his contemporaries and to successors the possibility to distinguish this molecule among all, opening on multiple pathways of research. In 1859, BERTHELOT another chemist, will show that cholesterin is an alcohol, and that three fatty acids join a molecule of glycerol to constitute what he will name *"triglycerides"*. WINDAUS will demonstrate that there are two chemical forms of cholesterol in plasma: one esterified, the other unesterified [2]. The term "cholesterol" will be commonly used in the middle of the twentieth century after the description of the tetracyclic backbone common to all sterols [6].

The nineteenth century was marked by a multiplicity of observations: chemical, pathological, histological and clinical, which reported the presence of cholesterin in various human or animal tissues, normal or pathological (blood, brain, liver, atheroma, tumors) and in various physiological or pathological circumstances (pregnancy, jaundice, diabetes, renal insufficiency, infections...) [2]. It is only at the turning point of 19th and 20th centuries that simple and reproducible quantitative measures of cholesterol will be used in clinical practice. Cholesterol conquered its aristocratic rank, it became a universal component of the living matter, by which it is synthesized; it took part in the function of certain vital glands like the adrenals and was highly abundant in the brain. In addition, it appeared to have antihemolytic and antitoxic properties. The clinical concerns of the time were mainly turned towards infectious diseases. On the basis of clinical observations that infected patients were hypocholesterolemic and that hypercholesterolemia marked convalescence, liquors containing cholesterol were proposed as tonics beneficial for health (*figure 1-1*). A notion after which a long tradition of cod-liver oil oral supplementation was proposed along the first half of the 20th century. Cholesterol had become a benefit for health.

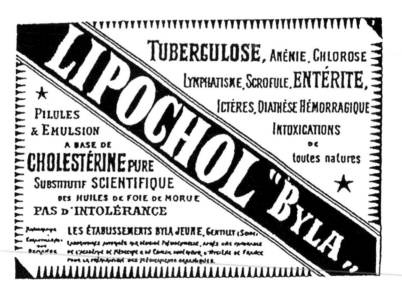

Figure 1-1. **Cholesterol Medicine**. Advertisement published in 1910 in "Le Nord Medical" [2]. "Pills and emulsion based on pure cholesterin. A scientific substitute for cod-liver oil. Well tolerated. May be used in tuberculosis, anemia, ..., scrofula, enteritis, jaundice, hemorrhage, intoxication of any nature".

The rehabilitation was only of short duration. In 1906, IGNATOWSKI in Russia, by studying the physiological and pathological effects of food, noted that the wealthiest subjects who consumed a diet rich in animal "proteins" (milk, eggs, meat) had aortic lesions of arteriosclerosis more advanced than those who did not [7]. He reproduced these effects in rabbits by allotting them to a diet rich in "animal" proteins. However, ANITSCHKOW showed in 1913 from experiments of dietary-induced atherosclerosis in rabbits, that this pathogenic effect was due to the only enrichment of cholesterol in their food [8]. Cholesterol consequently recovered a harmful role for health, which it did not depart from until our days. Thus, as emphasized by Michael BROWN and Joseph GOLDSTEIN in their lecture in distinction for the Nobel Prize of Physiology and Medicine [9], cholesterol is the small molecule, the most highly decorated (fourteen Nobel Prizes including theirs) but it is a "*Janus-faced*" molecule one may allot a harmful role or a beneficial role depending on where it resides. Thus the current controversies on "good" and "bad" cholesterol are the continuing story of the ambivalent role, which has marked its history, often reflecting the limits of the knowledge of the time.

1.1.2-   The Pathological Descriptions of the 19th Century

Towards 1740 CRELL, a German doctor distinguished from ordinary bone the calcified concretions in the arterial wall, which were denominated "*osseous plaques*" [7]. One century later in 1833 LOBSTEIN, a doctor in Strasbourg, gave the name "arteriosclerosis" to this particular hardening of the arterial wall. In 1843 VOGEL, a German pathologist described the presence of cholesterol in these lesions. In 1860 MERCHANT in Leipzig, gave the name of "atherosclerosis" to the presence in the arteries "*... of a yellowish matter, comparable with pea soup interposed between the internal tunic and the medium tunic, or infiltrated between strips of the latter...*" [2]. VIRCHOW, proposed in 1856, the first theory based on the "response to injury" hypothesis for atherosclerosis [10]. He described it as an inflammatory process leading to progressive degeneration of conjunctive tissues in the internal tunics of the artery, which favors "depositions" of substances circulating in the lumen, which in turn worsen the "lipidic degeneration" of the arterial wall. In 1912, HERRICK described the syndrome of myocardial infarction and established the relationship with atherosclerosis and arterial thrombosis [11]. At the same time studies reported the association of atherosclerosis with high blood cholesterol levels in various diseases.

It is also during the second half of the 19th century that xanthoma were reported. RAYER in 1835, a dermatologist made a very fine description of xanthelasma "*One observes sometimes, on the eyelids and in their vicinity, yellowish plaques similar to the color of chamois leather, slovenly projecting, soft, without heat nor redness, sometimes laid out in a rather symmetrical way* " [12]. Then descriptions grew rich in particular after the work of English doctors, ADDISON (1863), HUTCHINSON (1871), KAPOSI (1872) who recognized several clinical forms, then FAGGE (1873) and CHURCH (1874) who found their hereditary characteristic [13]. These lesions were described with a high degree of accuracy. For example xanthoma of the palmar folds were described by BRUCHET (1886) as follows: "*... On top of the folds which movements determine on the palmar face of the hand and fingers, the mode of grouping is not any more the same;*

the yellow lines, as if they had been traced with a brush, made of small humps or spots, laid out end to end, underline these various folds". However a great confusion reigns then on the pathogenesis of these lesions sometimes reversible sometimes permanent, observed at early ages or at oldest, under the most various conditions: chronic jaundice preceding death shortly, diabetes, and perfect health! BAZIN (1869) then CHAMBARD (1878) described the presence of cholesterol in these lesions, but the link with hypercholesterolemia was established in 1889 by CHAUFFARD who concluded that *"xanthelasma is with hypercholesterinemia what tophus is with the excess of uric acid in the serum of patients with gout"* [2,14]. TANNHAUSER, will establish in 1938, a classification of these lesions and their association with hypercholesterolemia [15]. MÜLLER gave evidence at the same time that these lesions could be associated with angina pectoris and sudden death from myocardial infarction [16].

1.1.3- From Lipido-Proteinic "Cenapses" to Apolipoproteins

The first authors had been puzzled by the particular nature of plasma cholesterol. In 1833, BOUDET was intrigued by small differences between the pure cholesterin of CHEVREUL and the *"problematic cholesterin"* of blood [17]. During the 19th century chemists sought of ways to develop the measurement of blood cholesterol. NERKING concluded in 1901 on the need for proteolysis before the organic extraction of cholesterol from blood because of its strong association with "albuminoid" substances [1,2].

In 1928, MACHEBOEUF of the Pasteur Institute had isolated by precipitation with ammonium sulfate of horse serum, a protein fraction particularly rich in lipids and yet water-soluble. He deduced that they were *"lipido-proteinic complexes"* which he called *"lipido-proteinic cenapses"* [18]. In 1941, BLIX recognized two classes by electrophoretic separation: the α–lipoproteins and the β–lipoproteins migrating respectively with α–1 and β–globulins [1,6]. This technique was applied in particular by NIKKILÄ to clinical studies. However, data from electrophoresis did not match with those of ultracentrifugation, which detected an unstable entity, the protein "X". GOFMAN and his group developed in 1949, a new technique of ultracentrifugation, which unraveled the heterogeneity of plasma lipoproteins, and provided a new way of isolating them. They proposed a new classification based on their density of flotation within specific density ranges: "High Density Lipoproteins" (HDL), "Low Density Lipoproteins" (LDL), "Very Low Density Lipoproteins" (VLDL). In addition, applied to clinical observations, data from preparative ultracentrifugation agreed with those of electrophoresis [19].

However these techniques were tedious and costly, making them hardly accessible to doctors and patients. In the early "60s" only measurements of plasma cholesterol and total lipids were routinely performed. Triglycerides became routinely measured in plasma at the end of the decade. The needs for a mass-diagnosis were more pressing since the results of investigations of large human cohorts like that of Framingham, brought first sound epidemiological evidence of causal relationships between high plasma cholesterol and cardiovascular disease. FREDRICKSON, LEVY and LEE proposed in 1965, a "classification of hyperlipoproteinemia" which integrated clinical data, with data from ultracentrifugation, electrophoresis and standardized quantitative analyses of blood lipids [20]. The five types of hyperlipidemia (from type I to type V), (see *figure 1-10*) offered the first classification, which established a phenotypic link with still arid and dispersed experimental data. A little later DE GENNES proposed a simplified clinical classification of the hyperlipidemia, comprising only three classes [21], based on serum decantation and plasma cholesterol and triglyceride measurements (see *table 1-2*). These classifications gave useful diagnostic tools to orient therapeutic decisions in clinical practice. Meanwhile, the fractionation of lipoproteins had revealed new entities: apolipoproteins that ALAUPOVIC proposed to classify in "A", if they were purified from α–lipoproteins (HDL), "B", if they were purified from β–lipoproteins (LDL) and "C" if they were purified from VLDL [1]. In 1975 UTERMANN purified a novel apolipoprotein, found in great excess in VLDL of subjects with type III hyperlipidemia, apo E [22].

1.1.4-   The Concept of Molecular Disease

This idea had emerged in England in 1901. GARROD had the intuition that the cause of several metabolic diseases causing the urinary excretion of abnormal products (alkaptonuria, cystinuria...), came from an enzymatic block with recessive inheritance [23]. The term "*gene*" was proposed initially by JOHANNSEN in 1911 to indicate the hereditary determinism of a trait. BEADLE and TATUM in 1941 defined this idea more precisely through the "*one gene-one enzyme*" concept [24]. This was substantiated by their observations on mutants of *Neurospora Crassa*, which became auxotroph for a given metabolite after they had lost a single enzymatic function. PAULING in 1949 brought evidence of this concept in humans by the identification of an abnormal pattern of electrophoretic migration of haemoglobin, in subjects suffering from sickle cell anemia. These results showed that the structural and functional anomaly of a single protein could be the cause of an inherited metabolic disease, through the deficit of a single gene [25]. This fundamental observation initiated a new era of intensive research on the molecular basis of "*inborn errors of metabolism*" still in progress.

The English school had noted as early as in the mid 19th century the inherited nature of xanthoma. OSLER, in 1897 had noted the familial aggregation of the arteriosclerosis. MÜLLER in 1938 approaches the concept of familial hypercholesterolemia while bringing together some disconcerting facts in Swedish subjects with xanthoma [16]. They all had died suddenly and prematurely. On pathological examination they presented lesions of myocardial infarction and had very advanced lesions of atherosclerosis. They often had hypercholesterolemia and had suffered from angina pectoris, both traits dominantly inherited in their families. KHACHADURIAN in 1964 showed in 10 Lebanese families that familial hypercholesterolemia was an autosomal dominant disorder (see *figure 2-1*). In this disease, a dose-effect of a single gene causes a severe and generalized form of the disease in homozygotes and a more attenuated form of the disease in heterozygotes [26]. For the first time the concept of an inborn error of metabolism was applied to one dyslipoproteinemia, which became a single clinical entity. Michael BROWN and Joseph GOLDSTEIN in early "70s" using newly developed cell-culture technology and analytical biochemistry, brought definite evidence that this disease resulted from functional defects of a novel membrane protein, which could take-up LDL out from the cell surface down into the cytosol: the LDL receptor.

Meanwhile, controversy was fed by results of experimentally induced atherosclerosis through dietary induced hypercholesterolemia, which thus became an "acquired" disease. ANITSCHKOW in 1913 had proven that a dietary enrichment in cholesterol could induce atherosclerosis in rabbit, however the same experiments failed in rat [8]. He concluded that the type of animal used was a necessary condition to induce atherosclerosis. After Second World War, large population samples taking the Framingham cohort as a paradigm, were conceived to study and tackle the first cause of mortality in the United States: cardiovascular disease [1]. Epidemiological results became robust. KEYS reported in 1957 a relationship between dietary enrichment in saturated fat and carbohydrates, and the frequency or severity of coronary arterial disease [27]. He also noted a progressive increase in the prevalence of the disease among Japanese emigrants to Hawaii, then to California along with them adopting a more "western" lifestyle and dietary habits of the host country [28]. Moreover the "susceptibility" to lifestyle and diet could vary from one individual to another. Therefore, "genetics" and "environment" were two culprits under one undeniable charge for only one offence: atherosclerosis. The consensus was obtained: atherosclerosis was recognized as a "multifactorial" disease.

Contradictions accumulated, for example type III hyperlipidemia a highly atherogenic dyslipidemia, with family inheritance could be corrected by an appropriate diet. In 1975, GOLDSTEIN & al. had undertaken in Seattle, the analysis of plasma triglycerides and cholesterol among 500 survivors of myocardial infarction [29]. In more than a third of these patients hyperlipidemia exhibited familial aggregation and in most cases it was monogenic. A new clinical entity was observed across the phenotypes established by FREDRICKSON: familial combined hyperlipidemia [30]. In these families, hyperlipidemia of variable type (IIa, IIb, IV) was dominantly inherited. With peptide sequencing, enzymatic activities analyses and the development of monoclonal antibodies, identification of the first mutants of apolipoproteins and their receptors, split the previous classifications in "types" and "phenotypes" of hyperlipidemia. New forms of familial dyslipoproteinemia were individualized as many novel molecular diseases.

In the late years "70" simplified techniques of molecular biology became applicable to the study of the human genome. They widened the breach, which opened the way to the cloning of novel genes of lipoprotein metabolism. Genetics of dyslipidemia could enter the field of clinical practice. One of the "suspects" being cornered it became possible to specify the role of its "accomplice".

1.2- LIPIDS: ESSENTIAL COMPONENTS FOR LIVING CELLS

Lipids are elementary components of the living cell. Characterized by their insolubility in water, they spontaneously delimit a closed space thereby insulating chemically and thermally the inner compartment, from the surrounding aqueous medium. This actually defines the concept of a living cell. This function would have contributed to the evolution of the first living species on earth, by confining a closed space in which primitive molecules could diversify faster their structural and catalytic properties by acceleration of their chances to interact with each other [9]. Two great classes of lipids are present in living cells: fatty acids derivatives, in particular including phospholipids and triglycerides, and isoprenoids, among which cholesterol is one of the most outstanding.

1.2.1-   Fatty Acids Derivatives

Free fatty acids are present in small quantities within cells. They are sources of elementary molecules like acetyl-Coenzyme A (acetyl-CoA), at the crossroads of multiple metabolic pathways, or may be used as modifiers to increase the hydrophobicity of other cellular compounds (proteins, carbohydrates or even cholesterol to form cholesteryl esters). They may be themselves modified into important signaling molecules like arachidonic acid, prostaglandins or other ecosanoids. However, they represent an important cellular source of energy through their oxidative degradation. They circulate in plasma, bound to albumin, or may be taken up by cells. Most of cellular fatty acids are generally compacted by esterification of glycerol to form phospholipids and triglycerides. Phospholipids are amphipathic, and under suitable conditions they may form a lipid bilayer on the water surface. They actually constitute about 50% of cellular membrane mass. Short or unsaturated hydrocarbon chains will give more fluidity to the membrane, and will keep it in liquid phase at low temperatures. However, less fluidity makes the membrane more water-repellent, limiting losses of hydrophilic molecules from intracellular compartments. Thus prokaryotes, plants, certain fungi or insects resist climatic changes by modifying the fatty acid composition of their extracellular membranes. Variations of membrane composition and fluidity build up a selective medium of integration for proteins. It modulates their activities and those of molecules associated with the inner or outer face of the plasma membrane. In the case of lipoproteins, the phospholipid composition modulates for example the activity of LCAT (Lecithin-Cholesterol Acyl Transferase), or of SR-BI (scavenger receptor B-1). Membrane phospholipids are also characterized by the nature of their polar groups: phosphatidylcholine and sphingomyelin are localized at the outer face of plasma membrane whereas negatively charged phosphatidylserine and phosphatidylethanolamine (Lecithin), are localized at the inner face creating a transmembrane gradient of electric charges. The cell controls the lipid and protein content of membranes or of lipoproteins, mainly in the endoplasmic reticulum where they are assembled. Certain phospholipids like the phospho-inosities, are less abundant (10% of membrane phospholipids), but essential in cell signalling.

Triglycerides or triacylglycerols are essential energy stores for cells. They form intracytoplasmic droplets, easily hydrolysable to release fatty acids. This property is used by cells specialized in the storage of fat, like adipocytes, "eggs", "ovocytes" or "seeds", to constitute a nutritive stock, essential at early stages of embryonic development. In addition to their role as structural components of membranous lipids, fatty acids are a major fuel for muscular activity and thermogenesis. Their hydrolysis provides two fold the energy of an equal mass of glucose. This "combustion" is as essential to the fly of insects as it is to myocardium contraction, or the maintenance of a constant body temperature.

1.2.2-   Isoprenoid Derivatives

Isoprenoids are unsaturated hydrocarbon molecules of the mevalonate pathway whose main end product is cholesterol [32]. The cascade of reactions along this pathway produces fundamental components of the cellular machinery (*figure 1-2*). This cascade begins with the condensation of three molecules of acetyl-CoA by the enzyme HMG-CoA (Hydroxy Methyl Glutaryl Coenzyme A) synthase. It is followed by the cleavage of the chemical bond with coenzyme A by HMG-CoA reductase producing the key product: mevalonate. Mevalonate is then modified to give isopentenyl

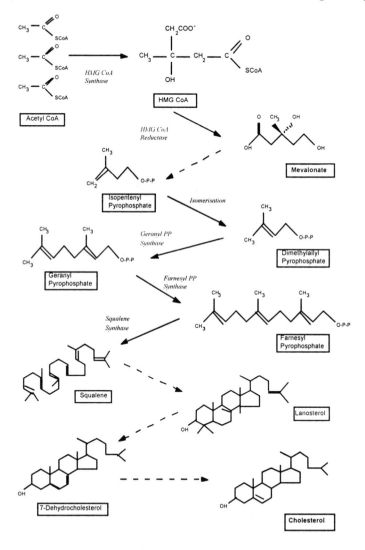

Figure 1-2. **The mevalonate pathway.** Three molecules of Acetyl CoA are condensed by HMGCoA synthase to give HMGCoA, which in turn will be reduced into mevalonate, by HMGCoA reductase. Follows a cascade of chemical reactions building up the carbohydrate backbone common to sterols, and finally cholesterol. Each reaction step produces intermediate compounds essential to living cells.

pyrophosphate (isopentenyl-PP). Maturation of several tRNA is achieved through the modification of adenine by isopentenyl-PP, thereby increasing anticodon specificity. Isopentenyl-PP combines with its dimethylallyl-PP isomer to give geranyl-PP, which combined with a new isopentenyl-PP gives farnesyl-PP. This compound and its derivative, geranylgeranyl-PP can be bound by a thioether bond to proteins and modify their structural and functional properties.

This function known as "prenylation" enriches the spectrum of post-transcriptional modifications of proteins [33,34]. A classical example is how farnesylation may confer transforming properties to the oncogene p21-Ras. The attachment of the hydrocarbon chain results as expected in a closer association of the protein to the plasma membrane. This was demonstrated in the case of lamins, proteins fixed at the inner face of the nuclear membrane. However farnesylation does not seem only to increase affinity for the lipid bilayer of membranes. In the case of p21 Ras, this modification is essential to switch the protein to an active form. This is the case for several pheromones in funghi, which become specific for their membrane receptor, after prenylation. Prenylation would change the conformation of proteins to facilitate interactions of their substrate (often GTP), with activating cofactors or other proteins in the vicinity of membranes. Thus many G proteins (including p21 Ras) are farnesylated or geranylgeranylated on their γ sub-unit which increases the affinity of the α sub-unit for GTP. GTP-binding proteins, which are associated with membranes of organelles, known as "Rab" proteins are geranylgeranylated. This modification ensured by a geranylgeranyl transferase, would control the fusion and the one-way assembly of vesicles and organelles. Defects of this enzyme expressed abundantly in the nervous system result in choroideremia, a recessive retinal degeneration linked to chromosome X [35]. Here, as sometimes observed in molecular pathology, a protein with a universal and fundamental role in cellular life proves to display mutations responsible for a monogenic disease of a single or a few organs in humans.

Isoprenoids are involved in the synthesis of factors of electron or ion transfer. The porphyrin core of heme-A is built starting from farnesyl-PP. There are two prenyl-transferases modifying farnesyl-PP. The trans-prenyl transferase provides "the all-trans" geranylgeranyl-PP for geranylgeranylation of proteins. This compound also constitutes the hydrocarbon chain of ubiquinone, an electron transporter at the inner face of the mitochondrial membrane. A transient or partial limitation in ubiquinone formation by HMGCoA reductase inhibitors, could contribute to muscle or liver intolerance of these compounds in humans. The cis-trans prenyl-transferase provides "the cis-trans" geranylgeranyl-PP, a basic unit constituting the dolichol-phosphate cofactor for protein glycosylation and the polymerization of several polysaccharides. Squalene synthase condenses two molecules of farnesyl-PP by their phosphorylated end to give squalene, the elementary backbone of all steroids. This function is archaic in eucaryotes, since funghal genes for squalene synthases (*Saccharomyces cerevisiæ and Saccharomyces pombe*) have strong homologies with the human gene and related enzymes in plants: phytoenes diphosphate synthases [36].

Cholesterol is the end product of this metabolic pathway. It is associated in a ratio near 1:1 with phospholipids of the cellular membrane in eucaryotes [31]. Mutant cells, auxotroph for cholesterol have greater membrane fragility and degenerate quickly in cell culture in the absence of cholesterol. Cholesterol brings polar groups of phospholipids closer to each other, and prevents the crystallization of acyl groups by interposition of its non-polar groups. This results in a reduction of membrane permeability to small water-soluble molecules, a better stability of the membrane structure, which at the same time becomes more flexible. Indeed, contrary to phospholipids, cholesterol can flip easily within the lipid bilayer, its polar extremity passing from the outer face to the inner face of the membrane, increasing membrane capacities of inflections and curving. In addition, cholesterol participates to the formation of caveolae at the cell surface, which are microdomains involved in membrane exchanges between intracellular and extracellular compartments (see page 26). In fact, eucaryotes are organized in multicellular systems of variable shape, and are equipped with organelles made of invaginated intracellular membranes that are able to exchange substances with the external medium by endocytosis and exocytosis. In addition to its structural role, cholesterol is the precursor of steroid hormones, vitamin D and bile acids. Moreover, cholesterol may be a crucial signaling molecule during embryonic development. Teratogens, including verartrum alcaloids from the plant jervine, induce cyclopia and holoprosencephaly when ewe would graze in fields where jervine grows. The morphogen *Sonic HedgeHog* (SHH) mediates covalent bonding with cholesterol by autocatalysis. In the presence of teratogens, *Patched*, a membrane protein cannot interact with the newly formed molecule by its cholesterol-sensing domain [37]. This prevents SHH from adopting an appropriate cellular topology, resulting in facial and brain malformations [38]. The recessive Smith-Lemli-Opitz syndrome manifesting with multiple neonatal malformations and death in infancy illustrates the vital role of cholesterol in humans [39]. The disease is caused by mutations of 7-dehydrocholesterol reductase, or delta-7 sterol reductase which catalyses the last step of cholesterol biosynthesis. Cholesterol is thus a molecule directly essential to the structure of cellular membranes, and directly or indirectly by its derivatives essential to cellular development, differentiation and metabolism. In vitro, HMGCoA reductase inhibitors added to cell lines in culture, disturb cellular multiplication by disrupting the cell cycle and interrupting DNA synthesis. They also disturb cellular differentiation and shape maintenance. In vivo, these drugs are inoffensive because intracellular balance of cholesterol concentrations is sustained by extracellular sources of cholesterol provided by lipoproteins.

Although all living cells are fully able to synthesize lipids that are essential to their survival, there exist complementary mechanisms of collecting lipids from the external milieu. Multicellular organisms have established along with the development of circulatory systems, lipid transport pathways for their delivery to the cell cytosol, saving the energetic cost of their biosynthesis. These pathways constitute the basic architecture of lipoprotein metabolism.

1.3- LIPOPROTEINS: SOLUBLE CARRIERS OF LIPIDS IN EXTRACELLULAR SPACES

1.3.1-   Structure and Functions of Lipoproteins

Lipoproteins are complex entities composed of a hydrophobic lipid core surrounded by a monolayer of phospholipids, unesterified cholesterol and amphipathic proteins, the apolipoproteins *(figure 1-3)*. Lipoproteins, produced by specialized organs for their assembly, ensure the transport of lipids in a soluble form in the plasma, and the extracellular compartments to direct them towards specific cellular targets. They can charge or discharge quickly from lipids, by modifying their conformation, without losing their affinity for lipids. The type and number of apolipoproteins associated with the particle fulfill these functions. The molecular element common to apolipoproteins is a motif of 11 amino-acid residues, forming a helix of 3.6 residues per turn (amphipathic α–helices). Polar amino acids are gathered on one face and non-polar amino acids clustered on the other face of the helix *(figure 1-4)*, conferring to the motif its amphipathic characteristics [41].

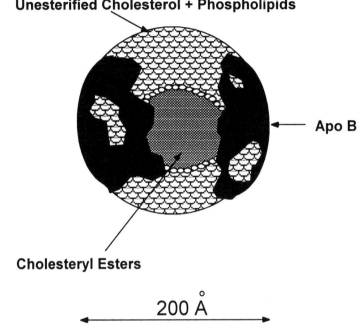

Figure 1-3. **Schematic representation of a lipoprotein: example of a Low Density Lipoprotein (LDL).** The hydrophobic core contains mainly cholesteryl esters and lipophilic compounds (e.g. vitamins); the surface is composed of a monolayer of phospholipids and unesterified cholesterol. Non-polar sides of molecules are oriented towards the inner core, whereas polar sides of molecules turn towards the surface. Apolipoprotein B, the main apolipoprotein constitutes the backbone of the particle and allows interactions with other lipoproteins, enzymes, and components of the extracellular matrix or cellular receptors.

Among millions of living species, it is not unexpected to observe a great diversity in the systems between species, and even within the same species from one strain to another. However one finds the continuity of the function of facilitated transport and accessibility of lipids to living cells, which has diversified during evolution.

-A-L-D-K-L-K-E-F-G-N-T-L-E-D-K-A-R-E-L-I-S-

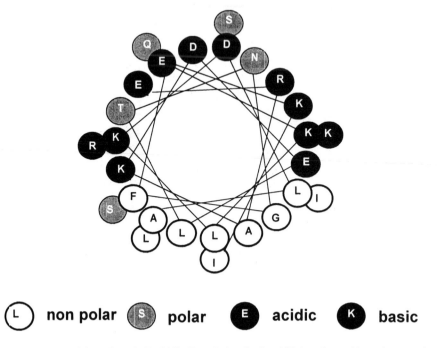

Figure 1-4. Amphipathic α-helix [41]. Top. Polar (hydrophilic) amino acids and non-polar (hydrophobic) amino acids (indicated by their letter code) are next to each other in the primary sequence. Bottom, in the α-helix secondary structure polar amino acids cluster on one face, whereas non-polar amino acids cluster on the other face of the helix.

1.3.2-   Lipoproteins Across the Evolution of Living Species

In plants, the lipid mass can reach up to a half of the total weight of a seed. Lipids are a nutritive reserve, and represent a protection against chemical threats, desiccation or freezing. Although they are deprived of an active circulatory system, plants accumulate lipids in their seeds, readily mobilized during germination. There are intracellular organelles the *"oil bodies"* consisted of a lipid core rich in triglycerides surrounded by a monolayer of phospholipids associated with an amphipathic protein, oleosin [42]. This protein, which accounts for 20% of the mass of these particles, has a molecular weight and a structure, similar to those of exchangeable apolipoproteins in animals. The cell controls the synthesis of *"oil bodies"* in the endoplasmic reticulum where they are assembled. By splitting intracytoplasmic lipid reserves into small droplets, oleosin increases the surface/volume ratio, and facilitates the accessibility of triglycerides for lipases during germination.

This function of lipid storage in the ovocyte is found in the assembly of the vitellus in animal species. Proteins of the vitellus ("yolk proteins", and vitellogenin) have highly conserved structures in all species, from nematodes to mammals. In insects, yolk proteins have significant domains of homology with lipases (see page 86), although they are deprived of their catalytic activity [43]. They bind lipids and a steroid hormone, ecdysone conjugated with fatty acids in inactive form within the vitellus. During embryonic development, fatty acids are hydrolyzed, thereby releasing the hormone in active form, driving a primordial stage of insect metamorphosis: the secretion of the larval cuticule. Thus proteins of the vitellus would take part in the chronology of embryonic development through the controlled release of the steroid hormone ecdysone.

Vitellogenins are large proteins associated with lipids that share structural and functional homologies with human apolipoprotein B [44]. They are secreted by a producing body, the intestine in the nematode, or the *"fat body"* an equivalent of liver in insects, and the liver in vertebrates, under the control of steroid hormones. They bind to a specific receptor at the surface of the ovocyte, and are internalized by a receptor similar to the VLDL receptor: the vitellogenin receptor (see page 123). In the ovocyte, these macromolecules (PM = 400 to 500 kDa, similar to that of apo B) are cleaved in smaller fragments to become vitellins, which will favor the mobilization of lipids during embryonic development.

In certain crab species, dense lipoproteins containing a protein related with vitellogenins ensure the shuttle in the hemolymph between sites of lipid absorption and tissues taking up lipids, in particular the ovocyte for the development of the vitellus [45]. **Insects** accumulate nutriments during the larval period to carry out their metamorphoses, and burn out energy in the adult stage to ensure their mobility and reproduction. Insects have an open circulatory system containing hemolymph (the equivalent of blood) together with a lipoprotein metabolism sharing many common points with that of mammal [46]. In these species, lipoproteins are named lipophorins. They have been particularly studied in the tobacco hornworm, *"Manduca sexta"* and in a migratory locust *"Locusta migratoria"*. There are two major classes of lipophorins: HDLp (High Density Lipophorins) and of LDLp (Low Density Lipophorins), of which the densities of flotation are similar with those of mammalian lipoproteins. The protein components of HDLp are the apolipophorins LpI and LpII. LpI is not exchangeable, and with LpII shares homologies with human apolipoprotein B. Lipophorins are assembled and secreted in the hemolymph by *the "fat body"* they collect absorptive lipids by the intestine and deliver them to muscular tissues. In case of intense energy demands, like those required by the movement of wings in the adult, a peptidic hormone, the adipokinetic hormone similar in sequence to glucagon, mobilizes fat and carbohydrates (*figure 1-5*). This hormone could be an activator of lipases in storage compartments, to mobilize diacyl-glycerol, the main source of carbohydrates used by these species. To accelerate lipid transport, a third apolipophorin, LpIII, associates with HDLp. This stabilizes larger and less dense particles, LDLp, which increase their lipid-loading capacities. After diacyl-glycerols are delivered to tissues, LpIII is detached from the particle and may circulate freely in the hemolymph. The gene for LpIII was cloned in these two species [47,48]. The sequence is highly conserved it presents α-amphipathic helices similar to those of human exchangeable apolipoproteins. Conversely, during the larval period a receptor homologous to human VLDL receptor with strong affinity for lipophorins, directs the movement of lipids, from the intestine where they are

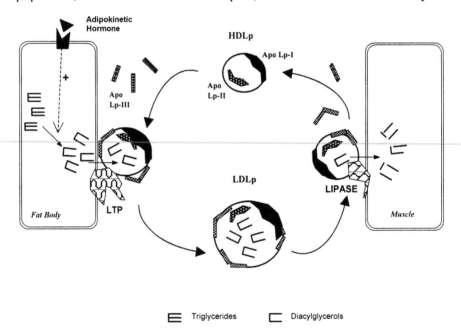

Figure 1-5. **Lipoprotein metabolism in adult insects** [46]. HDLp (High Density Lipophorins) contain non-exchangeable apolipophorins: apoLp-I on the surface, and apoLp-II on the inner side, binding lipids. In case of intense energy needs (wing muscles for flight), adipokinetic hormone is secreted and diacylglycerols stocks are mobilized from the fat body towards the lipophorin core, by a transfer protein: Lipid Transfer Protein (LTP). ApoLp-III is recruited to build up larger particles or lower density LDLp (Low density Lipophorins), from which it will detach after lipid delivery to the cell upon the action of lipases.

collected towards the "*fat body*" where they are stored [49]. Thus, from the early evolutionary stage of insects, the heterogeneity and the exchangeability of lipophorins, the specialization and the cooperativeness of apolipophorins, their interactions with cellular receptors, their modulation by lipolytic or hormonal factors, are foretelling the specific characteristics of lipoprotein metabolism in vertebrates.

Fish also have two stages of development, which condition their dietary behavior. They are poikilotherm and their circulatory system is a true closed cardiovascular system [50]. The cellular membranes and lipoproteins are rich in unsaturated fatty acids. It was shown in trout that their proportions change with the adaptation to lower temperatures. They possess major classes of apolipoproteins observed in mammals, however fish (like amphibians) would be deprived of apo E and would have only some of the apolipoproteins "C". They have lipases of the LPL family, which have the characteristic to be insensitive to salt, as opposed to mammalian lipases. Lipoprotein transport is mainly accomplished by particles of the HDL type, however it is prone to strong variations according to life cycles. During the larval period, lipoproteins of the VLDL and LDL type rich in cholesterol esters and triacylglycerols increase by 12 fold. At the beginning of the reproduction period, certain fish like the lamprey cease nourishing. They can loose 90% of their body mass during their migration like in salmon, leading in certain cases to the involution of the liver and of the intestine at the time of spawning. Fish are deprived of bile ducts, and do not have a system of excretion of cholesterol esters. Enormous accumulations of lipoproteins may be observed in adults, also considerable at the time of the development of eggs in the female, and under the dependence of estrogens. Even stranger fibrous plaques histologically identical to human lesions are observed in the arteries, these lesions are reversible after egg lying.

Birds have a lipoprotein metabolism very close to that of mammals with some specificity. For example, they store fat mainly in the liver. Moreover, the dietary lipids are forwarded directly from the intestine towards the liver via the portal vein in VLDL [45]. Birds have the main apolipoproteins but do not have apo E. In birds it is apo A-I, which has its ubiquity and its functions [51]. On fibroblasts, there is a LDL receptor of MW similar to human B/E receptor. Lipases are similar to mammalian lipases. However an apolipoprotein specific to birds, apo II is associated with particles rich in apo B and triglycerides: VLDL-II. Apo-II gene expression and VLDL-II secretion are controlled by estrogens. VLDL-II are taken up by endocytosis by an ovocyte receptor of MW 95 kDa, able to bind human LDL or β–VLDL, for the storage of fat during ovogenesis. Certain birds like the

White Carneau pigeon or the *Japanese Quail* are prone to dietary induced atherosclerosis.

In mammals, the metabolism of lipoproteins is highly diversified across species; however with an organization relatively close to that of human. It is intensively studied, because mammals represent useful experimental models of human pathology. The majority of animals would they be herbivorous, insectivorous, carnivorous, or omnivorous transport their fasting cholesterol in lipoproteins of the HDL type [45]. They have spontaneous circulating levels of cholesterol half or one third lower than those observed in humans *(figure 1-6)*. Under certain dietary or hormonal conditions, they produce

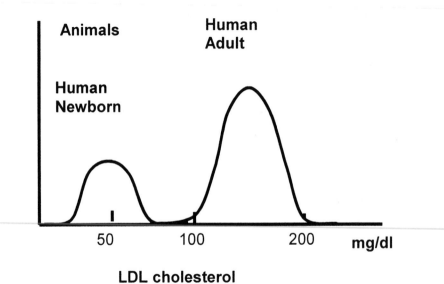

Figure 1-6. Distribution of fasting LDL cholesterol levels in serum [9].

lipoproteins of the VLDL or LDL type. It is only in primates, particularly monkeys of the Old World (cynomolgus, rhesus, and baboon) and in great monkeys (chimpanzee, gorilla), that lipoprotein metabolism is closest to that of human. In certain animals like the domestic pig, atherosclerosis may develop spontaneously [52]. Hypercholesterolemia is exceptionally observed in animals in the wild. For example, in more than 800 rhesus monkeys living in the wild, no rise in plasma cholesterol was noted [53]. Conversely hyperlipidemia and atherosclerosis can be induced in many sedentarized mammals, subjected to a diet rich in saturated fat and/or cholesterol.

In murine species, resistance to hyperlipidemia and atherosclerosis is generally observed, which regularly feeds criticisms concerning these animals as experimental models of human disease. In all species from the most sensitive like rabbit, pig, dog or primates, to the most resistant like murines, there exist strain-specific differences in susceptibility to dyslipoproteinemia and atherosclerosis. A genetic predisposition underlies this strain-specific sensitivity [54] whose caricature is represented by transgenic mice, in which extreme situations of resistance or sensitivity to atherosclerosis are produced by the modification of one or more genes of lipoprotein metabolism (see page 210). Gene polymorphisms are also described. Protein polymorphisms of apolipoprotein B are described in gorilla and pig [55,56]. In pig and Cynomolgus monkey, rare variants of apo B or of LDL receptors may be associated with severe hypercholesterolemia [57]. Authentic LDL receptor mutations were described, in Watanabe rabbit (a 12nt deletion in exon 4), in rhesus monkey (a nonsense transition G→A, at residue 284 of exon 6), and in pig [58-60]. Strains of cats or minks have been described with a single LPL gene mutation causing chylomicronemia [61,62].

The metabolism of lipoproteins represents an open metabolic system, genetically controlled to fulfill the cellular requirements of vital functions, but likely to display variable responses (physiological or pathological) according to environmental conditions. As a consequence of this adaptability there are obligate systems built up to sense interactions with the environment (gene-environment interactions), there is a variety of factors to multiply the possible set of responses (gene-genes interactions), and their polymorphism to modulate these responses from one individual to the other (evolutionary capacities).

1.4- THE METABOLISM OF LIPOPROTEINS

Absorption, synthesis, and transport of lipids towards cellular targets that store them, transform them or use them, are carried out by a complex network of inter-connected metabolic pathways. Within this network, the metabolism of lipoproteins ensures the traffic of lipids in plasma and extracellular compartments [7,63,64]. Lipoproteins exchange lipids or apolipoproteins with one another. Lipids may be taken up by tissues so that the various types of circulating lipoproteins permanently appear or disappear as the result of these exchanges. In addition, there are different pools of circulating lipids. The "turnover" of cholesterol and phospholipids is slower than that of triglycerides and of free fatty acids. For the sake of simplicity, the metabolism of lipoproteins has been conventionally divided into three main pathways: the exogenous pathway, the endogenous pathway and the reverse pathway of lipoprotein metabolism (*figure 1-7*).

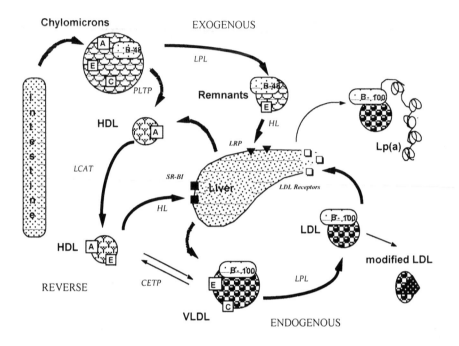

Figure 1-7. **Overview of lipoprotein metabolism**. It is organized into three main pathways: exogenous, endogenous and reverse.

• The exogenous pathway begins with the absorption of dietary lipids and the assembly of chylomicrons in the intestine. These large triglyceride-rich particles contain apoB-48 (non-exchangeable) and several other exchangeable apolipoproteins. Triglyceride hydrolysis by lipoprotein lipase (LPL) generates smaller particles: chylomicron remnants, which may undergo further lipolysis with hepatic lipase (HL). Apo E recognizes cellular receptors (LDLR, LRP) which mediate their endocytosis in the liver.

• The endogenous pathway begins with the secretion of VLDL in the liver. They contain triglycerides, cholesteryl esters and apoB-100 as the major non-exchangeable apolipoprotein. Lipolysis will generate the final product of the pathway, LDL, a major transporter of cholesterol in plasma. Modified LDL may result from a prolonged half-life in extracellular spaces or from other degradations (e.g. lipid oxidation). They are no longer recognized by LDL receptors and may be cleared by scavenger receptors. Lp(a) is also another form of modified LDL, resulting from covalent bonding between apo(a) and apoB-100 of LDL.

• In the reverse pathway native HDL take up cholesterol from peripheral cells (not shown) and return to the liver where specialized receptors (SR-B1) mediate the selective uptake of cholesteryl esters. Cholesterol is converted into bile acids, which will be secreted into bile. HDL shuttle between the other pathways exchanging lipids and apolipoproteins promoted by transfer proteins (PLTP, CETP).

1.4.1-   The General Architecture of Lipoprotein Metabolism

The exogenous pathway begins with the synthesis of chylomicrons - voluminous carriers of dietary lipids - in the epithelial cells of intestinal villosities. Chylomicrons are very large particles (10 to 100 times larger than HDL) whose lipid content is represented by 90% triglycerides *(table 1-1)*, and whose protein content represents less than 2% of the total mass. They are

Table 1-1. **Lipoprotein composition** [7]. CE : cholesteryl esters, UC : unesterified cholesterol, TG : triglycerides, PL : phospholipids, Prot. : proteins.

Lipoprotein	Density	% of total mass					Apoproteins
		CE	UC	TG	PL	Prot.	
Chylomicrons	< 0,98	3	1	90	4	2	B-48, C, E
VLDL	0,98 - 1,006	12	6	60	14	8	B-100, C, E
IDL	1,006 - 1,019	26	10	30	20	14	B-100, C, E
LDL	1,019 - 1,063	40	11	5	22	22	B-100
HDL	1,063 - 1,121	18	5	7	25	45	A, C, D, E
Lp(a)	1,050 - 1,120	33	9	3	22	33	B-100, (a)

secreted into the chyle, the fluid circulating in intestinal lymphatic vessels, whence their name. Reaching the blood stream chylomicrons acquire apo C-II from plasma HDL. Apo C-II is an activator of Lipoprotein Lipase, a lipolytic enzyme attached to heparan sulfate proteoglycans (HSPG) of endothelial surfaces. LPL hydrolyses triglycerides contained in the core of lipoproteins, releasing free fatty acids and glycerol taken up by cells as many energy sources for thermogenesis or muscular activity, or reassembled into triglycerides for storage. Lipolysis of chylomicrons induces the formation of smaller particles -chylomicron remnants- and the transfer of surface components such as phospholipids and small apolipoproteins (A-I, A-II and A-IV) towards the HDL pool -a process enhanced by PLTP (Phospholipid Tranfer Protein). The emergence of native HDL particles and the exchange of apo E from HDL towards chylomicron remnants, are direct consequences of chylomicron lipolysis. Apo B-48, the intestinally derived isoform of apo B, is not exchangeable. It constitutes the backbone of chylomicron remnants, whose clearance is ensured by hepatic receptors (LDL receptors, LRP, Low-density lipoprotein receptor Related Protein) recognizing apo E. The exogenous pathway is centered on the transport of dietary (exogenous) sources of lipids. Lipolysis, exchanges of small apolipoproteins and lipids with HDL, and the clearance of remnants by specific receptors, result in their adequate distribution in the organism.

The endogenous pathway begins in the liver by the synthesis of particles of low density, rich in triglycerides and cholesteryl esters: VLDL. Proteins constitute approximately 10% of their mass. Apolipoprotein B-100 the hepatically derived isoform of apo B constitutes their backbone, apolipoproteins C-II and E being the most abundant, and other exchangeable apolipoproteins. Triglycerides are hydrolyzed by LPL after its activation by apo C-II. Residual particles, or IDL (Intermediate Density Lipoproteins), are quickly evacuated from plasma by LRP or LDL receptors through apo E interaction. The remaining particles may undergo complementary lipolysis by LPL or by Hepatic Lipase attached to heparan sulfate proteoglycans (HSPG) of the space of Disse. Particles with relative core enrichment in cholesteryl esters become LDL. LDL further increase their core content in cholesterol, by net transfer of cholesterol esters from the core of HDL, facilitated by transfer proteins such as CETP (Cholesteryl Ester Transfer Protein).

LDL particles are rich in cholesterol. They contain one molecule of apo B-100 per particle, recognized by a ubiquitous receptor: the LDL receptor (see *figure 2-2)*. This receptor ensures the endocytosis of the whole LDL particle towards lysosomes, where it will be hydrolyzed releasing free cholesterol. The intracellular influx of free cholesterol down-regulates the *de novo* synthesis of cholesterol on the mevalonate pathway, the synthesis of LDL receptors, and stimulates cholesterol storage in the form of cholesteryl esters by the enzymatic system of ACAT (Acyl-coenzyme A: Cholesterol Acyl Transferase). LDL receptors ensure the daily clearance of two-thirds to three-quarters of circulating cholesterol in humans. Approximately 10% of circulating LDL are associated with a protein, apo (a) to form Lp(a). Lp(a) may be atherogenic and thrombogenic, however its physiological role remains unclear. When their residence time is prolonged in plasma or in extracellular compartments (like the extracellular matrix of arterial intima), native buoyant LDL can be deteriorated by oxidation, proteolytic degradation or other chemical modifications (glycation, acetylation, phospholipid hydrolysis, NO_2-, etc.), or may become denser. Modified, oxidized or dense LDL are no longer recognized by LDL receptors. They may be taken up by non-specific pathways enhanced by LPL bridging with HSPG or by receptors present at the surface of macrophages: scavenger receptors. Macrophages accumulate cholesterol indefinitely in their cytoplasm becoming foam cells (overloaded with cholesteryl esters), that are typical of the initial stages of atheroma. However, in physiological conditions, the main role of the endogenous pathway is to guarantee the delivery of triglycerides and cholesterol to tissues, whatever the dietary status (fed or fasting).

The reverse pathway of lipoprotein metabolism -a concept suggested by Glomset in 1968- ensures the return of excess cholesterol from peripheral tissues back to the liver, which is able to excrete cholesterol and its derivatives in soluble form: bile acids [65]. Cholesterol efflux is a major mechanism of reverse cholesterol transport. Most cells cannot keep excessive stores of cholesteryl esters in their cytosol and are unable to degrade the tetracyclic chemical backbone of sterols. Therefore, cells export 0.1% of their total cellular cholesterol contents every minute [66]. Cellular cholesterol efflux may be passive or enhanced by cyclodextrin from lipid poor domains of the plasma membrane [67]. However, active transport is rate-limiting and strongly dependent on HDL metabolism. Native HDL particles synthesized by the liver or produced by lipolysis of triglyceride-rich lipoproteins, are discoidal and consist in exchangeable apolipoproteins, mainly apo A-I (or apo E) surrounded by a monolayer of phospholipids and unesterified cholesterol [68]. They have a strong affinity for unesterified cholesterol, which being amphipathic resides on the surface of the particles, and of cellular membranes. Cholesterol efflux is controlled from within the cell to drive unesterified cholesterol to specific membrane microdomains enriched in cholesterol and sphingolipids: caveolae [69]. Caveolae are detergent resistant small membrane invaginations (or cave-like microdomains) in the plasma membrane. There are the site of many molecular exchanges and signaling pathways. Proteins binding cholesterol organize their structure: caveolins. Active cholesterol efflux from the cell membrane depends on the interaction of native HDL with membrane caveolae. A membrane protein, ABC1 (Cholesterol Efflux Regulatory Protein, or ABCA1 transporter), is a major regulator of cholesterol efflux. It mediates ATP-dependent transfer of unesterified cholesterol and the processing of caveolins from the trans-Golgi compartment to the plasma membrane [70,71]. It is also in caveolae that SR-BI mediate cholesteryl ester uptake by native HDL (see below) [72].

Cholesterol transport in HDL is also dependent on extracellular proteins like PLTP, which enhance HDL formation by surface remodeling of chylomicron-remnants. Moreover, reverse cholesterol transport depends on LCAT (Lecithin:Cholesterol Acyl Transferase). LCAT is bound to HDL in plasma (see *figure 2-28*). Apo A-I activates LCAT to esterify cholesterol. Hydrophobic cholesteryl esters integrate the core of the HDL particle, which by progressive swelling takes a spherical form. This conformational change of the lipoprotein particle slows down the activity of the enzyme, inducing a self-regulated mechanism of efflux and HDL release from the cell surface.

Circulating HDL particles now become spherical and cholesterol-rich (HDL$_3$). They may acquire additional exchangeable apolipoproteins, in particular apo E and apo C. They also serve as a shuttle for anti-oxidative and anti-inflammatory molecules (e.g. paraoxonase, apolipoprotein J, etc). They may as well exchange lipids with apo B-rich lipoproteins via transfer proteins: CETP (Cholesteryl Ester Transfer Protein) or PLTP (Phospholipid Transfer Protein). Large HDL then converted into particles of the HDL$_2$ type may be taken up by multifunctional liver receptors interacting with apo E. Alternately Hepatic Lipase (HL) may hydrolyze their triglyceride content in the liver space of Disse. HDL returned to the HDL$_3$ type of lipoproteins, are allowed to recirculate. Moreover, they use scavenger receptor class B type I (SR-BI) to deliver cholesteryl esters into hepatocytes through selective uptake, without HDL undergoing endocytosis, allowing unloaded HDL to cycle back to the general circulation [73]. In addition, SR-BI mediates bi-directional exchanges of cholesteryl esters with cell membranes. Depending on the cholesterol content of HDL SR-BI may favor cholesterol efflux or cholesteryl ester delivery to steroidogenic organs. More recently, HDL has been shown to undergo endocytosis in the kidney through multifunctional receptors: megalin and cubilin [74]. More than transporting cholesterol on a one-way "reverse" pathway, HDL ensure a permanent shuttle between lipoproteins and tissues. In keeping with more primitive transport systems they allow exchanges of lipids essential for cellular needs in close interaction with the metabolism of larger endogenous or exogenous triglyceride-rich lipoproteins. The 10 to 20 times faster in-vivo turnover of lipids transported in triglyceride-rich lipoproteins, compared to that of HDL apolipoproteins might be another argument in support of the archaic role of shuttle assigned to HDL [75].

1.4.2-   Regulations of Lipoprotein Metabolism

1.4.2.1-   Regulations by Exogenous Sources of Energy

Diets rich in saturated fat enhance and accelerate the synthesis of chylomicrons and VLDL and may lower plasma HDL [7]. Because carbohydrates reach the liver through the portal vein soon after food intake, diets enriched in carbohydrates preferentially stimulate liver biosynthesis and secretion of VLDL. In this line, alcohol, beside specific effects in the liver, acts as an external source of simple carbohydrates. Excessive number and size of secreted VLDL may in turn yield to high plasma LDL, provided that lipolysis is efficient. Thus many lipid-lowering diets used to control atherogenic hyperlipidemia are targeted at lowering lipoprotein biosynthesis in liver and intestine, through a limited intake of saturated fat and of simple carbohydrates. Moreover, polyunsaturated or omega-3 fatty acids by decreasing liver production of VLDL, and plant sterols by lowering plasma levels of LDL may be used in complement to lipid-lowering diets. In addition to dietary input, internal sources of energy, such as plasma glucose or free fatty acids, which may circulate in excess, in conditions such as insulin-resistance, diabetes or android obesity, are potent activators of liver production of lipoproteins [76]. They may also slower several mechanisms of HDL metabolism thereby contributing to atherogenic dyslipidemia often observed in these conditions.

1.4.2.2-   Regulations by Endogenous Sensors of Energy

Hormones have a major impact on lipoprotein metabolism as a whole [7]. Thyroid hormones influence cellular catabolism of lipoproteins: hypothyroidism increases plasma levels of LDL, HDL and sometimes VLDL, opposite effects being observed in hyperthyroidism. Glucocorticoids stimulate VLDL synthesis and inhibit LPL and Hepatic Lipase expression and/or activity. Sexual steroids have a major effect on lipoproteins [77]. Estrogens decrease the "LDL/HDL cholesterol ratio" in premenopausal women, whereas this effect is absent in men, or lost after menopause in women. During pregnancy, the production of VLDL is increased under the influence of estrogens and progesterone. Among pituitary hormones, in addition to those, which influence the above-mentioned peripheral hormones, growth hormone and other peptidic hormones also influence hepatic metabolism of lipoproteins [78]. Finally, most hormones regulating energy homeostasis, such as insulin, glucagon, or more recently leptin have great impact on many pathways of lipoprotein metabolism: lipoprotein synthesis and assembly, lipid exchanges between lipoproteins or cellular uptake of lipids and lipoproteins. Their effect may vary depending on the physiological or pathological conditions [79]. For example, high plasma insulin levels result physiologically in decreased liver apo B production, whereas in insulin resistance, the same insulin levels are associated with apo B and VLDL overproduction.

Cytokines and other circulating factors involved in the immune system (i.e. growth factors, the complement system) have a strong influence on lipoprotein metabolism [7,80]. Cachexia, viral or bacterial infections, acute or chronic inflammatory diseases may be marked by low plasma LDL and HDL cholesterol, sometimes associated with increased levels of plasma triglycerides. Apart from regulations depending on accelerated catabolism and increased energy needs for cell renewal, specific responses may be induced to enhance cellular signal transduction pathways mediated by NF-κB [81]. Cytokine-dependent cellular effects are also tissue specific. For example, Tumor Necrosis Factor-α, increases liver production of lipoproteins, decreases LPL expression and increases lipolysis in the adipose tissue, resulting in an overall raising effect on circulating triglyceride-rich lipoproteins. In addition there are proteins sharing immune and metabolic functions. Acylation-Stimulating Protein (ASP) is the desarginated form of complement component C3a (C3a-desArg). It is deprived of proinflammatory activity, however it is a potent stimulator of triglyceride synthesis and glucose uptake in adipocytes, which makes it a modulator of fat redistribution and fatty acid metabolism [82]. Aside from systemic regulations on global energy homeostasis, local immune reactions modulate the cellular handling of lipids particularly by the lymphocytic and monocyte/macrophage cell-lineages. Cellular and molecular components of the immune system play a crucial role in foam-cell formation in atherosclerosis (*see below*) and in extra-vascular cholesteryl esters deposits like xantomatosis [83].

Within cells the metabolism of lipoproteins is regulated at transcriptional and post-transcriptional levels. Regulatory sequence elements, upstream of genes for apolipoproteins, enzymes or receptors, are sensitive to multiple signals induced by hormones, cytokines, growth factors etc. Moreover, there are specific factors sensing intracellular lipid levels. Sterol Responsive Element Binding Proteins or SREBPs were recognized as major regulators of cholesterol homeostasis. They are integral membrane proteins of the endoplasmic reticulum, a pivotal compartment of intracellular pools of cholesterol [84]. When cellular pools of sterols decrease, a sterol sensitive SREBP proteolytic cleavage releases a sub-fragment belonging to the family of β-HLH transcription factors. Activated SREBPs stimulate the transcription of the LDL receptor and of several genes of the mevalonate pathway, resulting in increased cellular cholesterol levels. This crucial mechanism of feed-back regulation by intracellular sterols is at the basis of the cholesterol lowering effect of HMGCoA reductase inhibitors, or statins. Beyond their central role in cholesterol homeostasis, SREBPs monitor energy homeostasis, by regulating fatty acid biosynthesis, triglyceride synthesis or lipid uptake. Another group of major factors regulating lipid metabolism is the PPAR family of transcription factors [85,86]. Peroxisome Proliferator Activated Receptors (PPAR) belong to a family of nuclear hormone receptors. They bind to active derivatives of fatty acids (such as leucotriene B4, or prostaglandin J2) or to inducers of peroxisome proliferation (peroxisome proliferators) such as fibrates, in murine species. After dimerization with compounds of the RXR (9-cis Retinoic Acid Receptor) family of nuclear hormone receptors, they induce many ontogenic and regulatory responses at the crossroads of energy metabolism, cell proliferation and inflammatory responses. Through these universal functions and cell-specific expression, PPARs could sense energetic resources related with fatty acids that are essential for cell survival. Their identification has explained the classical pleiotropic lipid-lowering effect of fibrates in humans. The characterization of lipid specific families of transcription factors is only at its dawn. The first examples of SREBPs and of PPARs have demonstrated that lipids behave like signalling molecules relayed by potent and specific intracellular targets. Other regulatory pathways of intracellular lipid trafficking will be probably identified shortly, as many possible targets for future lipid-lowering treatments.

In addition to nuclear regulations of lipoprotein metabolism, intracellular factors controlling synthesis, assembly and catabolism of lipoproteins have been identified, so that the three-pathway architecture may become too schematic and limited in the near future. It reflects physiological facts however: the dietary induced synthesis and circulation of lipoproteins (transport of external sources of energy); the internal regulations of circulating lipoproteins (endogenous homeostasis of lipid metabolism); the "protective effect" of reverse transport of cholesterol by HDL (shuttle between internal and external influences challenging energy homeostasis).

The metabolism of lipoproteins has this originality in that it generates complex and transient entities, in permanent remodeling in all the compartments in which they circulate.

1.5- LIPIDS AND ATHEROSCLEROSIS: THE CAUSAL LINK

VIRCHOW a pathologist in the 19th century had suspected that an inflammatory process underlies lipid deposition within the arterial intima. The "lipid hypothesis of atherosclerosis" was confirmed early in the 20th century by ANITSCHKOW with experimentally induced atherosclerosis in rabbit. Since then, numerous supporting evidences -anatomical, epidemiological, clinical and experimental- have ever since confirmed this hypothesis, so that a causal link between disturbances of lipoproteins metabolism and atherosclerosis is now established.

Atherosclerosis could result from a chronic failure of the arterial wall to adjust its response to various injuries altering its integrity. These agents may be chemical: like toxic products resulting from tobacco smoking, or oxidized lipids from modified lipoproteins; infectious: viral like cytomegalo-virus or bacterial like chlamydiae; physical: like shear stress or high blood pressure or mechanical: like those induced by angioplasty balloons or local ischemia induced by adventice injury [87]. Even in the absence of a manifest injury, lipoprotein themselves may represent triggering factors if they remain trapped within the extracellular matrix [88], susceptible to undergo there structural modifications (degradation, oxidation etc.) upon the action of locally produced enzymes: phospholipases, metalloproteases. In any case, inflammatory cells will be recruited; inflammatory-mediating, haemostatic and growth factors will be secreted and additional lipoproteins will be attracted at the site of injury [89]. Lesions begin preferentially at junction points and at sites of mechanical stress in the arteries, where local endothelial metabolism is chronically challenged. At very initial stages one detects morphological changes characterized by a cytoplasmic accumulation of cholesterol esters, giving a typical aspect of "foam cells". At this stage of **fatty streaks**, lesions consist of a clustering of sub-endothelial or intimal foamy macrophages and of T lymphocytes retained in the extracellular matrix. Then a vicious circle establishes in which activated cells proliferate and recruit more smooth muscle cells from the media. Myocytes change from a contractile phenotype to a secretory phenotype, characterized by their proliferation and the secretion of a collagenous matrix and other tissue remodeling substances (i.e. proteases). Smooth muscle cells in turn become foam-cells. Because there exists no enzymatic pathway to degrade cholesterol within cells, when the rate of reverse cholesterol transport fails to compensate for excessive cholesterol ester storage, foam-cells eventually undergo apoptosis and cell-death [90]. Cellular death in turn may release as many substances in the extracellular matrix, further activating inflammatory, secretory and proliferative processes. In addition cholesterol eventually crystallizes on place. The cellular debris mainly made of cholesterol esters are encapsulated within a collagenous envelope (sometimes calcified), giving rise to the **fibrous plaque.** The process once installed develops, weakens the endothelial barrier, being particularly active at the margin of lesions. At this stage and at later stages, the lesion (narrowing the arterial lumen or not) is likely to ulcerate or break in periphery favoring a sudden thrombus formation. Most cardiovascular events result from the rupture of unstable lipid rich non-stenotic lesions, indicating that atherosclerosis is more an active process than a "sclerotic" process [89].

Pathological examinations have reconstituted the sequence of events for hypercholesterolemia-induced atherosclerosis. An elegant demonstration of the role of the hypercholesterolemia in the initial stages of the atherosclerosis, was brought by the examination in electron microscopy of coronary arteries from human hearts with ischaemic cardiopathy, compared to the same arteries in monkeys with experimentally induced hypercholesterolemia [87]. The margination of monocytes penetrating the endothelium, without any aspect of endothelial lesion begins soon after induction of hypercholesterolemia in primates. The same picture is found in human artery. The aspect is also the same at the stage of fatty streaks in which foam-cells hump underneath of the endothelium. Pathological studies on several thousands of young adults (age <30) have shown that the stage and number of lesions correlate very positively with pre-mortem plasma cholesterol [91]. Even more striking, fatty streaks may be detectable on the aorta as early as fetal development, in children expected by their hypercholesterolemic mother [92].

Descriptive and interventional studies in large human cohorts have defined the extent and the characteristics of the causal link between hypercholesterolemia and atherosclerosis in population at large [93]. The clinical and biological bi-annual follow-up of 2336 men and 2873 women living in the town of Framingham (Massachusetts) since 1946, has been the pioneer study showing that the rate of mortality from ischaemic cardiopathy is significantly correlated with chronic disturbances of lipoprotein metabolism [94]. Many other large human cohort studies have followed since then throughout the world further showing the multiplicative effect of hypercholesterolemia clustering with other cardiovascular risk factors. Moreover, at equal plasma levels of total cholesterol, low HDL cholesterol represents a risk equivalent to high LDL cholesterol (*figure 1-8*), [95]. The

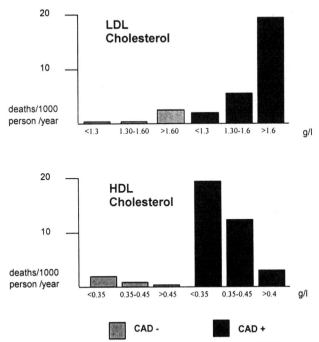

Figure 1-8. **Cardiovascular mortality rate as a function of LDL and HDL cholesterol** [95]. In 2541 men aged between 40 and 69 followed over 10 years, cardiovascular mortality rate is higher in subjects with patent coronary arterial disease (CAD+), or with high plasma LDL cholesterol. The risk profile is similar in subjects with low plasma HDL.

first studies by Keys & al. [28] had shown among Japanese emigrants that a progressive gradient of cardiovascular disease risk correlated positively with levels of plasma cholesterol together with the adoption of a North-American lifestyle. Other studies on migrants and studies on adoptees confirmed these data [96,97]. Between countries studies have also reinforced the lipid hypothesis. The 19 countries study had suggested that the relationship between cholesterol and atherosclerosis was universal, later confirmed by data from MONICA registries, a WHO initiative on cardiovascular disease risk monitoring in 38 populations [98,99]. Populations whose average plasma cholesterol was lowest also had an incidence of coronary disease among the lowest. The opposite was observed for populations with high average levels of plasma cholesterol. In addition, the tendency of the last decade to the decline in the incidence of premature cardiovascular death in Western populations followed a significant reduction in the average plasma cholesterol [100].

Interventional studies showed that plasma cholesterol lowering, by lifestyle changes and/or lipid-lowering drugs, prevented coronary events, in primary prevention and in secondary prevention [101]. The first angiographic longitudinal studies performed with cholesterol-lowering drugs, introduced an unexpected notion: the reversibility of atherosclerosis. These data had been suggested by animal experiments in which the extent and the nature of lesions could be directly and precisely measured, on anatomical examination. However, changes observed by angiography, appeared more modest in humans. Surprisingly, these interventions led to a stronger clinical benefit authenticated by a significant reduction of ischaemic events and total mortality rates. This paradox was explained by the fact that angiography did not assess directly a disease process, which takes place within the arterial wall: the active unstable atherosclerosis lesion [102]. The power to revert the atherogenic process by LDL-cholesterol lowering therapy was later confirmed by very large trials enrolling several thousands of patients, assessing morbidity and mortality rates as the study end-point in primary or secondary prevention of cardiovascular disease [103-105]. Their results introduced a new pathophysiological concept in atherosclerosis: lipid metabolism is not only a "lumenal" factor but intervenes at the root of atherosclerosis lesion development within the arterial wall [89].

Clinical facts in single patients also support the lipid hypothesis. Coronary diseased patients present in their great majority (*figure 1-9)*, at least one, qualitative or quantitative anomaly of lipoprotein metabolism [106]. Familial aggregation of lipid disorders and coronary heart disease suspected earlier in the 20th century, has been largely confirmed since [30,101,107]. In familial

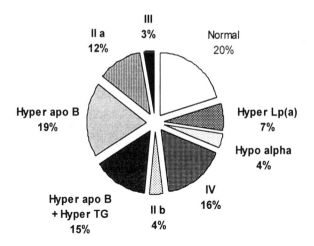

Figure 1-9. **Plasma lipid and lipoprotein profiles in coronary artery disease** [106].
Classical phenotypes of hyperlipidemia are observed, as well as other phenotypes of hyperapobetalipoproteinemia, hypo-alphalipoproteinemia or high plasma Lp(a). A minority do not
present any disorder.

hypercholesterolemia, a dominantly inherited disorder, atherosclerosis
progression and disease severity are a direct consequence of high plasma
LDL cholesterol levels. Heterozygotes who have 1.5 to twice higher than
normal plasma LDL cholesterol, experiment the first coronary event during
the fourth or fifth decade. Homozygotes who have three to four fold normal
levels have their first coronary events between the age of 10 and 20, and a
mortality rate close to 100% by the age of 30 [108]. In this disease xanthoma
and arterial lesions can regress under lipid lowering therapies. On the other
hand, heterozygotes for hypobetalipoproteinemia caused by apo B mutations
have half-normal LDL cholesterol levels [109], and a longevity syndrome.
Atherosclerosis, if present, is delayed. In dyslipidemia, secondary to thyroid
dysfunction, renal insufficiency or diabetes, lipid disorders are enhancing
factors for the macroangiopathy typical of these diseases.

Together with observational data in human, mechanistic explanations were
sought by **experimental studies** in vivo in animals, and in vitro in cell
cultures. In animals, whatever the way lipid disorders were induced; either
by drugs, diets, organ lesion (of endocrine glands, pancreas, kidney, liver
etc.), spontaneously inherited (naturally mutant strains) or genetically
engineered (transgenic or knock-out animals), arterial lesions may be
reproduced and develop in a similar way with those seen in humans. Thus
animal models are helpful for trials of future medical treatments. In cellular
models in vitro, it is possible to dissect the molecular mechanisms involved
in the atherogenic process. For example, foam-cell formation may be
induced after incubation of macrophages with oxidized or modified LDL

the normal artery, and secrete more cytokines, proteases, growth factors or oxidative products able to degrade even more LDL present in the cell-culture medium [87].

The vicious circle of atherosclerosis is manifest from large human populations down to the identification of its molecular actors within a single cell. However only a **causal link** is established with lipid metabolism, because atherosclerosis does not proceed from a single cause. It reflects a chronic failure of natural protective mechanisms of the artery to respond to injury. This dynamic process can be reverted by interruption of the chain of worsening events, each step challenging intracellular and extracellular lipid homeostasis.

1.6- HETEROGENEITY OF DYSLIPIDEMIA

The name dyslipoproteinemia indicate all the possible qualitative or quantitative disturbances in the metabolism of circulating lipoproteins. This definition has the merit to be general but the disadvantage to include a multitude of more or less defined entities. Lipoproteins are transient complex particles in permanent remodeling, so that a phenotype defined by their quantitative variations can be only heterogeneous. Moreover taking hormonal, dietary, immune, or cellular regulations, several hundreds of genes could more or less contribute to lipid homeostasis [111]. In fact, the majority of primitive dyslipoproteinemias have an inherited component [52]. Dyslipoproteinemias are thus by definition heterogeneous in their expression as much as in their genetic predisposition.

1.6.1-   The Phenotypic Heterogeneity of Dyslipoproteinemia

Clinical symptoms are rare, sometimes transient and except for planar xanthoma in homozygotes for familial hypercholesterolemia, not very specific. Lipoproteins even defined by their density are heterogeneous. Their content, their size and their function change in pace with modifications they may undergo. The nature of each lipoprotein species is controlled by apolipoproteins (A-I, A-II, B, C-I, C-II, C-III, D, E), enzymes (Lipases, LCAT, etc.), modifying proteins (CETP, PLTP), and cellular receptors (LDL, remnant, scavenger receptors etc.) with which they interact. The identification of these factors if possible by sophisticated techniques, is usually inapplicable in clinical practice. Classifications in phenotypes of dyslipidemia represent simple means, to orient diagnostic and therapeutic decisions [6,7].

Fredrickson's classification in five phenotypes [20] defined by electrophoretic profiles and fasting plasma lipid measurements, was a first approach which is still of a great practical utility (*figure 1-10*).

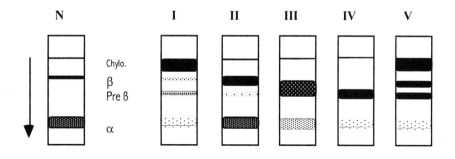

Type	Serum	Electrophoresis	Ultracentrifugation	Lipids	
I	lactescent	Chylomicrons	Chylomicrons	TC ↔↑	TG ↑↑↑
IIa	clear	ß lipoproteins	LDL	TC ↑↑	TG ↔
IIb	clear	ß and Pre ß	LDL, VLDL	TC ↑↑	TG ↑
III	opalescent	Broad ß lipoprotein	IDL	TC ↑↑	TG ↑↑
IV	opalescent	Pre ß lipoproteins	VLDL	TC ↑	TG ↑↑
V	lactescent	Pre ß, Chylomicrons	VLDL, Chylomicrons	TC ↑↑	TG ↑↑↑

Figure 1-10. **Fredrickson's classification in five types of hyperlipidemia** [20]. Top. Electrophoretic profiles of human serum run in agarose gel and non-specific lipid staining. Chylo: chylomicrons, ß: betalipoproteins, pre ß: pre-betalipoproteins, α: alphalipoproteins. Arrow: orientation of electrophoretic migration. Bottom: Lipoprotein and lipid phenotypes corresponding to each phenotype. TC: total cholesterol, TG: triglycerides.

Type I hyperlipidemia corresponds to an excess of circulating chylomicrons. It is characterized by massive hypertriglyceridemia in fasting plasma. The excess of circulating LDL characterizes **Type IIa** hypercholesterolemia. Excessive levels of both LDL and VLDL characterize **Type IIb** hyperlipidemia. In **type III** hyperlipidemia the presence of an abnormal lipoprotein, "broad-ß" lipoprotein, corresponds to the abnormal presence of IDL in fasting plasma. Excessive levels of VLDL characterize **Type IV** hypertriglyceridemia so that triglyceride levels exceed twice that of plasma cholesterol. Finally in **type V** hyperlipidemia both chylomicrons and VLDL circulate in excess in plasma. Thereafter, De Gennes [21] proposed a simplified classification in three classes including the phenotypes defined by Fredrickson (*Table 1-2*). Thus, "hypertriglyceridemia" where excessive triglyceride levels predominate, reflect disorders mainly dietary related,

Table 1-2. **De Gennes classification of hyperlipidemia** [21]. Correspondence with phenotypes of Fredrickson's classification and relative serum concentrations of total cholesterol (TC) and triglycerides (TG).

Class	Type	Serum	Lipoproteins	Lipids
Hypercholesterolemia	IIa	clear	LDL	TC/TG >2,5
Hypertriglyceridemia	I	lactescent	Chylomicrons	TG/TC >2,5
	IV	opalescent	VLDL	
	V	lactescent	VLDL, Chylomicrons	
Combined	IIb	clear	LDL, VLDL	TC/TG <2,5
Hyperlipidemia	III	opalescent	IDL	TG/TC <2,5

whereas "hypercholesterolemia", reflects disorders more related with endogenous regulations of lipoprotein metabolism, and combined hyperlipidemia reflect disturbances of both.

In clinical practice, the most common forms of hyperlipidemia correspond to three main classes, found in De Gennes classification. **Hypercholesterolemia** when permanent and significant (TC > 300 mg/dL) may manifest clinically by the presence of extravascular cholesterol deposits: corneal arcus or tendon xanthoma, sometimes a cause of inflammation on Achilles tendons. The progression and the severity of atherosclerosis are positively correlated with time exposure and intensity of hypercholesterolemia. Arterial lesions are most frequently observed on proximal segments of the coronary or peripheral arteries. Xanthomatosis reflects the exposure duration to high plasma levels of LDL cholesterol.

Hypertriglyceridemia include disorders sensitive to the type of food intake. They can be sensitive to carbohydrate/alcohol intake (type IV) or to fat intake (types I or V). They may complicate with acute pancreatitis, when plasma triglycerides exceed 1000 mg/dL. During these episodes, one notes sometimes liver and spleen enlargement, characteristic cutaneous and retinal signs: eruptive xanthomatosis and lipemia retinalis respectively. Hypertriglyceridemia may have a pro-atherogenic potential, usually reinforced by common risk factors like diabetes, insulin resistance, obesity and high blood pressure, with which they cluster frequently.

Combined hyperlipidemia of the IIb type, sometimes manifesting with xanthelasma and corneal arcus, is atherogenic. In type III hyperlipidemia, there is a specific and transient form of xanthomatosis: palmar folds xanthomatosis, and a plasma accumulation of intermediate lipoproteins. This hyperlipidemia is very atherogenic and sensitive to dietary modifications. A common feature in hypertriglyceridemia and in combined hyperlipidemia is that arteriosclerosis may appear more diffuse and distal on arterial segments than lesions resulting from pure excess of LDL (i.e. hypercholesterolemia).

Types and phenotypes of primary hyperlipidemia are not constant. Phenotypes may vary over time, or according to age, hormonal and dietary status in the same individual, or in different individuals from the same family. In addition because they are more reproducible, phenotypes of dyslipidemia were defined in plasma after 12-hour fasting. However post prandial lipoprotein profiles correspond to the physiological state found during 80% of a 24-hour cycle in "normally" fed humans. Moreover, classical phenotypes do not include other phenotypes such as those of **hypolipidemia** (hypoalphalipoproteinemia (hypo-α), characterized by low plasma HDL cholesterol, and hypobetalipoproteinemia (hypo-β), characterized by low plasma LDL cholesterol. These conditions may manifest with certain corneal deposits or by neurological signs, which will be described further. In the same line, disorders of lipoprotein Lp(a) are not included in these classifications. However classical phenotypes of hyperlipidemia are of precious help for the diagnosis of common lipid disorders, even if they encompass different pathogenic entities.

Other phenotypes based on lipoprotein density sub-fractionation by ultracentrifugation, or on apolipoprotein content analyzed by immuno-affinity-chromatography (LpAI, LpAI-AII, LpB-E), or on lipid content by HPLC or electrophoresis were proposed [112]. Certain lipoprotein sub-fractions are pro-atherogenic (dense LDL) or on the contrary may be protective (LpAI). Various technologies may be used for lipoprotein analysis, even nuclear magnetic resonance spectroscopy for the detection of their oxidative modifications. However, present routine techniques generally do not allow the identification of a single disease entity they at best localize a lipid disorder to one of the main three pathways of lipoprotein metabolism.

Heterogeneity of dyslipidemic phenotypes is thus the common rule. For example type IIa hyperlipidemia can result from disturbances at various checkpoints of LDL cholesterol metabolism [113]: VLDL production and secretion (*figure 1-11*), LDL production as the result of intermediate lipoproteins catabolism, LDL catabolism by cellular receptors. In addition, the latter, for example do not have a univocal cause: they can result from structural defects of LDL receptors, or from disorders of intracellular cholesterol homeostasis, which repress LDL receptor expression at the cell surface, or from a binding defect of the ligand (apo B-100 in LDL). Kinetic studies with tracer-labeled lipoproteins may solve part of these problems. Cell culture techniques coupled with biochemical analyses, or morphological analyses are also powerful tools in the search for molecular defects of

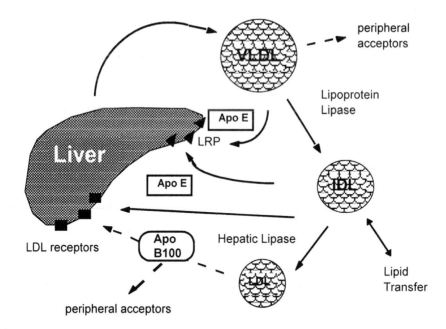

Figure 1-11. **LDL cholesterol metabolism** [113]. LDL may accumulate in plasma as the result of multiple mechanisms. They may either produce an excessive number of precursors (VLDL, VLDL remnants) or slower their catabolic rate in the liver or in peripheral tissues.

lipoproteins metabolism. These approaches have proven very fruitful in the identification of molecular mechanisms causing monogenic lipid disorders starting with familial hypercholesterolemia, more recently exemplified by Tangier disease [108,114-116]. Unfortunately, these methods are not easily transferable to routine clinical diagnosis. The absence of a simple relationship between "clinical phenotypes" of dyslipidemia and their molecular pathogenesis are the consequence of the multiplicity of factors controlling levels of circulating lipoproteins.

1.6.2-   Genetic Heterogeneity

The majority of atherogenic dyslipidemia have an inherited component [30,101]. Thus dyslipidemia may be considered as familial diseases, however with a far from univocal mode of inheritance. The first step thus consists in evaluating the familial aggregation of a lipid trait or associated cardiovascular disease over at least three generations. Indeed the type of dyslipidemia, its severity and age at discovery, those of its cardiovascular complications and its mode of inheritance may orient the analysis towards a monogenic or polygenic disease. In the first case, genetic predisposition depends on a single gene or locus. A single gene will determine most of the characters associated with dyslipidemia, which is generally more severe and premature. In the second case, several genes may interact to different extent, often modulated by environmental factors to give rise to lipid disorders, generally more variable in expression and of later onset.

Contrary to general knowledge monogenic dyslipoproteinemias are not rare [117]. They could account for 50 to 60% of dyslipidemia observed in patients with coronary heart disease [118]. Some of them result from defects on single genes known to encode for candidate genes of lipoprotein metabolism *(table 1-3)*. Other forms of dyslipidemia of unknown cause are defined by a set of inherited signs, segregating in families according to a mendelian mode of inheritance. Thus despite possible phenocopies, the search of causal genes -candidate genes- may identify novel pathogenic entities by reverse genetics, under the assumptions of single gene inheritance (see chapter III).

Table 1-3. **Familial dyslipoproteinemia and known disease causing genes** [117]. Frequencies were reported in Caucasian populations. AD: autosomal dominant, AR: autosomal recessive, ABC1: ATP-Binding Cassette transporter 1, APO AI: apolipoprotein A-I, Apo B: apolipoprotein B, Apo E: apolipoprotein E, CETP: Cholesteryl Ester Transfer Protein, CII: apolipoprotein C-II, LCAT: Lecithin:cholesterol Acyl Transferase, LDLR: LDL receptor, LPL: Lipoprotein Lipase, MTP: Microsomal triglyceride Transfer Protein.

Dyslipoproteinemia	Inheritance	Type	Locus	Frequency
Hyperlipidemia				
Familial Chylomicronemia	AR	I, V	LPL, C-II	$1/10^6$
Familial Hypercholesterolemia	AD	IIa	LDL R	1/500
Familial Defective apo B-100	AD	IIa	APO B	$1-5/10^3$
Autosomal Recessive Hypercholesterolemia	AR	IIa	ARH	rare
Familial Dysbetalipoproteinemia	Pseudo AD	III	APO E	$1/10^4$
Familial Combined Hyperlipidemia	AD	IIa, IIb, IV	1q, 11p	$3-5/10^3$
Familial Hypertriglyceridemia	AD	IV, V	?	$2/10^3$
Familial CETP Deficiency	AD	Hyper α	CETP	$<1/10^3$
Hypolipidemia				
Familial Hypobetalipoproteinemia	AD	Hypo β	APO B	$8/10^3$
Abetalipoproteinemia	AR	Hypo β	MTP	rare
Familial LCAT Deficiency	AR	Hypo α	LCAT	rare
Familial Apo A-I Deficiency	AD	Hypo α	APO A-I	rare
Tangier Disease	AR	Hypo α	ABC1	rare
Familial Hypoalphalipoproteinemia	AD	Hypo α	ABC1, 11q	$1-5/10^3$

In populations, biometric studies of blood lipids (twins studies, studies of adoptees, nuclear families or sib pair analyses), have often shown the predominant influence of genetic factors on the variance of these parameters

[111,119]. Complex analyses of segregation on large series of families showed that the model best fitted, was that of major loci modulated by other genes or environmental factors. Moreover, inter-individual variance of blood lipoprotein levels, the response to lipid-lowering therapy, or predisposition to atherosclerosis, necessarily involves genetic and environmental interactions modulating the expression of major loci. There are proven cases of multifactorial mechanisms: for example type III dysbetalipoproteinemia [120]. A specific common apo E allele combined with the effects other loci or dietary factors give rise to this particular phenotype of hyperlipidemia (*figure 2.27*). In this line, in large pedigrees of baboons two major loci could contribute to the regulation of apo A-I levels in response to various diets [121]. Last, authentic sporadic forms of dyslipidemia could account for 20 to 30% of dyslipoproteinemia observed in patients surviving from myocardial infarction [30,118]. The candidate approach gene applied to populations, without any presupposition on the underlying genetic model may identify modifier alleles contributing to susceptibility to frequent conditions such as myocardial infarction. Several applications of the candidate gene approach applied to populations will be presented in the third chapter.

Genetic predisposition to dyslipoproteinemia in keeping with the majority of multifactorial diseases is not univocal. Depending on the case, it may refer to a monogenic model, in which it is caused by rare mutations with a strong functional effect or to a polygenic model in which combinations of frequent alleles (polymorphisms) of different the loci contribute to the development of dyslipidemia. Whatever the genetic mechanism (monogenic or polygenic) environmental factors (physical exercise, diet, etc.) may modulate the expression of candidate genes. In sporadic forms, environmental factors would challenge a combination of allelic forms of gene regulating lipoprotein metabolism to predispose to dyslipidemia in a particular individual.

The adaptability of lipoprotein metabolism to physiological and environmental conditions is a consequence of the multiplicity and variability of the genes involved in its architecture. The majority of major genes controlling lipoprotein metabolism has been cloned some of them identified as causal genes for dyslipidemia. These candidate genes and the pathological consequences of mutations disturbing their physiological functions will be described in the next chapter.

II- GENES OF LIPOPROTEIN METABOLISM

The purification of proteins involved in lipoprotein metabolism and technologies of molecular biology, allowed the cloning of their structural genes in the mid "80s". Several groups sometimes carried out the identification of genes and of their chromosomal localization in sympathy [84,119]. Candidate genes of lipoprotein metabolism often appear to belong to gene families of diversified functions. The list of novel genes controlling lipoprotein metabolism enlarges every year, and is not expected ending soon *(table 2-1)*. However spontaneous mutations or common variants reported on loci characterized so far, have significantly improved our knowledge on the genetic predisposition to dyslipidemia in humans.

For more clarity, we chose to present candidate genes according to the main pathways of lipoprotein metabolism: the endogenous pathway, the exogenous pathway, the reverse transport pathway and other regulatory pathways.

2.1- GENES OF THE ENDOGENOUS PATHWAY

2.1.1- The LDL Receptor: a Gene for Familial Hypercholesterolemia

From the first clinical descriptions of xanthomas in the 19th century until the demonstration of their autosomal dominant inheritance a century later, familial hypercholesterolemia was gradually delineated as an inborn error of lipoprotein metabolism. Michael BROWN and Joseph GOLDSTEIN used cell-cultures of skin biopsy fibroblasts from homozygotes with familial hypercholesterolemia, to demonstrate the unsuspected existence of a membrane receptor specific for LDL endocytosis [123]. This receptor mediates the catabolism of circulating LDL and modulates intracellular concentrations of cholesterol [9]. BROWN and GOLDSTEIN's accomplishment (Nobel Prize winners in Medicine and Physiology) represent the paradigm of scientific progress. Guided by strong hypotheses they have combined fine observations of a naturally occurring disease and a judicious use of technological developments available then, to discover a universal regulatory mechanism in cell biology in general and in cholesterol homeostasis in particular.

Table 2-1. **Candidate genes of lipoprotein metabolism localised on human chromosomes.**
For conventional names and full spelling of gene loci refer to the index. nr: not reported.

Apolipoproteins	Chromosome	Enzymes	Chromosome
Apo A-I	11 q 23	ACAT1	1 q 15
Apo A-II	1 p 21	CEL	9 q 34
Apo A-IV	11 q 23	CYP 7	8 q 11
Apo (a)	6 q 26	CYP 27	2 q 33
Apo B	2 p 23	FASN	17 q 25
Apo C-I	19 q 13	HL	15 q 21
Apo C-II	19 q 13	HMGCoA Reductase	5 q 13
Apo C-III	11 q 23	HMGCoA Synthase	5 p 14
Apo C-IV	19 q 13	HSL	19 q 13
Apo D	3 p 14	LCAT	16 q 22
Apo E	19 q 13	LIPA	10 q 24
Apo H	17 q 23	LPL	8 p 22
Apo J	8 p 21	PNLIP	10 q 26
Apo L	nr	SCD	10
Receptors		**Others**	
ABC1	9 q 31	ABCG5	2 p 21
APOER2	1 p 34	APOBEC1	12 p 13
CD36	7 q 11	ARH	1 p 35
Cubilin	10 p 12	ASP (C3a desarg)	19 p 13
LDLR	19 p 13	CETP	16 q 21
LRP	12 q 13	FABP1	2 p 11
Megalin	2 q 24	FABP2	4 q 28
Scavenger SR-A	8 p 22	LMN A/C	1 q 21
Scavenger SR-B1	12 q 24	MTP	4 q 24
VLDLR	9 p 24	NPC1	18 q 11
		OCTN2	5 q 31
Transcription Factors		PLTP	20 q 12
PPARα	22 q 12	PON	7 q 21
PPARβ/δ	6 p 21	RAP	4 p 16
PPARγ	3 p 25	SAA	1 q 12
RXRα	9 q 34	SCAP	3
SREBP 1 (a,c)	17 p 11	SCP2	1 p 32
SREBP 2	22 q 13		

Familial Hypercholesterolemia (FH) is a frequent disorder worldwide, with a general prevalence of about 1/500 [108]. Autosomal inheritance of primary type IIa hypercholesterolemia and premature coronary heart disease characterize FH. An allelic dose effect determines its clinical presentation in families. In heterozygotes, serum LDL cholesterol concentrations reach about twice that of normal and the first cardiovascular events occur between the fourth and fifth decade (*Figure 2-1*). The penetrance of high plasma LDL cholesterol above the 90th percentile of the referral population is high (90%) in FH families. FH is naturally a prematurely life-threatening disorder. Data from the Simon Broome's Registry in UK, have shown that FH heterozygotes have a relative risk of cardiovascular mortality multiplied by about 48 in males and 125 in females below age 40 [124]. The homozygous form of FH is usually rare ($1/10^6$ in the population). Massive hypercholesterolemia (exceeding generally 3 to 4 times the normal) is already detectable during fetal life or at birth in cord blood. It is associated with typical skin deposits of cholesteryl esters: planar xanthomas. The first cardiovascular event usually occurs in childhood or adolescence. In the absence of lipid-lowering therapy, the mortality rate reaches 100% by the age of 30 in FH homozygotes.

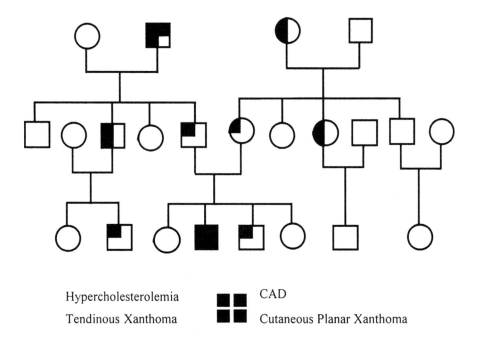

| Hypercholesterolemia | ■ ■ | CAD |
| Tendinous Xanthoma | ■ ■ | Cutaneous Planar Xanthoma |

Figure 2-1. **Familial Hypercholesterolemia** (FH). Hypercholesterolemia (LDL cholesterol >90th percentile) is dectetable from birth or soon after, in all carriers of both sexes. In heterozygous FH one copy of the LDL receptor gene is non functional. The disease may manifest with tendinous xanthoma by the age of 20 and coronary artery disease (CAD) before the age of 40. In homozygous FH, both copies of the gene are defective. Clinical symptoms are manifest from the first decade, cutaneous planar xanthomas being highly suggestive.

Familial hypercholesterolemia results from defects of the **LDL receptor**. This receptor ensures the endocytosis of LDL through a specific interaction with apolipoprotein B (*Figure 2-2*). LDL receptor may also bind and take up VLDL and β-VLDL, through specific binding with apolipoprotein E. It is a ubiquitous membrane receptor, expressed in all living cells. It is however more abundant in tissues with a high cholesterol demand: endocrine glands

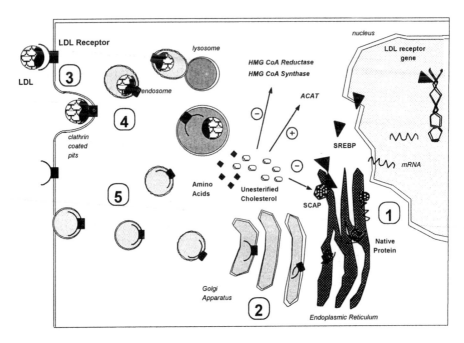

Figure 2-2. **The LDL receptor pathway.**
- LDL receptros are clustered at the cell surface in clathrin-rich coated pits. LDL ApoB-100 binding to LDL receptors triggers endocytosis of the whole complex within endosomes. Endosomal fusion with lysosomes by decreasing the pH of the vesicle lumen, dissociates the lipoprotein-receptor complex. Receptors are recycled to the cell surface ready to shuttle for another round of LDL endocytosis. LDL are degraded in lysosomes and free cholesterol is released within cellular compartments, inducing sterol-specific responses.
- A sterol sensor SCAP, integral to the endoplasmic reticulum (ER) membrane activates a proteolytic cascade on SREBP another microsomal membrane protein. Proteolysis releases an active SREBP subfragment homologous to the β-HLH family of transcription factors. In the nucleus SREBP activates biosynthetic enzymes of the mevalonate pathway and LDL receptor biosynthesis. The intracellular afflux of unesterified cholesterol down-regulates SCAP-meditated proteolysis and stimulates ACAT activity. Cholesterol biosynthesis is suppressed and cholesterol is oriented towards neutral storage in esterified form. On the contrary when cellular pools of sterols decrease SCAP activates SREBP and ACAT is down-regulated.
- Newly transcribed LDL receptor mRNA reach the ER for translation into a native protein of 120kD. The 160kD glycosylated mature protein is guided towards the cell-surface into clathrin coated pits. Numbers represent classes of functional defects observed on the LDL receptor. Class 1: defective synthesis. Class 2: defective transport and maturation. Class 3: defective binding. Class 4: defective endocytosis. Class 5: defective transport and recycling.

synthesizing and secreting steroid hormones, the white matter of the central nervous system, tissues with high cell-turnover and especially in the liver [125]. They usually cluster in clathrin-rich membrane domains, or clathrincoated pits, specialized in cellular endocytosis. After endocytosis, receptors are routed to lysosomes. The low pH of lysosomal lumen favors ligand-receptor dissociation. Lipoproteins are degraded, while receptors are rapidly recycled to the plasma membrane. In certain cellular types or certain physiological circumstances, LDL receptors may form dimers. Dimerization could diversify their functions either towards LDL transport through the cell, or towards LDL degradation to increase intracellular pools of cholesterol. It is a major pathway of LDL catabolism in humans. It has been estimated that about 2/3 to 3/4 of LDL cholesterol is removed daily from plasma by LDL receptors. Functional studies of LDL receptors on cell-culture fibroblasts have demonstrated that a dose effect of defective LDL receptors determines clinical and biological characteristics of FH.

The LDL receptor gene (LDLR) sequence is unique, in all genomes where it has been cloned. It is localized on the short arm of human chromosome 19 (19p13.2) [126]. It is 45kb long and organized into 18 exons. It is transcribed in a 5.3kb long mRNA of which 2.6kb encode for the mature protein. The LDL receptor is a monomeric protein of 839aa [127,128] (*figure 2-3*).

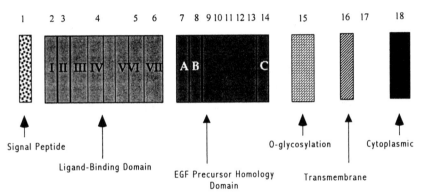

Figure 2-3. **Functional domains of the LDL receptor** [128]. Letters and roman numerals indicate repeated modules. Functional domains are organized into series of modules each encoded by a single exon (arabic numerals on top).

The first exon encodes for a signal peptide. Exons 2 to 6 encode for the 292 amino acids of the ligand-binding domain. It is composed of 7 repeated sequences of 40aa (repeats I to VII), rich in cysteines (LDL-A modules) involved in disulfide bridging. The folding of these structures is facilitated by the interaction of negatively charged amino acids with calcium [129]. These modules mediate interactions with basic amino-acid-rich binding domains of apolipoproteins. Exons 7 to 14 encode for the 400 amino acids of the second domain. This domain contains cysteine-rich repeated modules (EGF-like modules) homologous to those found in the EGF (Epidermal Growth Factor) precursor. Site directed mutagenesis experiments have shown that this domain is necessary for rapid recycling of receptors to the plasma membrane soon after endocytosis [130]. N-terminal cysteine-rich repeats of the EGF-like domain could also participate to ligand binding. The native protein has a 120kD molecular weight whereas it is 160kD in the mature protein. This difference is due to O-glycosylation. Exon 15 encodes for the O-glycosylation 58aa domain, which could strengthen membrane anchoring and confer antigenic specificity of the receptor. The single hydrophobic transmembrane domain (22 amino acids) is encoded by exon 16, and the cytoplasmic domain (50 amino acids) is encoded by exons 17 and 18. The cytoplasmic domain comprises a motif for 22 amino acids. In this domain a tyrosine at position 807 is necessary for endocytosis and interaction with clathrin. In epithelial cells LDL receptors have a polar distribution [131]. For example in hepatocytes they are present only at the basal side of cells and are totally absent from the apical side, which corresponds to the "biliary" face of cells. There is another 17-residue motif also including a tyrosine at position 824, which determines not only the clustering of newly synthesized receptors at the membrane surface, but also the destiny of endosomal receptors recently internalized [132]. Thus, most of the coding sequence of the LDL receptor contains essential functional elements. Transgenic mice overexpressing LDL receptors, or knock-out (KO) mice in which the LDL receptor gene was invalidated by homologous recombination have confirmed the fundamental role of receptors in LDL catabolism *in vivo*. Transgenic mice are resistant to atherosclerosis and rapidly clear LDL from their plasma; KO mice present a phenotype similar to that of familial hypercholesterolemia [133,134].

The primary sequence of the LDL receptor is highly conserved (70 to 86% of homology depending on domains) in many mammals (rat, rabbit, hamster, pig), in an Amphibian, Xenopus Laevis [135]. It also shares similarities with an insect lipophorin receptor [49]. That primary sequence of the LDL receptor was conserved suggests that a fundamental structure has been maintained to preserve important functions across the evolution of living species. The majority of exons of the LDL receptor gene encode for stretches of amino found in several hundreds of other proteins [136]. They are membrane proteins (cellular receptors for growth or differentiation factors, microorganism membrane receptors) or extracellular proteins (growth factors, components of the complement system and of the haemostasis cascade, extracellular matrix proteins, adhesion molecules etc.). However, only certain proteins contain several functional blocks of the LDL receptor. They constitute the LDL receptor gene family and include LRP (Low density lipoprotein Related Protein, Megalin, VLDL receptor, Apo E receptor 2, and several other LDL receptor related proteins (see *figure 2-27*, page 117). The LDL receptor gene family has diversified by exon shuffling leading to a set of proteins sharing at least some functions in common (e.g.

positively charged protein-protein interactions, single membrane anchoring domain, receptor-mediated endocytosis, clathrin binding etc.) [136].

LDL receptor gene expression is controlled transcriptionally and post-transcriptionally. There is usually only one LDL receptor transcript in all types of cells. However during pregnancy, LDL receptors express in trophoblast in a particular way [137]. Two mRNA species are present: one is of usual length (5.3 kb), the other is smaller of 3.7 kb in length, at the expense of untranslated regions. This unusual example of differential gene expression for LDL receptors is probably the consequence of the high metabolic demand of the placenta. Hormones (in particular insulin, growth hormone, and steroid hormones) and certain growth factors or cytokines increase LDL receptor gene expression through several pathways of intracellular signaling, independent from sterol-mediated regulation [125].

However the most powerful regulation is mediated by intracellular sterol concentration [9]. Negative feedback regulation by sterols was one of the red lines followed by BROWN and GOLDSTEIN to identify the existence of LDL receptors. The LDL receptor gene promoter includes several binding sites for activating factors of the Sp1 type (a family of housekeeping transcription factors) in a region of 144 bp upstream from the initiation codon [138]. There are also regulatory sequence elements mediating sterol negative feed back: Sterol Regulatory Elements (SRE) [139]. These sequence elements are also found upstream of many other genes controlling intracellular energy homeostasis (see page 169). SREBPs binds to SRE to activate gene transcription in conjunction with other factors interacting with Sp1 elements. SREBP are resident membrane proteins of the endoplasmic reticulum (ER). When cellular levels of cholesterol drop below a certain threshold, SCAP (SREBP Cleavage Activating Protein) senses this change in ER cholesterol concentration and activates SREBP cleavage. The activated proteolytic fragment of SREBP detaches from the ER membrane and moves into the nucleus to induce LDL receptor gene transcription (*Figure 2-2*). When intracellular cholesterol (and 25-hydroxycholesterol) concentration increases, SRE remain unbound, turning down receptor synthesis to baseline levels. Thus, both functions of a transcription factor and of a sensor of intracellular sterol levels combine in the ER to mediate concerted regulations on all SRE upstream of genes controlling cholesterol homeostasis. SREBP gene targets include HMGCoA reductase, HMGCoA synthase, caveolin 1 and the LDL receptor. This regulation has major physiological and pathological consequences. LDL receptors are abundantly expressed in the liver. Any increase in dietary intake of cholesterol may suppress their expression thereby increasing plasma LDL cholesterol owing to reduced catabolism. This mechanism was verified in primates under a diet rich in cholesterol or saturated fat, the number of LDL receptors decreased on hepatocyte surface as the result of suppressed mRNA synthesis [140,141]. Negative feed back regulation by sterols also explains the dose effect of defective alleles of the LDL receptor gene in familial hypercholesterolemia. The normal LDL receptor allele in FH heterozygotes, in not overexpressed on cultured skin fibroblasts, to compensate for defective LDL endocytosis [142]. This indicates that appropriate levels of intracellular cholesterol are maintained over time by concerted regulation of several pathways, at the expense of LDL circulating in the blood stream. Negative regulation by sterols is the target mechanism accounting for the strong and specific LDL cholesterol lowering effects as well as the overall anti-atherogenic effects of HMGCoA reductase inhibitors (statins). The transient suppression of intracellular cholesterol *de novo* synthesis activates SREBP signaling, increasing the number of LDL receptors at the cell surface. Moreover, the mevalonate pathway generates essential components modulating the inflammatory, proliferative and cell-activating processes taking place in the atherosclerotic plaque. Therefore these molecules represent at the moment a specific preventive treatment of cardiovascular complications in FH [143].

The LDL receptor gene is highly **polymorphic**. Several flanking short tandem repeats and SNPs (Single Nucleotide Polymorphisms) have been identified in introns or as neutral mutations in the coding sequence [144,145]. By comparing allelic frequencies in 7 different populations for 1298 alleles, recombination rates could be estimated to be low within the locus, although more pronounced in *Alu* repeats containing regions [146]. Based on these estimations an optimal diagnostic strategy using these polymorphisms was defined. The indirect diagnosis based on linkage of LDL receptor gene polymorphisms was used in early studies of FH in several populations: in Finland [147], Germany [148], Israel [149], Italy [150], in Japan [151], United Kingdom [152] and in Switzerland [146]. By segregation analysis, it was possible to establish that the LDL receptor locus was the locus for FH in 60 to 95% of families. Remaining cases were either excluded (20 to 25%) or not informative at this locus. The analysis of polymorphic markers has identified a founder effect for LDL receptor gene mutations with a high prevalence in genetic isolates (see below) or the existence of recurrent mutations across different populations [136,153].

Naturally occurring mutations of the LDL receptor gene (LDLR), cause familial hypercholesterolemia. Allelic mutations are very heterogeneous from one subject or one population to another [154]. Several hundreds of different molecular defects have been identified worldwide. Databases accessible on the "world-wide-web" are regularly updated for novel LDLR mutations [145,155,156]. Mutations may occur on any of the functional domains of the gene (promoter, exons, splice sites and their flanking sequences). Their identification has contributed to the molecular dissection of LDLR structure-function relationships. The direct access to the primary cause of the disease had also significant diagnostic and therapeutic implications. The study of LDLR mutagenesis and of mutation frequency in human populations has increased knowledge on its evolution.

Five classes of functional defect [154] result from LDL receptor gene mutations *(Figure 2-2)*. *Class 1 defects* (or null alleles) correspond to an absence of receptor synthesis (absence of mRNA or protein). They generally result from large gene deletions, frameshift, nonsense, or splice site mutations. These mutations may remove transcription initiation sequences, or produce abnormal mRNA or proteins, quickly degraded before they reach the cell surface. *Class 2 defects* correspond to a defective maturation or transport of receptors within the endoplasmic reticulum and Golgi apparatus, leading to a slower rate of synthesis. They may result from specific mutations of highly conserved residues in the ligand binding and EGF homology domains. Receptors have an abnormal conformation that slows down their maturation more or less severely. In the Watanabe rabbit, a classical animal model of familial hypercholesterolemia, a 12-nucleotide in-frame deletion disturbs LDL receptor's conformation [58]. *Class 3 defects* correspond to defective binding of receptors to lipoproteins. They may disrupt binding to all lipoproteins or only to apo B or to apo E. These defects result from in-frame gene rearrangements, or missense mutations within the first two domains. *Class 4 mutations* produce endocytosis defective receptors. The receptor binds ligands normally but the complex does not enter the cell cytosol. These defects are rare and result from mutations located in the last three exons. The critical role of a Tyrosin residue at position 807 once suspected by naturally occurring mutations in humans with endocytic defective receptors was later confirmed by site directed mutagenesis in vitro. *Class 5 defects* correspond to recycling-defective receptors retained within endosomal membranes after endocytosis. Here again, as predicted by site directed mutagenesis experiments, these mutations are mainly found in exons encoding for the EGF homology domain. Conversely, data of site directed mutagenesis were not confirmed for deletions of exon 15, which encodes the entire O-glycosylation domain [157]. In vitro these modifications do not seem to cause a major defect whereas in vivo exon 15 mutations have been described in patients with moderate forms of familial hypercholesterolemia [147,158]. Moreover the VLDL receptor (see page 123), undergoes naturally an alternate splicing which removes this domain. The significance of the functional role of the O-glycosylation domain in vivo remains to be elucidated. Thus the concept of the LDL receptor gene consisting of a mosaic of exons each encoding for a precise function is confirmed by its natural mutagenesis [128,154].

The mechanisms favoring exon shuffling during the evolution were also suspected by the study of **major gene rearrangements** [9,128]. The non-coding sequences are rich in repeated *Alu* sequences. There are approximately 900 000 copies of these stretches of 300 nucleotides in non-coding domains of the human genome and in certain primates (gorilla, chimpanzee). They are particularly highly represented in chromosomal regions that are rich in genes: R bands [159]. Their role is not fully understood however during meiosis these sequences can pair and mediate interchromatid exchanges. When one examines more carefully the breakpoints of LDL receptor gene large rearrangements, they generally coincide with *Alu* repeats (*Figure 2-4*). The significant number of intronic *Alu* repeats could account for the heterogeneity LDLR gene rearrangements and for the gene structure organized in homologous blocks [129]. Similar cases of mutagenesis have been reported at other loci, supporting a possible role for *Alu* repeats in meiotic recombination [160]. This mechanism was shown for certain deletions or duplications. One 4kb deletion between exons 7 and 8 observed in a Dutch patient had its counterpart in an insertion of the same length in a patient from UK [161,162]. This strongly suggests the existence of unequal crossovers during meiosis, driven by repeated sequences located in introns 6 and 8. This mechanism may not be the only

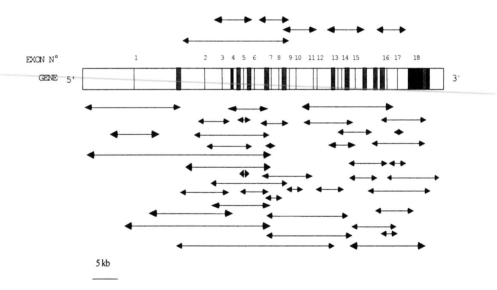

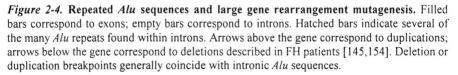

Figure 2-4. **Repeated *Alu* sequences and large gene rearrangement mutagenesis.** Filled bars correspond to exons; empty bars correspond to introns. Hatched bars indicate several of the many *Alu* repeats found within introns. Arrows above the gene correspond to duplications; arrows below the gene correspond to deletions described in FH patients [145,154]. Deletion or duplication breakpoints generally coincide with intronic *Alu* sequences.

one involved in the generation of large gene rearrangements. For example, the Helsinki mutation [163] does not result from such a recombination. In prokaryotes genetic defects known as "mutators" and " antimutators " may control mutation rates [164], some of them predisposing to short repeated sequences mismatch. Similar mechanisms may be involved in human mutagenesis [165]. It is not known whether they contribute or not, to large gene rearrangements of the LDL receptor gene. However large gene rearrangements do not represent the majority of LDL receptor gene mutations, not exceeding 5 to 8% of mutations causing familial hypercholesterolemia in panmictic populations [154].

The spectrum of LDLR mutations includes several hundreds of gene defects. Some occur on CpG dinucleotides, which are hot spots for recurrent mutations [166]. Cytosines may be methylated at these dinucleotides (*figure 2-5*). They can undergo a deamination, inducing a nucleotide substitution for a thymine, so that the DNA repair machinery is deluded by this transition. This mechanism could account for approximately 25% of naturally occurring genomic mutations. These mutation hotspots explain cases of recurrent mutations. Mutations have been identified in unrelated subjects, carrying the same nucleotidic variation at the same codon, however they carry a different set of flanking polymorphic markers, or haplotype at the gene locus. This indicates that mutagenic events occurred independently in different populations, defining a recurrent mutation. Cases of recurrent mutations have been described [153,154,167,168]. For example, the Glu207Lys mutation is worldwide recurrent, and results from a transition on a CpG dinucletide [145]. Other mutagenic mechanisms have been described. Small rearrangements involving one or several nucleotides could result from DNA slippage between homologous sequences in the vicinity of insertion or deletion breakpoints [169]. Thus LDL receptor gene mutations are numerous

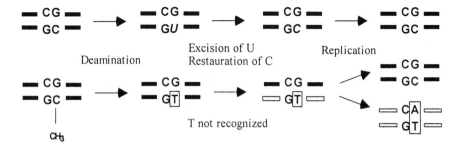

Figure 2-5. **Mutagenesis on CpG dinucleotides.** Nuclear DNA is frequently methylated on cytosines in CpG dinucleotides. A 5-methyl-cytosine deamination gives rise to a thymine, which remains unrecognized by the DNA repair machinery. The following replication cycles will stabilise the transition.

and highly heterogeneous from one individual and one population to another. Because of high conservation of gene sequences and of secondary structure, one can predict that nearly any amino acid of the coding sequence could be a possible site for FH mutations.

In certain populations representing genetic isolates specific mutations are highly prevalent (1/100 to 1/200). The high level of inbreeding has favored the spreading over generations of a small number of mutations, which may cause more than 90% of cases with familial hypercholesterolemia as the result of a **"founder effect"** [170-175]. It is the case in Finland (3 deletions: exons 7 to 10, exon 15, exons 16 to 18); in Afrikaners of Dutch origin in South Africa (Asp154Asn, Asp206Glu, Val408Met); in French Canadians (2 deletions: 5' upstream region to exon 1, exons 2 and 3, and point mutations: Trp66Gly, Glu207Lys, Tyr468Stop, Cys646Tyr); in Sephardic Jews (Asp147His) or Ashkenazy Jews (deletion of Glycin at 197); in certain populations of North Africa (complex frameshift mutation in exon 10, in central Tunisia) and of Middle-East (Cys660Stop, in Lebanon). Founder mutations allowed dating some of them back to at least several centuries ago [176]. LDL receptor gene mutations are however highly heterogeneous in the majority of human populations [177-181]. Known mutagenic mechanism do not explain the frequency of the disease worldwide (1/500 on average) [9,154]. The rate of neomutations is not precisely known, and mutations in flanking or noncoding gene sequences are seldom described [154,145]. In FH, mutations prevent neither normal development, nor reproductive fitness (homozygous women may carry out normal pregnancies and delivery). Therefore FH mutation events were allowed to drift up to a significant frequency across the evolution of human populations. We will discuss further the implications of such observations for the comprehension of the genetic basis of multifactorial and late-onset diseases, which are the hallmark of dyslipemia (see chapter III).

Despite the great allelic heterogeneity of mutations, significant clinical implications arose from their identification. **The variability of disease expression** depends on the type of molecular defect in homozygotes and to some extent in heterozygotes [108]. Subjects carrying mutations that completely abolish expression (null alleles or class 1 defect), generally have a more severe disease (higher plasma LDL cholesterol levels, CAD more frequent) with earlier onset than those carrying mutations for which a residual receptor activity is preserved (defective alleles) [154,182,183]. This has also been observed for the severity and prematurity of coronary arterial disease [184]. However minor mutations may be associated with almost normal plasma cholesterol in heterozygotes [154,177,185]. Moreover, response to lipid-lowering drugs may vary according to the type of allele [186-188]. Studies have suggested that the type of mutation could contribute for a significant proportion (20-30%) of the variance of the response to lipid-lowering drugs.

However, if the type of mutation determines most of the phenotype in homozygotes, in heterozygotes **disease expression is more variable** depending on diet or on epistatic effect of other genes. For example in several populations who consume diets low in saturated fat (China, Tunisia, Cuba), FH heterozygotes may have plasma cholesterol concentrations lower than in western countries and delayed or absent cardiovascular complications [189-191]. Genetic factors may also modulate the expression of LDL receptor mutations (see page 212). A pioneer example was that of a large Porto-Rican family reported by Hobbs et al. [192] in which a cholesterol-lowering effect of an unknown gene compensated for the effect of a Ser156Leu FH mutation. In vivo LDL kinetic studies in this family showed that this gene decreases the concentration of LDL precursors, and increases LDL clearance [193]. A locus for a cholesterol-lowering gene has been recently localized on chromosome 13q, however, the nature of the gene and its functions remain unknown [194]. A case with double heterozygosity for hypobetalipoproteinemia and a LDL receptor gene mutation led to normal levels of plasma cholesterol [195]. Conversely double heterozygotes for FH and Familial Defective apo B-100 (see below) have enhanced hypercholesterolemia by additive effects of two plasma LDL cholesterol-raising mutations. Interestingly, the phenotype was less severe than that of FH homozygotes [196,197]. Common gene variants of apo E or LPL may shift the lipoprotein profile in FH heterozygotes towards higher risk profiles [198,199].

The identification of genetic mutations of the LDL receptor has contributed to understand their pathogenic effect and to define the place of this disease among other forms of dominantly inherited hypercholesterolemia. The identification of this important pathway of LDL cholesterol homeostasis has contributed to the development of targeted therapies represented by HMGCoA reductase inhibitors [200]. These drugs are able to change the natural course of the disease in heterozygotes by preventing premature cardiovascular complications. In homozygotes, the extreme severity of the disease could justify efforts of gene therapy. Issues opened by the first experiments in this domain will be discussed in the next chapter.

Molecular genetics of familial hypercholesterolemia represent a paradigm, which has opened the way to other studies on the genetic basis of hyperlipidemia and to the genetic diagnosis of monogenic dyslipidemia in common practice.

2.1.2 -   Apolipoprotein B, a Single Locus for Opposite Diseases

Apolipoprotein B (apo B) plays a major role in the regulation of lipoprotein metabolism. This apolipoprotein, which distinguishes itself in many points from other apolipoproteins, had remained a mystery because of its high molecular weight and of its great hydrophobicity [201]. The first monoclonal antibodies directed against apo B opened the way to gene cloning, which was carried out simultaneously by several groups [202-204]. Essential knowledge in physiology and pathology has emerged from these studies.

Apolipoprotein B is a large glycoprotein with a molecular weight of 549kD, whose primary sequence comprises 4536aa. It serves as a structural backbone (one molecule per lipoprotein particle) for the assembly and secretion of lipoproteins, and as a regulatory protein for their catabolism (lipoprotein-receptor interaction) [205]. Its sequence comprises several β-sheet domains and α-helices weakly homologous to those of exchangeable apolipoproteins, however tightly bound to lipids and necessary for intracellular assembly of lipoproteins. These domains present homologies with invertebrate vitellogenins [206]. Several thrombin-cleavage sites are found on the primary sequence. Anchoring domains for carbohydrate chains are exposed in VLDL *(figure 2-6)*. They interact with heparin and heparan sulfate proteoglycans, favoring their binding to the endothelium and

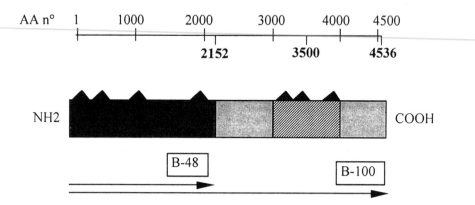

Figure 2-6. **Domain map of apolipoprotein B.** Top: a scale of amino acid position. Bottom: arrows indicate the length of apo B-48 (intestinally derived) and of apo B-100 (hepatically derived). Hatched area includes the receptor-binding domain. ▲: Heparan sulfate binding sites.

the hydrolysis of core triglycerides by lipases [207]. The resulting core enrichment in cholesteryl esters and the reduction in size of the particle, induce apo B conformational changes that enhance its interaction with LDL receptors [208]. Epitopes of 40 monoclonal antibodies purified by several groups, have delineated a C-terminal domain located between residues 2835 and 4189, which would include the binding site for LDL receptors [209].

The gene for apolipoprotein B (APOB) also has a singular structure. It is unique, localized at the short arm of human chromosome 2, 2p24 [210]. Apo B mRNA is mainly expressed in the liver and the intestine. It is also expressed at lower levels in heart myocytes, which could represent a way for myocytes to unload excess fatty acids no longer required for fuel [211]. Apo B transcript is very long, 14kb, a length comparable with that of transcribed dystrophin, 16kb [212]. The gene is about 43kb long, organized in 29 exons and 28 introns [213]. The originality holds in the fact that the 25 first exons are of small size (175bp on average), whereas exon 26 is 7572bp long and exon 29 coding sequence is 1906bp long *(figure 2-7)*. This 3' domain has a crucial role, since it encodes for the receptor interacting domain and for regulatory sequences of apolipoprotein B intestinal maturation (see below).

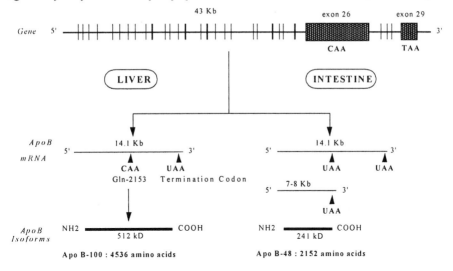

Figure 2-7. **Apo B gene organization and editing**. Top: exon/intron gene organization. Bottom: liver and intestine apo B editing. A cytidine deamination modifies codon CAA of the genomic sequence into UAA in apo B mRNA, generating a STOP codon. Alternative polyadenylation sites also generate shorter mRNA species. The protein translated in the intestine (apo B-48) will be only 2152aa long, representing 48% of the full-length apo B-100 (4536aa) expressed in the liver.

Regulatory sequences upstream from apolipoprotein B transcription start site are not methylated in the liver and in the intestine, which confirms apo B tissue specificity. These unmethylated domains could be exposed in the nucleus, and would form two loops anchored to the nuclear matrix: one corresponding to 2kb of regulatory sequences, the other loop 45kb long corresponding to the protein encoding region [214]. Promoter sequence elements interact with several nuclear factors including some with liver and/or intestine specificity, also found upstream of other apolipoprotein genes [215]. Factors belonging to the family of steroid hormone nuclear receptors: ARP-1, (Apolipoprotein AI Regulatory Protein-1), EAR2 and EAR3, (v-ErbA Related human receptor) interact 2kb upstream of the transcription initiation site. Certain regulatory sequences do not have the same effect in hepatocytes or enterocytes, depending on the nature and the relative abundance of other nuclear factors [216-219]. Moreover sequences binding to hepatic transcription factors HNF-1 and HNF-2 (Hepatic Nuclear Factor) are found within introns 2 and 3 enhancing apo B transcription [220-221]. Thus apo B expression is finely controlled, though in complex and variable ways depending on its tissue specificity. For example, transgenic mice overexpressing a genomic fragment spanning the entire gene (including 19kb upstream of the coding sequence and 17.5kb downstream), express apo B mRNA and protein in the liver, while remaining undetectable in the intestine [222]. The intestine specific regulatory sequences have been later found to reside between 33kb and 70kb upstream of apo B transcription start site [223].

Cholesterol is not a main regulator of apo B gene expression [141]. Cholesterol-enriched diets do not change the rate of hepatic apo B transcription whereas that of LDL receptors is lowered by 50%, accounting for most of the plasma LDL-raising effect. Apo B intestinal transcription is however increased by 30%, as the result of an increased synthesis of apo B-48 necessary for chylomicron assembly.

Indeed the apolipoprotein B gene encodes for two distinct isoforms: apolipoprotein B-100 synthesized by the liver and apolipoprotein B-48 synthesized by the intestine in human [224]. Apolipoprotein B-48 protein sequence corresponds to the first 2152 amino acids, N-terminal of full-length apolipoprotein B (or 48% of apo B-100 total mass, or 4536aa). Its synthesis results from an original post-transcriptional mechanism, named "**RNA editing**" [225]. In the intestine, an in frame C→U substitution at position 6666 of the mRNA, changes a codon Gln (CAA) for a STOP codon (UAA), inducing translational arrest at position 2153 (*figure 2-7*). This transition is absent from genomic DNA. The mechanism is post-transcriptional, taking place together with splicing and polyadenylation. A multicomponent enzyme-complex ApoBEC (apo B Editing Complex) deaminates cytidine 2153 to transform it into uracil. This reaction is highly controlled by specific interactions of these proteins with native apo B mRNA. Sequences flanking both sides of Cytidine [6666] are highly conserved across evolution [226]. A necessary "mooring" sequence of 11 nucleotides (6671 to 6681) and alternate polyadenylation sites have been identified downstream of Cytidine[6666]. Complementation experiments have allowed the molecular cloning of APOBEC-1 (Apo B Editing catalytic Component-1), the 229aa catalytic subunit, containing phosphorylation sites and "leucine-zipper" motifs [227,228]. This protein contrary to bacterial cytidine deaminases, restores apo B editing in chicken ovocytes (which do not edit apo B naturally), requires interaction with other stimulatory factors [229]. Apo B editing may be controlled by hormones, several dietary components such as fatty acids or ethanol [230], and could change along with development and aging [231]. Apo B editing must be finely tuned. Transgenic mice or rabbits overexpressing high levels of APOBEC-1 may develop liver dysplasia and carcinoma [232]. Protein editing seems universal: mRNA of chloroplasts in plants, mitochondrial tRNA in marsupials, viral RNA (hepatitis virus δ, the HIV "Tat" protein), glutamate-dependent neuronal calcium channels, WT-1 gene of susceptibility to Wilms tumour, undergo protein editing.

This mechanism is species-dependent. Whereas murine species and many mammals (horse, rabbit, dog) express apo B-48 at similar levels in liver and in intestine, only several primates and humans express Apo B-48 only in intestine. Transgenic mice for human apo B perform a less efficient editing than on murine apo B [222]. Chylomicrons with apo B-48 as a molecular backbone are secreted directly in the general circulation, which bypasses the liver for rapid peripheral tissue delivery of dietary triglycerides. Apo B-48 is deprived from the domain interacting with LDL receptors. Whereas fatty acids derived from apo B48-containing lipoproteins may be used by peripheral tissues, cholesterol will return back to the liver where it is taken up by apo E-specific receptors. Apo B editing seems to have coincidently appeared with apo E expression across the evolution of animal species [225]. Birds are deprived both of apo E and of apo B editing. Moreover mammals, which use HDL as the main pathway for cholesterol transport, edit apo B-48 in the liver, whereas those lacking apo B editing in liver use LDL as major cholesterol transporters [233]. Indeed in the latter species triglyceride-rich lipoproteins produced by the liver contain only apo B-100, the end product being LDL, with a longer residence time in plasma. Species editing apo B in liver produce endogenous lipoproteins, which use a short cut back to the liver. Alternate expression of apo B editing was demonstrated to result from additional regulatory elements driving liver expression of ApoBEC-1 in species editing Apo B in the liver. Thus, stimulation apo B editing in the liver has been considered as a possible target to prevent the formation of atherogenic lipoproteins, by reorienting cholesterol transport towards HDL [232].

Apo B mRNA has a prolonged half-life within hepatocytes (16 hours), which is not in favor of rapid transcriptional regulation [234]. Another modulatory mechanism of protein expression was identified for apo B. The native protein undergoes a particular type of cistern transfer towards the endoplasmic reticulum [235]. Apolipoprotein B when delipidated, is highly hydrophobic and rapidly precipitates in water [236]. Soon after the beginning of translation the signal peptide guides the entry of the native protein towards the endoplasmic reticulum. Along with protein chain elongation, protein translocation occurs stepwise: the pause-stop protein transfer. At intermediate stages of protein translation, the native protein would form loops inside microsomal membranes without integrating the lipid bilayer. This mechanism would leave the possibility for the protein to from a backbone able to bind lipids and to stabilize a water-soluble micellar complex, passing gradually into the lumen of the endoplasmic reticulum [124]. It seems that there are several intracellular pools of native lipoproteins [237]. A microsomal protein, MTP, ensuring lipid transfer within the ER lumen, towards the core of native lipoproteins, was identified as being defective in abetalipoproteinemia (see page 67) giving a molecular basis to a disorder of apo B-containing lipoprotein assembly. Translocation of native apolipoprotein B may be interrupted, thereby exposing the protein to ubiquitination and degradation into the cytosol by the proteasome [238,239]. Conversely a microsomal sterol-sensitive block in ubiquitin conjugation reverts apo B translocation arrest and prevents it from proteasomal degradation [240]. Thus translational regulations operate by a control on apo B translocation: they could accelerate the degradation of the protein whose transfer is in "pause" or on the contrary accelerate apo B transfer to hasten mature lipoprotein secretion.

Apo B synthesis and maturation are closely related with lipoprotein assembly. All components of apoB-rich lipoproteins (lipids, apolipoproteins) are synthesized synchronously [215]. In vitro, the synthetic rate of cholesteryl esters, or triglycerides derived from fatty acids and glycerol, drives the production of apoB-rich lipoproteins. However they do not act in the same way. For example, glucose stimulates the secretion of VLDL by increasing triglyceride synthesis, whereas oleate stabilizes native particles by slowing down intracellular degradation of apo B. Phospholipids also modulate the rate of assembly and the number of secreted particles [241]. In contrast, small exchangeable apolipoproteins do not seem to interfere. The type of fatty acid and the source, from which it reaches the liver, has differential effects on apo B rich lipoprotein synthesis and secretion [242]. These properties have therapeutic implications. Dietary unsaturated fatty acids, particularly omega 3 fatty acids slower lipogenesis, thereby reducing liver production of lipoproteins. Late maturation mechanisms also modulate lipoprotein secretion at the stage of native particle transfer between the endoplasmic reticulum and the Golgi apparatus where further glycosylation takes place. Apo B-rich lipoprotein assembly is thus a very complex process, finely controlled at multiple stages. Its elucidation has opened interesting prospects in the molecular basis for dyslipoproteinemia and in drug development, to decrease hepatic or intestine lipoprotein production.

Apolipoprotein B is polymorphic. This polymorphism was discovered in the early 60s by the presence of antibodies directed against LDL in plasma from blood-transfused subjects [243]. Bütler & al. [244] were described five allotypes Ag(c/g), Ag(a₁/d), Ag(x/y), Ag(h/i), and Ag(t/z), later related with gene polymorphisms of the apo B (*table 2-2*). An extensive study of these markers in 1464 subjects originating from 13 different populations showed that this system was neutral in population evolution [250]. Moreover, a comparison between these haplotypes and those observed in hominid monkeys confirmed the "out of Africa" hypothesis for human origin, and the consecutive migration of populations towards other continents [251] in agreement with previous observations carried out with others polymorphic protein systems [252]. Another interesting application to these findings is that the Ag(c/g) polymorphism may be detected by monoclonal antibody MB19 [253]. As opposed to small apolipoproteins, apolipoprotein B is not exchangeable between lipoprotein species. The study of heterozygotes with this antibody detecting either of apo B isoforms has highlighted the allelic specificity of circulating rates of apo B containing lipoproteins. It was possible to trace two pools of apolipoprotein B in circulating LDL: one, of whose the circulating rates correspond to the maternal allele, the other determined by the paternal allele [254]. It was shown that each apo B allele determined either differential synthetic or catabolic rates when detected by antibody MB19. However the mechanism accounting for these differences could not be elucidated. Moreover, in familial defective apo B-100 (see below), this polymorphism provided evidence that relative proportions of

Table 2-2. Protein polymorphisms of apolipoprotein B.

Antigen	AA substitution	NT #	Exon #	NT substitution	RFLP	Allele	Ref.
c/g	Ile71 Thr	421	4	ATC→ ACC	Apa LI	1/2	245
a₁/d	Val591 Ala	1981	14	GTT→ GCT	Alu I	1/2	246
x/y	Leu2712 Pro	8344	26	CTA→ CCA	Mae I	2/1	247
h/i	Gln3611 Arg	11041	26	CAG→ CGG	Msp I	1/2	248
t/z	Glu4154 Lys	12669	29	GAA→ AAA	Eco RI	2/1	249

plasma LDL particles containing the wild type (Arg3500) or the mutant allele (Gln3500) of apo B were 0.3/0.7 respectively, as the result of slower catabolic rate of the mutant allele [255,256]. Finally the use of MB19 antibody in hepatic transplant recipients showed that after a fat meal, triglyceride-rich lipoproteins containing apo B-100 were of hepatic origin, confirming the specificity of hepatic secretion of apo B-100 and of intestinal expression of apo B-48 in humans [257].

Apolipoprotein B gene is also highly polymorphic [258], in addition to protein polymorphisms, there are many informative SNPs and short tandem repeat sequences (STRs) along its sequence *(table 2-3)*. The combination of these markers gives a large number of theoretically possible haplotypes, providing with extended possibilities of genotyping. Ludwig & Mac Carthy [259] have found complete informativity of 22 haplotypes constructed with these polymorphic markers in 8 North American families. These markers are in strong linkage disequilibrium in particular in the 3' sequences of the gene, which are poorer in introns than 5' sequences [260]. Biometric studies in large families or in twin pairs have shown the implication of major loci controlling plasma lipids and lipoproteins [261,262]. Numerous association

Table 2-3. **Apo B gene polymorphisms.** This table is not exhaustive, for a complete description see refs 245,247,249,260,265-270. NC: non-coding. STR: short tandem repeat.

Marker	Location	AA change	Allele Frequency
Ava II	-4kb, 5' NC	-	0.20
(TG)$_n$	-3256nt, 5' NC	-	STR, 7 alleles
Msp I	-265, promoter	-	0.21
Ins/del 9bp	exon 1, signal peptide	del Leu^{-16}-Ala-Leu^{-14}	0.43
Int3 G/T	intron 3, G$\rightarrow$T	-	0.47
Apa LI	exon 4, Ag(c/g)	Thr^{7}Ile	0.40
Hinc II	intron 4, 3' ex 4	-	0.14
Pvu II	intron 4, 5' ex 5		0.07
Alu I	exon 14, Ag(a$_1$/d)	Ala^{591}Val	0.47
Bal I	intron 20, 5' ex 21	-	0.50
(TTTA)$_n$	intron 20, *Alu*	-	STR, 7 alleles
T$\rightarrow$C	exon 26	Neutral Asp^{2285}Asp	-
Xba I	exon 26	Neutral Thr^{2488}Thr	0.46
Mae I	exon 26, Ag(x/y)	Leu^{2712}Pro	-
Msp I	exon 26, Ag(h/i)	Arg^{3611}Glu	0.11
Eco RI	exon 29, Ag(t/z)	Glu^{4154}Lys	0.20
A$\rightarrow$G	exon 29	Asp^{4311}Ser	0.21
HVR	3' NC	-	STR, 17 alleles

studies have used apo B polymorphisms to test whether it could be a candidate gene for common metabolic disorders and atherosclerosis [263]. Polymorphic markers of apo B do not induce detectable functional mutations by themselves except for the insertion/deletion polymorphism of the signal peptide, which could decrease apoB secretion rate [264]. Studies have included numerous populations of diseased or healthy subjects: coronary artery disease, peripheral arteriopathy, plasma levels of LDL, of apo B etc (for review see [263]). If results have been consistent to identify associations with plamsa lipid levels, they failed to show associations with any forms of atherosclerosis. Moreover, in subjects carrying the "susceptibility" allele of the neutral polymorphism XbaI, for high plasma LDL cholesterol and apo B, a systematic screening of possible genetic variations was performed by SSCP, completed with genomic sequencing of domains encompassing this polymorphism [271]. No mutation was found. However, the possibility of linkage disequilibrium with other functional genomic variations is not excluded. As has been demonstrated by studies in transgenic mice, elements located more than 33kb apart from apo B transcription start site have a strong regulatory effect on intestinal gene expression [223]. Any variation in remote chromosomal regions could modulate levels of circulating apo B-rich lipoproteins. However, these functional variations are still unkown. It is also possible that preferential associations have no physiological relevance, but result from small size or selection bias of the study population.

Naturally occurring mutations of the apolipoprotein B gene are associated with two opposite forms of inherited dyslipidemia: familial hypobetalipoproteinemia and familial defective apo B-100. Their study has contributed to identify the physiological role of apolipoprotein B in humans.

Hypobetalipoproteinemia is a disease with autosomal dominant inheritance characterized by low levels of plasma LDL cholesterol and apolipoprotein B. Its prevalence was estimated to be 0.8% based on plasma apo B measurements [272]. Homozygotes present with a serious disease characterized by intestinal malabsorption and neurological disorders [273]. Heterozygotes are usually asymptomatic although their plasma LDL cholesterol and apo B are lowered down to half of normal. Linkage analyzes had pointed the apo B gene as the predisposing locus [274]. Several gene mutations (*table 2-4*) were identified on the apo B gene causing familial hypobetalipoproteinemia. They are nonsense or frameshift mutations expressing as abnormal truncated proteins circulating in plasma [109,273,302]. The shortest may separate in the HDL fractions, while heavier forms are found in VLDL and LDL. These "mini" apo B are named after the ratio of the size of the truncated protein to full-length apo B: B-27.6, B-31, B-37, B-39 etc. They are still able to bind lipids and to form the

Table 2-4. **Apo B gene mutations causing hypobetalipoproteinemia.** (-) no apoB in plasma. Abnormal apo B found in HDL (H), VLDL (V) or LDL (L) lipoprotein fraction.

AA Position	Location	Mutation	Expression	Type	Ref
-	Intron 5	G→T	(-)	5' splice	267
412	Exon 10	C→T, 412	B 9 (-)	Nonsense	267
1085	Exon 21	Δ exon 21	B 25 (-)	*Alu*-repeats	275
1254	Intron 24	C→T	B 27.6 (H)	5' splice	276
1306	Exon 25	C→T, 4125	B 29 (-)	Nonsense	277
1425	Exon 26	Δ1 nt, 4480	B 31 (H)	Frameshift	278
1450	Exon 26	C→T, 4548	B 32 (H)	Nonsense	279
1474	Exon 26	T→G, 4631	B 32.5 (H)	Nonsense	280
1728	Exon 26	Δ4 nt, 5391	B 37 (HVL)	Frameshift	281
1745	Exon 26	Δ1 nt, 5444	B 38.9 (H)	Frameshift	282
1755	Exon 26	C→T, 5472	B 38.7 (VL)	Nonsense	283
1768	Exon 26	Δ14 nt	B 38.95	Frameshift	284
1795	Exon 26	Δ1 nt, 5591	B 39 (VL)	Frameshift	277
1829	Exon 26	Δ2 nt, 5693	B 40 (HVL)	Frameshift	285
1986	Exon 26	C→T, 6162	B 43.7 (HV)	Nonsense	286
2014	Exon 26	Ins 11 nt	B 44.4	Frameshift	287
-	Exon 26	C→T	B45.2 (V)	Nonsense	288
2058	Exon 26	C→T, 6381	B 46 (HVL)	Nonsense	289
2166	Exon 26	Δ1 nt, 6627	B 48.4 (HVL)	Frameshift	290
2251	Exon 26	C→T, 6963	B 50 (V)	Nonsense	291
2355	Exon 26	Δ5 nt, 7273	B 52 (L)	Frameshift	292
2362	Exon 26	Δ1 nt, 7295	B 52.8 (VL)	Frameshift	280
2384	Exon 26	Δ1 nt, 7359	B 52.8 (VL)	Frameshift	280
2486	Exon 26	C→T, 7665	B 54.8 (VL)	Nonsense	293
2492	Exon 26	C→T, 7692	B 55 (VL)	Nonsense	294,287
2772	Exon 26	Δ37 nt, 8525	B 61 (VL)	Frameshift	295
3040	Exon 26	Δ1 nt, 9327	B 67	Frameshift	296
3184	Exon 26	Ins 1nt, 9760	B 70.5 (VL)	Frameshift	297
3387	Exon 26	Δ1 nt, 10366	B 75 (VL)	Frameshift	298
-	-	-	B 76	-	293
3734	Exon 26	C→A, 11411	B 82 (VL)	Nonsense	280
3750	Exon 26	C→A, 11458	B 83 (V)	Nonsense	299
3896	Exon 26	Δ1 nt, 11840	B 86 (VL)	Frameshift	300
3978	Exon 29	Δ1 nt, 12032	B 87 (VL)	Frameshift	301
4039	Exon 29	Δ1 nt, 12309	B 89 (VL)	Frameshift	285

backbone of circulating lipoprotein particles. These natural mutants seem to behave in the same way as those induced by site directed mutagenesis. An artificial apo B-17 expressed in liver cell lines is still able to bind lipids and to form discoidal particles, which are secreted in the culture medium [303]. Moreover the study of mutants of larger size (B-60 to B-94) have shown that there is a linear relationship between the lipid content of the secreted lipoprotein, its diameter, its density and the length of apo B [304]. This confirms the structural role of apolipoprotein B, in apo B-rich lipoproteins. When mutations occur beyond nucleotide 6666, the abnormal apo B and normal apo B-48 may circulate in plasma [285]. A lower lipid content and secretion rate of apoB rich lipoproteins explain hypolipidemia, but it is not the only mechanism. A particular behavior was noted for truncated proteins generated by mutations of the C-terminal domain: they have increased affinity for cellular receptors, which accelerates LDL catabolism and further decreases plasma LDL [305]. This supports a conformational role for C-terminal domains in ligand-receptor interactions.

In homozygous familial hypobetalipoproteinemia the disease is more severe and its onset earlier when apo B is undetectable in plasma [267,275]. Homozygotes who express longer forms of truncated apo B in their plasma seem less affected. They may present with chronic diarrhea and certain women could have normal pregnancy and child delivery despite a partial steroid-hormone deficiency [306]. Heterozygotes for familial hypobetalipoproteinemia are generally asymptomatic, however patients may have steatosis and increased biliary synthesis. They may present with longevity and delayed onset atherosclerosis [307]. Therefore a lipid-lowering therapy, which would mimic heterozygous hypobetalipoproteinemia would confer protection against atherosclerosis. Interestingly, no apoB missense mutations werer reported in familial hypobetalipoproteinemia, and more particularly in the domain comprising amino acids 3300 to 3700 *(figure 2-8)*. The only mutation found in this domain corresponding to the LDL receptor binding domain is the cause of an opposite phenotype: **familial defective apolipoprotein B-100 (FDB)** [308-311]. A G→A mutation substitutes a glutamine for an arginine at amino acid position 3500, which disrupts apo B-100 binding to LDL receptors, resulting in hypercholesterolemia.

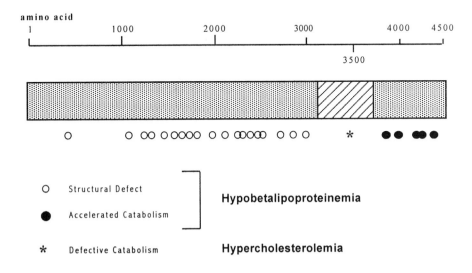

Figure 2-8. **Naturally occurring mutations of the apo B gene.** Depending on their location they may cause opposite phenotypes: hypobetalipoproteinemia or hypercholesterolemia.

The clinical presentation appears similar to that of heterozygous familial hypercholesterolemia resulting from LDL receptor defects (*Table 2-5*).

Table 2-5. Comparison of characteristics of two dominantly inherited hypercholesterolemias: **Familial Defective apo B-100 and Familial Hypercholesterolemia** [108,308-311]

Characteristics	Familial Defective Apo B-100	Familial Hypercholesterolemia
Frequency	1/200 – 1/1200	1/200 - 1/500
Inheritance	Autosomal Dominant	Autosomal Dominant
Chromosome	2	19
LDL Receptor	Normal	Deficient
Apo B-100	Deficient	Normal
Mutations	Arg3500→ Gln ± few others	Several hundreds (>700)
Cholesterol deposits	(age >18)	(age >18)
Tendinous Xanthoma	30-40%	33%
Atherosclerosis (age >40)		
Coronary	20-40%	> 50%
Peripheral	+	+
Hyperlipidemia (age >18)	IIa	IIa
Total Cholesterol (mean)	~300 mg/dL	~370 mg/dL
LDL Cholesterol (mean)	~235 mg/dL	~290 mg/dL
Variability	+++	+
Lipid-Lowering Response	++	++
Diet + lipid-lowering drugs		

Dominantly inherited type IIa hypercholesterolemia is detectable from childhood and may complicate in adults with extravascular cholesterol deposits (tendinous xanthoma, corneal arcus). Cardiovascular complications are 7 times more frequent in comparison with age and sex matched subjects of the general population [312]. A significant lipid-lowering response is obtained with LDL-cholesterol lowering drugs [311]. Although the clinical presentation of FDB is similar with that of FH caused by LDL receptor mutations there are some distinctive features. Plasma total and LDL cholesterol levels may be more variable within and between individuals and lower on average even in childhood [313,314]. Cardiovascular events are slightly delayed by about 5 to 10 years [309-312]. Moreover, homozygous FDB is surprisingly associated with moderate hypercholesterolemia, in contrast with homozygous FH, in which severe hypercholesterolemia is a hallmark. It was precisely documented in one case [315]. Kinetic studies of lipoproteins have shown that clearance of LDL precursors is maintained in these subjects, in particular through the binding of apo E to peripheral or liver receptors (LDLR, LRP) *(figure 1-11)*. However defective catabolism of LDL led to plasma accumulation of dense LDL. These lipoproteins are more susceptible to oxidation and thus more atherogenic, in agreement with clinical observations in heterozygotes [316].

Because the FDB mutation was identified as unique the screening of the Arg3500Gln mutation was easily undertaken throughout the world. The mutation is mainly found in populations originating from central Europe. The prevalence of the disease has been estimated between 1/200 in Switzerland to 1/1200 in Denmark in the general population. FDB could account for approximately 3% of patients with primary type IIa hypercholesterolemia. However this mutation was not detected in Norway, Finland, Russia, Israel, Japan and it seems rare in Italy.

This mutation occurs on a CpG dinucleotide, a known hotspot for recurrent mutations, (see page 51*)*. Cases of recurrent mutations were described in Europe and China [317-319]. However, in most individuals, the mutation has spread in European populations as the result of a founder effect. Indeed, a unique haplotype constructed with 10 apo B polymorphic markers was associated with this mutation in populations or European ancestry [259]. The fact that the mutation is absent in Finland and its associated founder effect elsewhere in Europe suggests that this mutation occurred rather recently. Strikingly, this prevalent mutation appears as a single cause for FDB in contrast with the several hundreds of mutations described on the LDL receptor gene causing FH. Another mutation disrupting the same codon, the Arg3500Trp (C→T) mutation has been reported in Scotland, Germany and Asia, though it is rare [311,320]. Other mutations have been described in the receptor-binding region of apo B or in its close vicinity: an Arg3531Cys and a Glu3405Gln mutation. The functional and pathogenic significance of these mutations remains debated [311,312]. More recently, an interesting observation was made from site directed mutagenesis experiments. Transgenic mice overexpressing recombinant apo B modified on basic amino acids between residues 3359 and 3369 are defective in binding to receptors and to proteoglycans [321]. Mice had a delayed LDL clearance and hypercholesterolemia however LDL failed to bind to heparan sulfate proteoglycans. They could represent an interesting animal model in which atherosclerosis could be prevented by a lack of retention of lipoproteins within the extracellular matrix. If mutations of this kind were to be found in humans, it would appear as a model of primary hypercholesterolemia, but somewhat associated with resistance to atherosclerosis.

Molecular genetics of apolipoprotein B have shown that this locus plays a central role in lipoprotein assembly and catabolism particularly in humans.

2.1.3 - Microsomal Triglyceride Transfer Protein (MTP), a Locus for Abetalipoproteinemia

In 1950 Bassen and Kronzweig described the case of a brother and sister born from consanguineous union, presenting with a syndrome associating a spino-cerebellar degeneration, retinitis pigmentosa, peripheral neuropathy, chronic diarrhea and abnormally shaped erythrocytes (acanthocytes) [322]. **Abetalipoproteinemia** was later identified as a recessively inherited disorder characterized by the absence of intestinal and hepatic secretion of apo B containing lipoproteins in plasma [273]. Total cholesterol, LDL cholesterol and triglycerides are extremely low in plasma, most of cholesterol esters being transported in HDL. Apo B-rich lipoproteins are not detectable in plasma. Fat malabsorption explains diarrhea and acanthocytosis. Malabsorption of vitamins A, E, and K contribute neurological complications. On tissue biopsy, liver and intestine present with cytoplasmic lipid accumulation leading to liver steatosis, which may evolve into cirrhosis. Contrary to homozygous hypobetalipoproteinemia, in which only few apo B mRNAs are detected, apo B mRNAs are abundant on hepatic or intestinal biopsies from patients with abetalipoproteinemia the [323,324].

Although apolipoprotein B is absent in plasma, segregation analysis had shown that abetalipoproteinemia was not resulting from a primary defect of the apo B gene [325]. A previously unknown defective protein was identified [326]. **Microsomal triglyceride transfer protein** or MTP, allows triglyceride enrichment of the lipid core of lipoproteins during their assembly in hepatocytes and enterocytes. This protein is a heterodimer comprising a disulfide isomerase, PDI (Protein Disulfide Isomerase) of 58kD and a larger sub-unit of 97 kD associated with PDI (*Figure 2-9*). PDI is an ubiquitous enzyme that contains specific amino-acid "KDEL" motifs for microsomal membrane proteins, which facilitate the formation disulfide bridges on lately translated polypeptide chains. It has no lipid transfer activity. This activity is mediated by the large sub-unit, which proved to be defective in abetalipoproteinemia [327,328]. The MTP gene (MTP) encoding the large sub-unit was localized on the long arm of chromosome 4, band q24 [329]. It spans 50-60kb of genomic DNA organized in 18 exons. The MTP cDNA is 3120 nucleotides in length, including 24 noncoding in the 5' end and 550 to 621 nucleotides of a 3' untranslated sequence. The mature protein contains 894 amino acids, and is highly conserved (86% of homology with the bovine MTP). It presents homologous regions with the phospholipid-binding domain of vitellogenins and for cholesterol ester

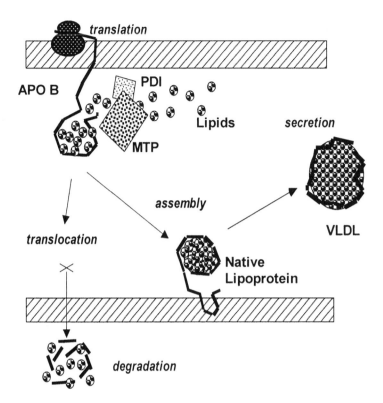

Figure 2-9. **Microsomal triglyceride transfer protein (MTP) and apoB-rich lipoprotein assembly.** MTP is a dimer composed of a small subunit (PDI) anchored in the ER membrane and of a large subunit (MTP) with lipid transfer properties. ApoB translation and maturation are coordinated with lipoprotein assembly. PDI could favor disulfide bridging for proper folding of apo B. MTP favors lipid-core enrichment of the native lipoprotein and prevents apoB translocation arrest. MTP protects native apoB from degradation in the cytosol by the ubiquitin/proteasome pathway.

binding region of CETP [328], confirming its role as a lipid transfer protein. MTP expression is upregulated by cholesterol, whereas insulin downregulates MTP transcription [330].

The first gene mutations identified in patients with abetalipoproteinemia, were homozygous nonsense or splice site mutations. The undetectable and inactive MTP resulting from these mutations authenticated the MTP gene as the predisposing locus for the disease. Although the disorder is rare, other cases with abetalipoproteinemia have been described with MTP mutations worldwide [331-336]. Interestingly a naturally occurring missense mutation Arg540→His identified this residue as necessary for complex formation with PDI [333]. Another interesting example of mutagenesis was a case of abetalipoproteinemia resulting from uniparental maternal isodisomy of chromosome 4q [334]. In this case reduction to homozygosity results from the inheritance of both chromosome homologues from a single heterozygous parent carrying the mutation. It may be unmasked in humans by recessive disorders. UPD was first identifed as a cause of recessive dyslipidemia by a case of complete paternal isodisomy of chromosome 8, causing lipoprotein lipase deficiency [337]. Uniparental isodisomy (UPD) may also reveal regions of genomic imprinting, a mechanism by which each parental allele undergoes differential expression [338,339]. In imprinted regions UPD may disturb growth and development. In this case and in the case of chromosome 8 UPD, the absence of developmental abnormalities in addition to the classical signs of abetalipoproteinemia or LPL deficiency, suggested that maternal chromosome 4q and paternal chromosome 8 were not imprinted. A short tandem repeat $(CA)_n$ was identified in intron 10. In addition a promoter variant –493G/T was described to modify transcriptional activity in vitro. Associations with plasma lipoprotein profiles remained so far inconsistent from one study to the other [340-342].

The identification of MTP has increased knowledge on **apoB-rich lipoproteins assembly** (*Figure 2-9*) in the liver and in the intestine [223, 343,344]. The primary structure of apo B is essential for lipoprotein assembly. However truncated forms of apo B, such as those causing familial hypobetalipoproteinemia, do not prevent lipoprotein secretion. In abetalipoproteinemia apo B is normal, but it is retained and degraded in the proteasome before any further lipoprotein maturation. In vitro and in vivo experiments have demonstrated that the presence of MTP in the ER is critical for lipoprotein assembly and that it prevents Apo B degradation in the proteasome [345,346]. Thus apo B maturation and lipoprotein assembly are coordinated. MTP lacks a specific anchoring motif to the endoplasmic reticulum membrane. MTP dimerization with PDI could allow the formation of disulfide bridges essential for proper folding of apo B, along with MTP anchoring to the ER membrane. The large sub-unit would facilitate lipid transfer and enrichment of core of the native particle. If lipid biosynthesis lowers or if MTP is defective, native apo B will be directed to the cytosol where it will be degraded in the proteasome. On the other hand, any situation favoring high cellular lipid influx (dietary or other causes of increased lipogenesis) or enhanced MTP expression (transgenic mice) leads to accelerated apo B rich lipoprotein assembly and secretion [347]. MTP seems to have a more significant effect in mice as compared with humans [348,349]. MTP gene invalidation in line with apo B knock out, is embryonic lethal in mice as the result of defective lipoprotein synthesis in the yolk sac. Moreover, heterozygotes express a phenotype of lowered plasma apoB-rich lipoproteins, which is not the case in human abetalipoproteinemia. Finally, the major regulatory role of MTP in lipoprotein assembly has been explored in therapy. MTP inhibitors dramatically reduce levels of circulating LDL when given to Watanabe rabbits, a model of homozygous familial hypercholesterolemia [58,350]. The draw back was the development of liver steatosis. However, this demonstrated the relevance of the lipoprotein assembly pathway as a promising target for the design of future lipid-lowering therapies.

Other factors may modulate maturation and secretion of apo B rich lipoproteins. Rare syndromes associated with defective secretion of apo B rich lipoproteins or low circulating levels of apo B have been described [273]. Anderson disease is a particular form of hypobetalipoproteinemia characterized by absent chylomicron secretion by the intestine, however with no intracellular lipid accumulation, a feature observed in chylomicron retention disease. Thus the identification of MTP together with identifying a cause for a rare and severe disorder of lipoprotein metabolism has uncovered a crucial pathway of intracellular maturation and assembly of lipoproteins.

2.1.4-   Apolipoprotein (a), a Mysterious Candidate for Atherosclerosis

Soon after its first descriptions [351, 352], lipoprotein Lp (a) was associated with a higher risk of cardiovascular disease. However despite intensive research and significant progress in understanding its pathophysiology, this particle at the crossroads of lipoprotein metabolism, inflammation and haemostasis, still remains a mystery [353,354].

Lipoprotein Lp(a) is formed by the assembly of one LDL particle with one protein named apo(a) (*Figure 2-10)* [355]. The primary sequence of apo(a) presents strong homology with plasminogen [356]. Therefore, apo(a) belongs to a family of circulating serine-proteases involved in haemostasis, including plasminogen, prothrombin, tPA (tissue Plasminogen Activator), factor XII, and other liver specific growth factors (Hepatocyte Growth Factors). Contrary to other members of the gene family, apo(a) is deprived of proteolytic activity, as a consequence of nucleotidic substitutions in the cleavage site for tPA. However, its specificity resides in multiple repetitions (12 to 51) of cysteine-rich domains named "Kringle" after their resemblance of their secondary structure with Danish cakes [357]. Kringles are highly glycosylated, inducing charge modifications at the surface of Lp(a). Moreover, they mediate protein-protein interactions starting with apo B in LDL, plasminogen receptor, fibrin or components of the extracellular matrix, contributing to its atherogenicity [358]. Apolipoprotein (a) is mainly

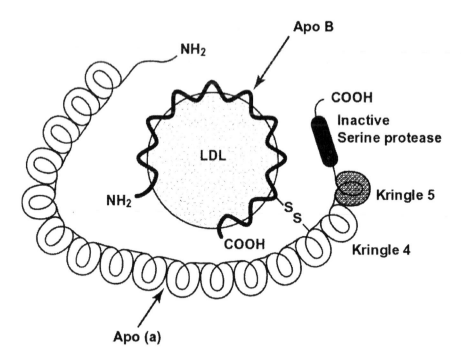

Figure 2-10. **Structure of Lp(a) particle.** Apo(a) associates by disulfide bridging with apo B-100 of LDL to form lipoprotein(a). Apo(a) shares strong homology with members of the plasminogen family. However it is unique by the high degree of repetition of Kringle IV (4) motifs and by the absence of serine-protease activity.

synthesized by the liver [359] and is mainly secreted as free apo(a) [360]. Soon after apo(a) secretion apo B-100 captures apo(a) and associates by disulfide bridging in its C-terminal part to form Lp(a) [361-363]. Noncovalent interactions between apo(a) and apo B-100 could also stabilize the cohesion of Lp(a) particle and contribute to apo(a)-apo B-100 covalent bridging. Apo(a)-apoB-100 bonding is species dependent. Human apo(a) expressed in transgenic mice circulates unbound in plasma, failing to associate with murine apo B-100, whereas infusion of human LDL or crossbreeding with transgenic mice expressing human apo B induces Lp(a) assembly in vivo [364,365].

The physiological role of Lp(a) is unknown. Lp(a) represents only a minority of circulating LDL (7-10%) [366]. The large number of Kringles suggests that apo(a) could be a competitive inhibitor for plasminogen [367] and for ligands of the vascular endothelium, thereby having thrombogenic properties [368,369]. Moreover, Lp(a) would inhibit the in vivo activation of TGFβ induced by plasmin, which in turn could activate several atherogenic cascades in the arterial wall [370]. Lp(a) favors activation and foam-cell formation of cellular components (i.e. macrophages, myocytes) of the atherosclerotic plaque *in vitro*. Apo(a) and apo B-100 colocalize in human lesions of atherosclerosis [371,372]. Moreover in retrospect of more than three decades of observations in humans, there is clinical evidence for a positive correlation between plasma levels of Lp(a) and the risk of cardiovascular disease [373]. The role of apo(a) in the development of atherosclerosis was shown in transgenic mice for human apo(a), which develop more extended and more advanced arterial lesions than their control littermates [374]. Lp(a) could direct LDL particles towards sites of injury into the arterial wall, contributing to early development of atherosclerosis.

The gene for apolipoprotein (a) (LPA) was localized on the long arm of chromosome 6 (6q26-q27) [375,376] soon after the identification of its cDNA [356]. There is near complete identity (98-100%) of the 5' noncoding and signal peptide sequence between apo(a) and plasminogen. The large size of the gene (> 200kb) due to the Kringle repeats domain, was a handicap for the characterization of the gene organization and of regulatory sequences. Moreover, there are homologous genes and pseudogenes [377] in other genomic regions and within the apo(a) locus. The analysis of large genomic fragments showed that the apo(a) gene is part of a cluster approximately 400 kb in length. It includes apo(a) and plasminogen, together with two other genes homologous to apo(a): APO(a)RG-B (Apo(a) Related Gene) and APO(a)RG-C, one of them being a pseudogene of plasminogen (*Figure 2-11*) [378,379]. Although there is a Stop codon within the third Kringle of APO(a)RG, gene transcripts were detected in the liver. The genes for apo(a) and plasminogen are oriented in opposite directions; their 5' ends being

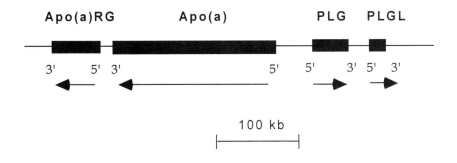

Figure 2-11. **Gene cluster of plasminogen** [378-80]. Apo(a): apolipoprotein(a), Apo(a)RG: apo(a) related gene, PLG: plasminogen, PLGL: plasminogen pseudogene (plasminogen-like). Arrows indicate transcription orientation.

separated by a 50kb intergenic sequence [380]. The apo(a) gene promoter comprises response elements for interleukin 6, which might mediate stimulatory responses of apo(a) expression in inflammatory conditions [381]. There are several binding sites for liver specific nuclear factors (HNF1α, C/EBP, Liver Factor A1). The first exon encodes for a signal peptide. It is separated from the sequence of the first kringle by a 14kb long intron. In the Kringle repeat domains, blocks approximately 6kb-long and containing 2 exons each encoding for ~114 aa, and two introns 4.2 and 1.3 kb long respectively, are repeated more than 10 times [382,383].

The apo (a) gene is polymorphic in sequence and more particularly in size [384]. The number of repeated Kringles is not identical from one individual to the other, and not all plasminogen-like Kringles are repeated. The first and the last eight Kringles are unique [356]. Apo(a) size polymorphism was more precisely defined for its size polymorphism resulting from the repetition of two Kringles differing only by 3 neutral nucleotides (Kringles IV-A and IV-B) *(Figure 2-12)*. At least 40 alleles are found in humans with a heterozygosity index exceeding 95% [385]. A similar phenomenon was reported in the gene for complement receptor type 1, in which a 70aa motif is repeated more than 30 times [386]. In this case as in that of apo(a) this polymorphism is expressed in the mRNA [387] and in the protein [382,383,388]. Therefore autosomal codominant transmission of apo(a) size isoforms is readily detectable in plasma [389]. Numerous sequence polymorphisms (SNPs) were also described in exons, introns and within the promoter sequence, where a pentanucleotide repeat $(TTTTA)_n$ was described [385]. The pentanucleotide did not induce any change in promoter activity in vitro, whereas the +93C/T polymorphism created an alternative translational start site reducing translational activity in vitro.

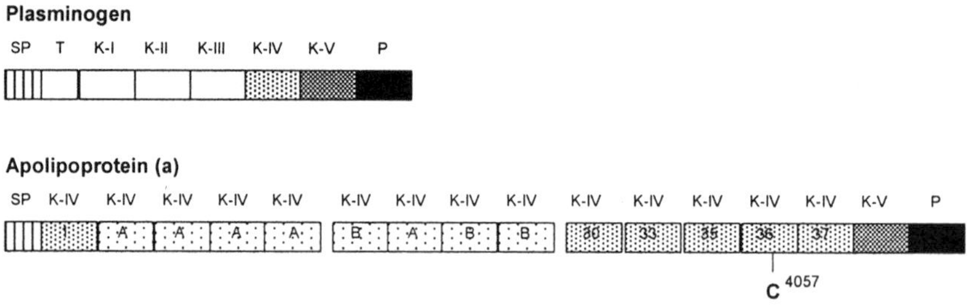

Figure 2-12. **Domain maps of plasminogen and apolipoprotein(a)** [356]. Apo(a) and plasminogen signal peptides are homologous. The kringle repeat domain is composed of repeated Kringles (type A and B) homologous to kringle IV (K-IV) in plasminogen. It was repeated 37 times in the gene initially cloned. The K-IV number may vary from 12 to 51 repeats from one individual to the other. One K-IV before the last contains the binding site for disulfide bridging with apo B-100. The 3' region contains Kringle V and serine-protease homologous domains, although the latter is deprived of catalytic activity in apo(a).

The particular structure of the gene cluster and the high degree of polymorphism of the apo(a) gene suggest that this genomic region has evolved through successive gene conversion events. In fact recombination events have been observed by independent segregation of SNPs with apo(a) isoforms [390]. Through this combined approach a meiotic recombination event was identified within a large family (376 meiotic events analyzed), creating a new apo(a) allele in an offspring, which was absent from the parental chromosomes [384]. The possibility of intragenic recombinations was also supported by the presence of specific sequences within introns separating both exons encoding for Kringle IV-A. The first, a sequence "*Chi* (χ)" (GGTGGTGG), is found in E Coli to mediate recombination events in the λ phage and in human genes undergoing conversion events (immunoglobulin or β-globin gene clusters) and in the generation of several VNTR. The second, is a short tandem repeat $(CA)_n$ reported at breakpoints of gene or chromosomal rearrangements.

Apo(a) polymorphism would remain anecdotal if it were not universally correlated with plasma concentrations of Lp(a) [373] (*figure 2-13*). In all populations there is a preferential association of the highest plasma Lp(a) levels with the lightest apo(a) isoforms and an increased risk of coronary or carotid atherosclerosis. However, there are interethnic differences in the distribution of plasma Lp(a) concentrations [354,385]. In Caucasian and Asian populations the distribution is skewed towards the lowest levels, whereas in black populations the distribution tends to be gaussian, increasing mean and median Lp(a) levels. Twins studies have shown the strong genetic determinism of Lp(a) concentration [391]. Family studies, combining

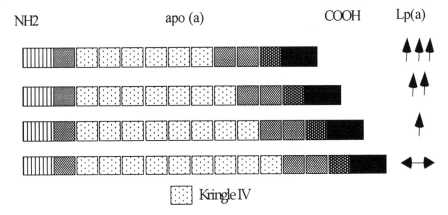

Figure 2-13. **Inverse relationship between apo(a) size polymorphism and plasma Lp(a).** The shorter the isoform the higher the plasma concentration may be found. However, the relationship may vary between individuals of the same family or between populations of different ethnic backgrounds.

analyzes of apo (a) allele size and plasma Lp(a), have shown that the apo(a) locus controls a significant proportion of plasma Lp(a) variance [389]. These observations were confirmed in the Cynomolgus monkey, in which apo(a) mRNA size and levels in liver determine 58% of Lp(a) concentration in plasma [392]. However, the relationship is not linear, particularly from short to long size apo(a) isoforms and may vary from one population to the other, or from one individual to the other within a family.

Some apo(a) isoforms are not expressed in plasma, after which they were named "null alleles" [354,385]. It was suggested that the largest isoforms would be those undetectable in plasma. Indeed, large isoforms are processed more slowly in the ER, because they are more difficult and longer to synthesize than light isoforms. Their mass and the higher number of disulfide bridges necessary to fold the secondary structure of Kringles could explain this difference. However, apo(a) allele size determination has shown that larger alleles were not over-represented in subjects who did not express apo(a) in plasma, suggesting that secretion rate of Lp(a) was independent form apo(a) size. However, both facts may be reconciled. The sequence of the longest isoforms could be more likely susceptible to point mutations interrupting translation (i.e. nonsense, frameshift or splice site mutations). Indeed a splice site mutation prevalent in Caucasians (6%) was described as a cause of apo(a) "null" mutation [393]. Missense mutations particularly on cysteine residues, could also cause intracellular apo(a) degradation as the result of improper folding of Kringle motifs. In addition, when expressed in the appropriate genetic context, the +93C/T variant was shown to lower Lp(a) levels. With possible chromatid exchanges during meiosis, particularly in the Kringle repeats domain, point mutations could have "spread" to all isoforms over several generations. Therefore, sequence polymorphisms could introduce an additional factor of variability in plasma Lp(a) concentration. Despite a relative homogeneity within individual or within family between circulating Lp(a) concentration and genetic apo(a) size isoforms, there is great variability between individuals or populations, even for a given size isoform. This suggests that beside genetic determinants of plasma Lp(a), protein expression may be modulated.

Cis-regulatory factors in the apo(a) gene were reported to modulate Lp(a) independently from apo(a) isoforms [390]. Sequence polymorphisms were found upstream of the apo(a) transcription start site [882]. Although interindividual levels of Lp(a) may vary by a 1000-fold, plasma concentration of Lp(a) are rather constant within a single individual [373]. Lp(a) concentrations are independent of age, sex and very little on diet. They are not always associated with premature cardiovascular disease, high levels of Lp(a) being found in centenarians [394]. *Trans*-acting regulatory factors may modify apo(a) gene expression. Apo(a) has been proposed as an acute-phase reactant [395]. Lp(a) levels may increase in moderately inflammatory conditions, such as the acute phase of myocardial infarction, rheumatoid arthritis or in HIV positive patients [396,397]. On the other hand, aspirin reduces liver apo(a) expression [398]. Interestingly, major inflammatory conditions such as severe burns, sepsis or acute hepatitis in humans and high doses of acute-phase inducers in mice profoundly decrease Lp(a) levels [399-401]. Therefore depending on physiological or environmental conditions Lp(a) may be oppositely regulated. Growth hormone treatment increases plasma Lp(a) in humans and in transgenic mice expressing apo(a) [402,403]. Steroid hormones lower plasma Lp(a) concentrations [404,405]. Danazol (an anabolic steroid) given to premenopausal women [406], and Tamoxifen (an anti-estrogen) also lower plasma Lp(a) [407]. In transgenic mice expressing human apo(a), sexual steroids have a negative effect on apo(a) expression after sexual maturation [403]. Omega-3 fatty acids are moderately lowering Lp(a). In keeping with the fact that Lp(a) may be low in peroxisomal disorders, this suggests that regulatory mechanism are mediated by intracellular fatty acid derivatives [408]. Controversial data were reported on Lp(a) lowering effects of N-acetylcysteine (a disulfide bridge cleaving molecule) in-vivo [409,410], however a lysine derivative 6-aminohexanoic acid is able to act as an intracellular molecular chaperone enhancing apo(a) secretion by the liver. Certain drugs (nicotinic acid, colestyramin) modestly lower plasma Lp(a) [355].

Kinetic studies have shown that Lp(a) levels are primarily controlled by its secretion rate and to a lesser extent by its catabolism [411,412]. The role of LDL receptors as a catabolic pathway for Lp(a) has been a long lasting dispute [373]. In FH patients Lp(a) concentration was reported to be high. Moreover in transgenic mice overexpressing LDL receptors, LDL clearance was accelerated and plasma LDL and Lp(a) were lower than in controls [413]. However HMGCoA reductase inhibitors, which upregulate LDL receptor expression, do not have a major effect in humans [414]. Moreover, in pedigrees of rhesus monkeys with a naturally occurring LDL receptor mutation, in patients with familial hypercholesterolemia, or in patients with familial defective apo B-100, there was no difference in Lp(a) concentration between carriers of the same apo(a) isoform, either hypercholesterolemic or not [415-417]. Several in vitro and in vivo experiments have shown that Lp(a) is assembled in extracellular spaces [360]. A high plasma LDL-cholesterol level could represent *per se*, a condition favoring higher number of Lp(a) particles, as the result of increased substrate availability for Lp(a) assembly, though independent of its catabolism. Lp(a) catabolism was recently shown to be mediated by a receptor belonging to the LDL receptor gene family: megalin [418]. Mouse embryonic cell lines were used to demonstrate that megalin is able to bind and take up Lp(a) with much greater affinity than the LDL receptor. Megalin is expressed in the kidney. These data could reconcile with the physiological notion that Lp(a) may be cleared in the kidney and with clinical observations reporting increased Lp(a) levels in kidney diseases [419]. However, Lp(a) is a large particule which cannot be ultrafiltrated in primary urine. Together with this specific mechanism, lipoprotein lipase could stimulate the cellular catabolism of Lp(a) through a receptor independent pathway [420]. This mechanism could contribute to the moderate effect of fibrate derivatives on Lp(a) concentrations in-vivo [421].

The physiological role of apo(a) remaining unknown, it is still problematic to identify putative Lp(a) modifying drugs. Many animal species (including several mammals) have neither Lp(a), nor a structural gene for apo (a) [359]. Its rather recent occurrence should be posterior to the divergence between New World and Old World monkeys, dating back to 40 millions years. Only the hedgehog has been found so far with circulating Lp(a) [422,423]. Interestingly in the hedgehog apo(a) gene Kringle-III is repeated instead of Kringle-IV. However, an additional cysteine residue allows the formation of a disulfide bridge with apo B-100, like in humans. Therefore, a parallel but convergent evolution of similar genes has driven the formation of lipoprotein-complexes (i.e. Lp(a)) with similar functions in primates and in hedgehog. Plasma Lp(a) has high interindividual variability. Although apo(a) size alleles are normally distributed, in several populations, plasma Lp(a) distribution is highly skewed towards low levels, which is suggestive of environmental or other selective pressure. On the other hand, apo(a) gene and protein are highly polymorphic, suggesting a low selective pressure on the locus. The evolutionary advantage provided by Lp(a) circulation in plasma thus remains very obscure. The majority of small rodents do not have Lp(a), though transgenic mice overexpressing human apo(a) may develop atherosclerosis under a saturated fat diet [374]. It may act as a factor enhancing recruitment of LDL at the site of tissue injury, with an expression strongly regulated by hormonal and inflammatory states. Selective pressure might have contributed to higher levels of plasma Lp(a) in a specific environment found in Africa, or conversely it could have suppressed apo(a) expression in non-African populations. High levels of Lp(a) are found in all human populations, and in approximately 30% of myocardial infarction survivors, innovative therapies modulating apo(a) metabolism could represent a significant step forward in the prevention of atherosclerosis.

2.1.5 - HMGCoA Reductase, the Rate Limiting Enzyme of the Mevalonate Pathway

HMGCoA (3-hydroxy-3-methylglutaryl coenzyme A) reductase is the rate-limiting enzyme for intracellular cholesterol biosynthesis: the mevalonate pathway [32]. This pathway of basic metabolism is essential for living cells (see page 11). HMGCoA reductase is a glycoprotein found in membranes of the nucleus and of endoplasmic reticulum [424]. This monomeric protein has several transmembrane domains (7 or 8) in all living species including the most primitive [425]. The catalytic site is cytosolic. Transmembrane domains contain sterol-sensing domains that regulate enzyme degradation (see page 162).

The gene for HMGCoA reductase (HMGCR) was localized on the long arm of chromosome 5 (5 q12-q13.1), [426]. A 4471bp long mRNA, which encodes for a 888aa protein, also includes 5' and 3' non coding sequences. The gene is highly conserved [427]. In the hamster it comprises 20 exons, of which the first and the last are noncoding [428]. In the human gene sequence, there is no TATA box; alternative transcription start sites are used, a feature common to houskeeping genes. HMGCoA reductase has several transcripts some being stabilized by the presence of specific sequences near the polyadenylation site [429]. Two RFLPs ScrFI/BstNI and HgiAI were described [430,431].

HMGCoA reductase **gene expression** is complex. Transcriptional regulation is partly under the control of SREBPs (see page 169). However negative regulations by sterols may be uncoupled from those of other SREBPs sensitive genes by regulatory sequences 5' upstream of the gene [432]. Mevalonate and intranuclear calcium concentrations also exert a powerful control on gene transcription [433]. Post-transcriptional regulations that accelerate enzyme degradation, are mediated by transmembrane sterol-sensitive domains [434]. Nonsterolic products of the mevalonate pathway also control enzyme translation and degradation in the cytosol, modulating 70 to 80% of its activity [435]. Finally there is a phosphorylation site on Serine 871. Phosphorylation by an AMP-sensitive kinase would reversibly inhibit enzyme catalytic activity when cellular ATP concentrations decrease [436]. This reaction is coupled with the inhibition of fatty acid synthesis, which would represent a mechanism for sparing cellular energy resources.

There are no natural mutations described for HMGCoA reductase in humans. Pharmacological data brought by HMGCoA reductase inhibitors (statins) suggest that partial enzyme defects would result in permanently lowered concentrations of plasma cholesterol. Homozygosity for a severe impairment of enzyme activity would be probably lethal, when one considers the number and the vital importance of products derived from the mevalonate pathway. This has been observed in vitro [34]. Moreover, mutations on the very last step of cholesterol biosynthesis cause a very severe malformative disease, the Smith Lemli Opitz syndrome in humans (see page 13). One could also imagine models of activating mutations in transmembrane domains or in sterol sensitive elements (SRE) of the gene promoter [437]. However, in Chinese hamster ovary cells insensitive to negative control by sterols, intracellular cholesterol levels were unchanged as the result of accelerated enzyme degradation in the cytosol [438]. Therefore chances are low that this type of mutation would have a detectable effect on circulating lipoproteins.

Despite the modest role of HMGCoA reductase as a candidate gene for human dyslipidemia, it has been a pioneer example of an excellent therapeutic target of lipoprotein metabolism and atherosclerosis. A past decade of clinical and biological progresses in the prevention of atherosclerosis with HMGCoA reductase inhibitors has provided the sound and ultimate evidence [105].

2.1.6-   Scavenger Receptors

2.1.6.1-   Scavenger Receptors Class A: Molecular Flypapers for Modified Lipoproteins

Early studies had suggested that receptors for modified LDL were necessary to initiate the pathogenesis of atherosclerosis [492]. These macrophage-specific receptors ensure unlimited endocytosis of many extracellular substances including modified LDL, whence their name of "scavenger" receptors. Macrophages indefinitely accumulate cholesteryl esters in their cytoplasm, becoming foam cells typical of primitive lesions of atherosclerosis and of xanthomas [110]. Their existence was suspected by the persistence of acetylated LDL endocytosis by macrophages, in completely LDL receptor deficient patients [440].

Scavenger receptor Class A (SR-A) bind oxidized, acetylated, maleylated LDL or modified HDL, having no particular affinity for native LDL [441]. In addition to their affinity for modified LDL, these receptors specifically recognize various types of negatively charged ligands. These may be modified proteins (including advanced glycation end products found in diabetes, or β-amyloid fibrils found in Alzheimer's disease), acid phospholipids (phosphatidyl-serine), polysaccharides, polyribonucleotides (poly-I, poly-G), exogenous components like asbestos or bacterial endotoxins. They may also take up whole cells like "Gram +" bacteria through the binding with lipoteichoic acid, apoptotic cells and damaged red blood cells [442-445]. Moreover scavenger receptors confer properties to macrophages to adhere on inert supports like plastics made of polyvinyl sulfate [446]. Therefore scavenger receptors class A, have been identified as molecular "flypapers" at the crossroads between atherosclerosis and immunity [447]. Genetically modified mice for scavenger receptor class A have proven that these receptors are involved in host defense, tissue scavenging and atherosclerosis [448-450]. Mice overexpressing SR-A have increased uptake of modified LDL in macrophages and rate of foam cell formation [451]. Surprisingly, when crossbred with atherosclerosis susceptible strains of mice, SR-A transgenic mice appeared resistant to atherosclerosis, indicating other regulatory functions of MSR on macrophage activation. Conversely, SR-A knockout mice have increased susceptibility to infection and endotoxic shock, they have decreased uptake of modified LDL in macrophages however, their lipoprotein profile and plasma clearance of modified lipoproteins are normal [452]. However, when crossbred to apo E deficient mice – a genetic model of atherosclerosis- lesion formation is significantly reduced as compared with control apoE knockouts.

The genomic sequence (MSRE) is unique spanning 80 kb long. It includes 11 exons and 10 introns (*figure 2-14*), located 11 cM telomeric to the LPL gene on chromosome 8p22 [453]. Polymorphisms detected by Hind III, Bam HI and Msp I were reported. An alternative splicing generates two mRNAs encoding for type I and type II class A scavenger receptors [454,455]. A third truncated splice variant could act as a dominant negative isoform over both, by reducing cellular uptake of modified LDL by type I or type II receptors [456]. Type I and type II cDNAs are highly conserved in several mammals (mouse, rabbit, or bovine) [457]. Scavenger receptors type A form trimers made of either type I or type II monomers, however heterotrimers made of both types have been described. The various combinations obtained and a higher order of oligomerization could increase the affinity of scavenger receptors for a higher number of ligands or for ligands of larger size.

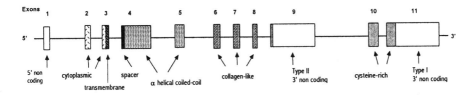

Figure 2-14. Domain map of scavenger receptors class A type I and type II.

Six functional domains compose the primary chains. Exon 1 is noncoding. There is no signal peptide. Exons 2 and 3 encode for the N-terminal cytoplasmic domain. This 50aa long domain is mediating receptor-ligand endocytosis and receptor clustering into clathrin-coated pits, however through residues distinct from those of LDL receptor [457]. Phosphorylation sites modulate the rate of ligand endocytosis. The transmembrane domain comprises 26 hydrophobic amino acids. Exons 4 and 5 encode for an extracellular "spacer" domain (32 aa), which possess N-glycosylation sites. They also encode for the α–helical coiled-coil domain (163 aa). In this domain 23 heptad repeats (7 amino acids) are likely to from right handed amphipathic α-helices. Interestingly, there is a skip domain at the junction between exons 4 and 5 generating a distortion between to clusters of coiled-coil domains. Coiled-coils assemble into 2 or 3-stranded bundles to form multimers. Disulfide bridges and a 7-residue trigger sequence stabilize the multimeric structure in this domain [458,459]. Multimers are pH sensitive and easily dissociate in acidic compartments after endocytosis. Exons 6 to 8 encode for the collagen like domain, rich in positively charged residues [454]. This domain offers a great surface for ligand attraction, remaining selective however [447]. Site directed mutagenesis experiments confirmed ligand binding properties of the fifth domain [460-462]. Collagen-like motifs are conserved in other species, and in a family of proteins belonging to complement component C1q, antigen CD5, lung surfactant apoproteins, conglutinin (binding mannose), lectins involved in host defenses and clearance of nonself substances [450,463]. Exon 9 encodes for the last 6 amino acids of type II receptors, and comprises a long 3' noncoding sequence. The sixth domain of type I receptors, is encoded by exons 10 and 11, the latter also comprising a 3' noncoding sequence. It is composed of 109 amino acids C-terminal, rich in cysteines.

Upstream regulatory sequences are also conserved among species [464]. There are no TATA boxes, but A+T rich motifs, which could be used as alternate transcription initiation sites [453]. In monocytic cell lines, SR-A expression is restricted to macrophage [465]. Transcripts and mature protein are found in different cellular compartments of non-monocytic and in monocytic non-macrophagic cells; however, the receptor is present at the surface of the cells and in endosomes only in macrophages at a very late stage of cell differentiation. PDGF induces scavenger receptor expression in smooth muscle cells and fibroblasts [466-468]. As expected, oxidized-LDL and phorbol esters are strong inducers of SR-A expression [469,470]. Stimulation by phorbol esters can be inhibited by retinoic acid and dexamethasone [464]. There is an interleukin 2, sensitive binding site upstream of the gene, consistent with the expression of SR-A in a cytokine-responsive cell. Interferon γ, a macrophage-activating lymphokine suppresses SR-A expression [471]. As expected there are no SREBPs sensitive elements upstream of the gene. Interestingly, PPARγ (nuclear receptors mediating modified fatty acids intracellular signaling, see page 171) down-regulate SR-A expression by antagonizing AP-1, STAT or NFκ-B signalling pathways usually involved in inflammation [472,473]. Vitamin E (α-tocopherol) also downregulates AP-1 signalling pathway and suppresses in a dose dependent manner macrophage SR-A activity [474]. These observations further support the hypothesis of necessary molecular interactions between lipid metabolism and the immune system in the development of atherosclerosis.

No naturally occurring mutation has been described on scavenger receptor class A in human dyslipidemia or atherosclerosis. Their ancestral function of polyanions attraction mediate by collagen-like domains is found in phagocytic cells in *Drosophila* [441]. Scavenger receptors remain however a privileged target in the prevention of atherosclerosis.

2.1.6.2-   CD 36, A multifunctional Receptor for Fatty Acids and Thrombospondin, Scavenging Oxidized Lipoproteins.

CD36, or thrombospondin/collagen I receptor, or GPIIIb, or FAT (fatty acid translocase) was initially recognized as an antigen expressed on human platelets, enhancing platelet adhesion to collagen. CD36 deficiency is defined by the absence of immunodetectable CD36 on human platelet membrane [475]. However, lack of CD36 in platelets does not induce significant haemostatic disorders. CD36 deficiency is prevalent in Japan (2-3%), in Thais and in sub-Saharan Africa where it may favor severe forms of cerebral malaria [476]. It is less frequent in populations of European ancestry (0.2-0.3%). Later CD36 was identified as a receptor belonging to another class of scavenger receptor (scavenger receptor class B) expressed on the surface of monocytes/macrophages, on erythrocytes and in tissues active in fatty acid metabolism: heart, muscle, adipose tissue and intestine [477]. CD36 is a multiligand receptor, which could serve as an adhesion molecule to collagen, thrombospondin, erythrocytes infected with plasmodium falciparum or shed photoreceptor outer segments [478]. Moreover, it could be a transporter of long-chain fatty acids, anionic phospholipids in heart and adipose tissue. Although it is able to bind HDL and mediate cholesteryl-ester uptake, contrary to SR-B1 (a close relative) it does not enhance selective cholesterol uptake from HDL particles [479]. It avidly mediates the uptake of oxidized LDL particularly LDL locally modified by the myeloperoxidase-hydrogen peroxide-nitrite system (NO_2-LDL) produced by macrophages [478]. It contributes to the scavenging functions of macrophages by cooperating with $\alpha_v b_3$vitronectin for the phagocytic clearance of apoptotic cells.

The gene for CD36 (CD36) has been identified. It extends over 32kb, organized into 15 exons, localized to chromosome 7q11.2 [480,481]. It encodes for an 88kD protein encoded by 472aa, sharing homologies with other transmembrane receptors: SR-B1 (*Figure 2-15*) and another multifunctional protein LIMPII (for Lysosomal Integral Membrane Protein II) [479,482]. They have 2 transmembrane domains and a large extracellular domain containing a cysteine-rich region. Extracellular domains are N-glycosylated and cytoplasmic domains are palmitoylated. SR-BI and CD36 are found in membrane caveolae. The promoter includes a response element for PEBP2/CBF (Polyomavirus enhancer-binding protein2/core-binding factor) a myeloid specific transcription factor, which controls most of CD36 expression in macrophages [483]. CD36 gene expression is induced by IL-4, Macrophage Colony Stimulating factor, NO_2-LDL, intracellular cholesterol and PPARγ ligands [484-486]. On the other hand interferon γ, lipopolysaccharide and vitamin E down regulate CD36 expression [487]. The broad tissue-specificity suggests that CD36 might play a role on levels of circulating lipoproteins and atherosclerosis. Transgenic mice overexpressing CD36 in muscle have enhanced ability to oxidize long-chain fatty acids [488]. They have lower fasting blood concentrations of triglycerides, fatty acids and VLDL depleted in TG. In addition they appear resistant to dietary-induced hyperlipidemia. Fasting glucose was higher, however associated with normal glucose tolerance curves. Consistently, knockout mice for CD36 had combined hyperlipidemia with increased levels of free fatty acids and lowered glucose in serum [489]. They had higher adipose mass in both sexes and tended to gain more weight than

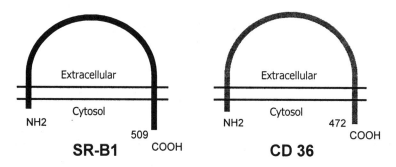

Figure 2-15. **Structural homologies of scavenger receptors class B** [479]. N-terminal and C-terminal domains are cytoplasmic. Extracellular domains contain N-glycosylation sites. Both ligand domains mediate selective uptake of lipids, however the cytoplasmic tail in SR-B1 contains specific regions for enhanced HDL cholesteryl-esters selective uptake.

control littermates under western diet. Despite this apparently atherogenic lipoprotein profile, CD36 KO mice were protected against dietary or genetically induced atherosclerosis [490].

This contrasts with results of quantitative linkage mapping experiments which had identified CD36 as the disease causing locus in the NIH-strain of Spontaneously Hypertensive, insulino-resitant, hyperlipidemic Rat (SHR) [491]. This strain was found with multiple mutations in CD36 primary sequence. However, the original Japanese strains of SHR rats were not carriers of such mutations [492]. Moreover, complementation experiments showed that the SHR phenotype was associated with another locus in close linkage with CD36, but independent from CD36 mutations [493]. Therefore CD36 appears to be more specialized in the regulation of cellular fatty acid uptake in peripheral tissues and oxidized lipoproteins in monocyte macrophages. In this line, a prevalent Pro90Ser mutation for CD36 deficiency in Japan is associated with defective uptake of long-chain fatty acid in the myocardium of carrier patients and could be more prevalent in hypertrophic myocardiopathy [494-496]. Besides, macrophage uptake of oxidized LDL was also reduced [497]. In vivo observations in humans or animal models suggest that CD36 may significantly contribute to plasma lipoprotein profile and that its local overexpression in macrophages may contribute to initiation and progression of atherosclerosis. These characteristics are good prospects to qualify CD36 as a candidate gene for atherogenic dyslipidemia.

Other membrane receptors were identified and able to bind modified or oxidized LDL [498]. Another member of scavenger receptor class A, MARCO (macrophage receptor with collagenous structure) mediates phagocytic removal of pathogens [499]. Scavenger receptor B-1 has been identified to play a major role in HDL metabolism (see page 156). The LOX-1 receptor (for Lectin-like Oxidized-LDL receptor-1), which belongs to the C-type lectin family of membrane receptors, is expressed in endothelial cells and binds specifically oxidized LDL [500-502]. Interestingly, it is activated by Angiotensin-II establishing a link between hypertension and atherosclerosis [503,504]. Macrosialin (CD68) a mucin-like scavenger receptor also belongs to this "functional" family of receptors [456,457].

2.1- GENES CONTROLLING TRIGLYCERIDE-RICH LIPOPROTEIN METABOLISM

2.2.1- Lipases

2.2.11- Lipoprotein Lipase, the Rate-limiting Enzyme for Triglyceride-rich Lipoproteins

It was reported in the early fifties that an enzymatic factor released by heparin injection was able to clear lipemic serum from humans or animals in vivo and in vitro. The serum "clearing factor" was later identified to be **Lipoprotein Lipase** (LPL) [505]. LPL is an enzyme belonging to a family of serine-proteases [506,507]. It is a homodimer bound noncovalently to heparan sulfate proteoglycans at the surface of capillaries. LPL uses apo C-II as an activating cofactor on lipoprotein surface, to hydrolyze triglycerides from the core of lipoproteins. Muscles or adipose tissue rapidly take-up the released free fatty acids and glycerol. The relative cholesterol enrichment of lipoproteins generated by triglyceride removal stimulates lipid transfer and exchange of surface components (apolipoproteins, phospholipids) towards HDL and generates LDL *(figure 2-16)*. Lipoprotein lipase was confirmed experimentally as a rate-limiting enzyme for triglyceride-rich lipoprotein metabolism in genetically modified mice. Mice overexpressing LPL have lower plasma triglycerides and higher HDL cholesterol than controls [508]. Conversely knock out mice die soon after birth as the result of massive chylomicronemia in the suckling period. Thus LPL regulates essential steps of cellular energy delivery coming from dietary sources of triglycerides and rates of circulating lipoproteins.

In addition to its role of plasma hydrolase, LPL has a tissue-specific action attracting lipoproteins towards the sub-endothelial matrix and cell surfaces facilitating their endocytosis [509,510]. Indeed, it was shown that LPL fixed at cell surfaces or in the extracellular matrix favors the bridging of apo B-rich lipoproteins with LDL receptors and/or LRP or with heparan sulfate proteoglycans. In addition, small apolipoproteins (Apo C-I, C-III or apo E) may modulate lipoprotein interactions with LPL prolonging or preventing their retention in the extracellular matrix. Lipoprotein retention is an essential step to extracellular lipoprotein modifications and foam-cell formation. Thus, LPL plays a pivotal role in atherosclerosis, modulating tissue retention or clearance of atherogenic lipoproteins.

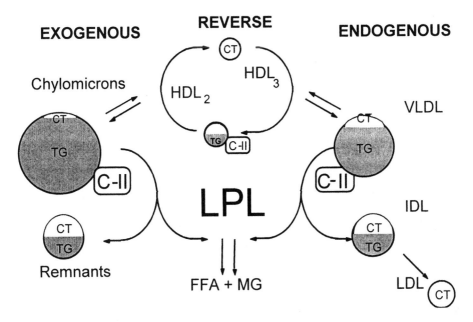

Figure 1-16. **Lipoprotein lipase, a pivotal role in lipoprotein metabolism** [507]. LPL hydrolyzes the TG-content of lipoproteins releasing free fatty acids (FFA) and mono-acylglycerol (MG) rapidly taken-up by cells. The relative changes in lipid content stimulate surface component exchanges and the generation of novel lipoprotein particles. CT: cholesterol, TG: triglyceride, CII: apo C-II

The gene for lipoprotein lipase (LPL) spans approximately 30kb [511,512]. It is single copy and localized to the short arm of chromosome 8 (8p22), [513]. Lipolysis is an ancient function, found in primitive lipases of certain fungi, *Rhizomucor Miehei* or *Geotrichum candidum* fulfilled by a catalytic triad (Ser-Asp-His) within a hydrophobic pocket of highly conserved conformation [514]. LPL is homologous to yolk proteins, which regulate lipid stocks in eggs, however, yolk proteins are deprived of lipoplytic activity [44,515]. LPL belongs to a family of lipases (hepatic, pancreatic or salivary lipases) including LCAT [511,516]. The LPL gene is highly conserved in various animal species, sharing more than 87% homology in mammals (bovine, rats, mouse, guinea pig), and 76% with chicken. This homology is also present in 5' flanking regions suggesting concerted regulatory mechanisms. Moreover, *Alu* repeats are observed in similar introns of other lipase genes suggesting that the lipase gene family originates form an ancestral gene (*figures 2-17 and 2-20*) by successive exon shuffling [512,517].

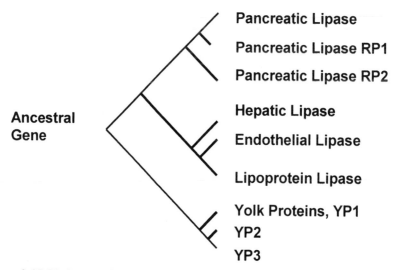

Figure 2-17. **Phylogeny of lipases** [516]. Sequence alignments and comparisons of primary sequence helped to reconstruct evolutionary links between members of the gene family.

The gene comprises 10 exons, encoding for a 448 amino acid protein of MW 50kD [518,519]. The first exon includes a 5' noncoding region, a signal peptide and the first two amino acids of the mature protein (*figure 2-18*). The protein is encoded by exons 2 through 9. The last base of the termination codon is encoded by exon 10 followed by a 3' noncoding sequence of 1948 bp. Exon 2 contains a N-glycosylation site on Asn43 essential for catalytic activity [520]. Exon 3 would contain residues involved in homodimer stabilization [521]. Exons 4 and 5 encode for Ser132 and Asp156 of the catalytic triad and for the highly conserved hydrophobic lipid-binding domain. The catalytic site is closed by a "lid" encoded by residues 217 to 238, forming a loop, which determines specific activity, and the affinity for lipid substrates [510,522]. Residues Lys147-148 also encoded by exon 4 would be significant for apo C-II binding [523]. Exon 6 comprises His241 of the catalytic triad and a domain rich in positively charged amino acids interacting with negatively charged heparan sulfates [524]. There would be a site for N-glycosylation in exon 8 and basic domains in exon 9 interacting with heparan sulfates and membrane receptors. In exon 8, Lys407 would be critical for LRP binding, and residues Trp393-394 would interact with membrane receptors. Moreover, there are cysteine residues (264-275 and 278-283) encoded by exons 6 and 9 involved in disulfide bridge formation for a proper folding of the active site [520,525]. The C-terminal region downstream of Lys428 is likely to be critical for LPL secretion [526]. Several polyadenylation sites in the 3' noncoding sequence of exon 10,

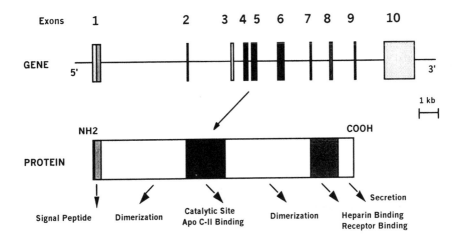

Figure 2-18. **Domain map of lipoprotein lipase.** Correspondence of gene domains with protein domains.

generate transcripts of different sizes (3.8kb, 3.3kb) fairly tissue-specific [528]. LPL is expressed in cells of mesenchymal origin, mainly in muscles (cardiac, smooth and skeletal) and adipose tissue. In addition to tissue specificity, LPL expression strongly depends on hormonal, metabolic (cytokines, growth factors) and nutritional conditions. Sequence elements sensitive to nuclear factors are found in the promoter: Sp1, acting synergistically with SREBP, PPRE, Oct-1 (octamer binding motif), cAMP, glucocorticoid and AP-1 like [527-529]. Growth hormone upregulates LPL, whereas estrogens or TNFα suppress LPL expression. Moreover, PPAR are strong activators of LPL expression, a reason why LPL is recognized as a typical and early gene of adipocyte differentiation [530,531]. Practically this regulatory mechanism explains part of the plasma triglyceride lowering effect of fibrates (PPARα activators) [85].

Lipoprotein lipase activity may be regulated in opposite ways between muscle and adipose tissue [507]. When energy demand increases LPL activity increases in the muscle and decreases in adipose tissue. In the post-prandial state, the opposite effect is observed. Regulations can be uncoupled; for example, TNFα (Tumor Necrosis Factor) also called cachectin, is a strong suppressor of LPL transcription and activity in adipose tissue whereas its effect is modest in muscles [526,532]. Tissue specific expression of LPL is genetically driven. The resemblance in LPL activity in adipose tissue is higher among monozygotic twins compared with dizygotic twins [533]. Moreover, cellular expression of LPL may change depending on pathophysiological conditions. For example, basal expression of LPL in macrophages is low, however human macrophages have high LPL synthetic rates in atherosclerotic plaques [534] and in atherosclerosis susceptible mouse strains compared to resistant stains [535].

Post-transcriptional regulations modulate LPL activity. Trimming of carbohydrate residues after N-glycosylation is an essential step for complete maturation and protein secretion [536]. The protein is active as noncovalent homodimers, stabilized by heparin and heparan sulfates, which prolong LPL activation. Heparan sulfate proteoglycans (or perlecan, versicans etc.) are extracellular matrix or cell surface proteins involved in many biological processes: embryonic morphogenesis, angiogenesis, neuronal growth and tissue repair [537]. The N-terminal domain of these macromolecules has complete homologies with the ligand-binding domain of LDL receptors. There are also binding sites for glycosaminoglycans: heparan sulfates [538,539]. These domains would facilitate apo B rich lipoprotein binding and LPL activation. HSPG attract and bind lipoproteins representing a high capacity compartment for lipoprotein remodeling and clearance in the liver. Moreover, these macromolecules could mediate LPL catabolism through cellular endocytosis and or its plasma secretion [540]. In the liver, LRP mediates LPL endocytosis if present at the surface of triglyceride-rich lipoproteins [441].

The LPL gene locus is highly polymorphic [541]. Several dozens of SNPs were described together with several short tandem repeats. Systematic DNA sequencing of a 9.7kb genomic region encompassing exons 4 to 9 in 71 individuals from 3 different populations has shown at least 69 variable sites [542]. All were neutral changes on primary protein sequence. Several gene polymorphisms were in strong linkage disequilibrium, which brings additional evidence of the great stability of the LPL locus organization across evolution [512]. However, mutation hot spots and non-random patterns of recombination were observed, challenging association studies with one or several neutral SNPs. These findings ruled out previous inconsistency found between association studies of lipoprotein traits or atherosclerosis with neutral SNPs of the LPL gene. Several RFLP markers were also used for haplotyping LPL gene mutations causing familial chylomicronemia. A founder effect was shown for the Gly188Glu, in French Canadians and in South Africans of Indian origin [542]. This mutation as well as Pro207Leu, and Asp250Asn mutations explain 97.3% of cases with LPL deficiency among French Canadians [544]. These mutations were found in families originating from France [545]. However, the most prevalent mutation in French Canadians (Pro207Leu) was associated with a haplotype distinct from that found in French patients, however similar to a haplotype found in a German patient. A Ser259Arg founder mutation has been found in

Moroccans of Berber ancestry [546]. In contrast, a recurrent Ile194→Thr mutation was found in subjects originating from South Africa and Holland [547].

Homozygous LPL deficiency is the cause of familial chylomicronemia [506,548]. Havel and Gordon demonstrated in 1960, that the metabolic basis of familial chylomicronemia was an impaired catalytic activity of LPL [549]. Thereafter, the disorder was distinguished from other familial hyperlipidemias on the basis of it's recessive mode of inheritance, the absence of circulatory inhibitors, the detection of a functional apo C-II and most important, a low LPL activity in plasma after intravenous injection of heparin. LPL deficiency is observed worldwide. It manifests early in life (80% before age 20) with recurrent abdominal pain, failure to thrive in infants, fasting chylomicronemia or acute pancreatitis. Symptoms (hepatosplenomegaly, eruptive xanthomatosis and lipemia retinalis) and risk of acute pancreatitis have been shown to evolve as a function of the intensity of hypertriglyceridemia (10-100 folds normal) and of chylomicronemia. They are strongly inducible by dietary fat intake.

Naturally occurring mutations of the LPL gene causing familial LPL deficiency are summarized in *table 2-6*. The great majority of patients with homozygous LPL deficiency have plasma LPL activity below 10% of normal. Therefore, levels of plasma LPL activity appear strongly predictive of an underlying LPL gene mutation. Nonsense mutations (premature stop codon, frameshift, splice site mutations or large gene rearrangements) occur all along the gene sequence. However, 78% missense mutations (single amino-acid change) reported to cause complete LPL deficiency, were found within exons 4, 5 or 6. Structure-function studies and the high degree of sequence conservation in these exons predicted that a defective lipolysis in plasma would result from mutations located within this region. The combined study of enzymatic mass and activity in plasma after heparin injection has defined three classes of functional defects in-vivo [550]. Class I mutations correspond to the absence of the enzyme in plasma. Class II defects correspond to absent catalytic activity, and class III defects to absent heparin or heparan sulfate binding. Studies of natural mutants have confirmed functional domains of the lipoprotein lipase previously located by site directed mutagenesis in-vitro: dimerization [551], apo C-II interaction [523] and heparin binding [524]. Nonsense mutations appear to express earlier in life, than missense mutations. This is reminiscent of the early onset of the disease in LPL knock out mice (an animal model of LPL null allele), in which the absence of LPL is lethal at birth, soon after suckling [552]. A residual ability to clear some of circulating triglyceride-rich lipoproteins may account for a delay in disease manifestation in carriers of missense alleles. In

Table 2-6. Naturally occurring mutations of the LPL gene causing familial chylomicronemia.

Mutation	Location	Type	Defect	Population	Ref
G→C	Intron 1	5' Splice	Class I	Italy	553
Del 3nt	Intron 1	3' Splice	Class I	Austria	554
Del 11 nt	Exon 2	Ala^{17} (Fs)	Class I	Spain	555
Ins A	Exon 2	Glu^{35} (Fs)	Class I	France	337
G→A	Intron 2	5' Splice	Class I	Japan	506
G→A	Intron 2	3' Splice	Class I	France	506
Del. 6kb	Exons 3-5	-	Class I	UK	506
T→A	Exon 3	Tyr^{61}→STOP	Class I	Japan	506
G→A	Exon 3	Trp^{64}→STOP	Class I	USA	506
G→C	Exon 3	Val^{69}→Leu	Class II	Holland	506
Del GT	Exon 3	Val^{69} (Fs)	Class I	France	556
Del GC	Exon 3	Ala^{71}(Fs)	Class I	France	557
C→A	Exon 3	Tyr^{73}→STOP	Class I	Germany, Ireland	521
A→C	Exon 3	Arg^{75}→Ser	Class II-III	Germany, Ireland	521
T→C	Exon 3	Trp^{86}→Arg	Class II	UK, USA	506
T→G	Exon 3	Trp^{86}→Gly	Class II	Europe	560
A→G	Exon 3	Thr^{101}→Ala	Class II	France	557
Ins 5 bp	Exon 3	Lys^{102}(Fs)	Class I	Malaysia	506
C→T	Exon 3	Gln^{106}→STOP	Class I	Germany, Poland, UK, USA	506
Del AACT	Exon 4	Asn^{120}(Fs)	Class I	France, USA	556,558, 559
A→G	Exon 4	His^{136}→Arg	Class II	USA	506
G→A	Exon 4	Gly^{139}→Ser	Class II	Spain	506
G→A	Exon 4	Gly^{142}→Glu	Class II	USA, Europe	506
G→A	Exon 4	Gly^{154}→Ser	Class II	Holland	506
A→G	Exon 5	Asp^{156}→Gly	Class II	Holland, Turkey	506
G→C	Exon 5	Asp^{156}→His	Class II	France	561
G→A	Exon 5	Asp^{156}→Asn	Class II	Italy	506
C→G	Exon 5	Pro^{157}→Arg	-	Holland	506
G→A	Exon 5	Ala^{158}→Thr	Class II	Europe	560
C→G	Exon 5	Ser^{172}→Cys	Class II-III	India	506
G→A	Exon 5	Ala^{176}→Thr	Class II-III	USA, black	506
C→G	Exon 5	Asp^{180}→Glu	Class I	Italy	562
C→G	Exon 5	His^{183}→Asp	Class II	Russia, Switzerland	563
C→G	Exon 5	His^{183}→Gln	Class II	Europe	560
G→A	Exon 5	Gly^{188}→Arg	Class II	France	560
G→A	Exon 5	Gly^{188}→Glu	Class II	Worldwide recurrent	506
C→G	Exon 5	Ser^{193}→Arg	Class II	Europe	560

Mutation	Location	Type	Defect	Population	Ref
T→C	Exon 5	Ile^{194}→Thr	Class II	Europe, South Africa, USA	506
G→A	Exon 5	Gly^{195}→Glu	Class II	USA Hispanic	551
C→G	Exon 5	Asp^{204}→Glu	Class II	Japan	506
T→G	Exon 5	Ile^{205}→Ser	Class II	Spain	506
C→T	Exon 5	Pro^{207}→Leu	Class II	Quebec, USA, Europe	506,545 560
T→A	Exon 5	Cys^{216}→Ser	Class II	Italy	506
Del G	Exon 5	Ala^{221}(Fs)	Class I	Japan	506
T→C	Exon 5	Ile^{225}→Thr	Class II	Holland	506
A→C	Exon 6	Cys^{239}→STOP	Class I	Japan	564
G→T	Exon 6	Arg^{243}→Leu	Class II	Holland	565
G→A	Exon 6	Arg^{243}→His	Class II	USA, Italy, China, Holland	506
C→T	Exon 6	Arg^{243}→Cys	Class II	France, Germany	506
T→A	Exon 6	Ser^{244}→Thr	Class I	France	506
G→A	Exon 6	Asp^{250}→Asn	Class II	Europe, Quebec, New Orleans	506,545
C→G	Exon 6	Ser^{251}→Cys	Class II	Holland	506
C→G	Exon 6	Leu^{252}→Val	Class II	Taiwan, China	566,567
T→G	Exon 6	Leu^{252}→Arg	Class II	China	567,568
T→A	Exon 6	Leu^{253}→STOP	Class I	Europe	560
A→G	Exon 6	Ser^{259}→Gly	Class II	Germany	569
T→A	Exon 6	Ser^{259}→Arg	Class II	Morocco	546
G→A	Exon 6	Ala^{261}→Thr	Class II	China	506
C→A	Exon 6	Tyr^{262}→STOP	Class II	Germany	570
T→C	Exon 6	Tyr^{262}→His	Class II	USA	506
T→C	Exon 6	Leu^{286}→Pro	Class II	France	557
Del A	Exon 6	Asn^{291}(Fs)	Class I	Japan	571
T→C	Exon 6	Met^{301}→Thr	Class II	Europe	560
C→A	Exon 6	Tyr^{302}→STOP	Class I	Sardinia	572
T→C	Exon 6	Leu^{303}→Pro	Class II	Europe	560
Duplic.	Exon 6	-	Class I	Europe	506
C→A	Intron 6	3' splice (-3)	Class I	Austria	573
G→A	Exon 7	Ala^{334}→Thr	Class II	Japan	574
C→G	Exon 8	Leu^{365}→Val	Class I	Italy	553
G→A	Exon 8	Trp^{382}→STOP	Class II-III	Japan, Caucasian	506
A→T	Exon 8	Glu^{410}→Val	Class II	Egypt	575
G→A	Exon 8	Glu^{421}→Lys	Class II	South Africa	576
Del 2 kb	Exon 9	Intron 8-*Alu* Intron 9	Class I	France	517

contrast, pancreatitis has a similar prevalence in carriers of nonsense or missense mutations. This suggests that LPL mass is not a protective factor in this case. However, pregnancy stands out as a risk factor for acute pancreatitis [506,548]. More than 2/3 of LPL-deficient women suffer from their first episode of pancreatitis during the last trimesters of pregnancy. Interestingly, pregnancy may predispose as well, to bouts of pancreatitis in heterozygous carriers of mutations of the LPL gene [568,576,577]. Moreover, mass assessment in conjunction with mutation identification may be useful in exploring the relationships between LPL deficiency and development of atherosclerosis. It was shown recently that LPL catalytic activity was not necessary for lipoprotein binding and HSPG interaction [578]. The development of peripheral atherosclerosis and of coronary artery disease was described in patients aged over 50 with complete LPL deficiency [557,579,580]. Some of them were carriers of missense mutations, with low to normal levels of LPL mass in plasma. These findings appear to contradict the classical assumption that LPL deficiency would not predispose to atherosclerosis, since these patients are lean, follow a low fat diet and since plasma levels of LDL cholesterol are low. However, together with profound disturbances in lipoprotein metabolism (delayed clearance of triglyceride-rich particles, impaired HDL metabolism), which may in turn activate thrombogenesis and lipoprotein modifications, the presence of an inactive enzyme may favor the binding and retention of lipoproteins within the arterial wall, thereby inducing foam-cell formation [509]. Indeed chances were low to observe cardiovascular disease in complete LPL deficiency because the disease is rare and manifests early in life. However with life-long low fat diet (a preventive measure against pancreatitis), several patients may reach and exceed their fifties giving more opportunities to observe such situations.

Consanguinity was also considered a predominant feature in LPL deficiency. Indeed, a founder effect has been reported for a small number of mutations in inbred population with a high prevalence of homozygous mutations [544,546]. However, except for mutations Gly188Glu, Ile194Thr, Arg243His and Asp250Asn found in patients of different ancestries, mutations usually differ from one family to the other [506,548]. Moreover, cases of recurrent mutations have been reported, the Gly188→Glu mutation being most caricatural since it has been found in numerous subjects from every continent. The majority of recurrent mutations result from a transition on a CpG dinucleotide (*figure 2-5*). Therefore, the disorder has greater allelic heterogeneity than initially predicted. However, there is a striking difference between the high rate of polymorphisms at the LPL locus in the non-coding sequences, compared with the confinement of disease causing mutations to specific domains of the gene. This gives further evidence of the selective pressure, which operated on the evolution of lipases.

Beside the identification of the molecular basis of a rare lipoprotein disorder (complete LPL deficiency), the most remarkable data were brought by analysis of partial LPL deficiency in heterozygous carriers and in carriers of common gene variants (*Table 2-7*). Heterozygous LPL deficiency was classically regarded as silent in humans, fasting lipid profile and LPL activity being usually normal. However, the possibility to distinguish between carriers and non-carriers by direct genotyping has changed this view. Consistent observations in several thousands of subjects worldwide, suggest that partial LPL deficiency either caused by rare (<1% in the general population) or common variants are significant modifiers of the lipoprotein profile and of the predisposition to atherosclerosis [581,582]. Common protein variants are particularly prevalent in Caucasian populations. A common promoter variant (-93T/G) is highly prevalent in populations of African origin. It is in strong linkage disequilibrium with the Asp9Asn mutation in Caucasians. However lipoprotein profile modifications associated with the -93T/G promoter variant were less consistent between populations, than for protein variants. Despite the fact that common protein variants do not cause familial chylomicronemia in homozygous carriers, site-directed mutagenesis experiments have demonstrated that they modify LPL activity in vitro. Consistently a low HDL-high TG trait is found in carriers compared with non-carriers of the Asp9Asn, Asn291Ser and Gly188Glu mutations [583-585]. Moreover after oral fat load carriers accumulate atherogenic triglyceride-rich lipoproteins and may have increased levels of small dense LDL. In keeping, this trait may be worsened to frank combined hyperlipidemia or to hypertriglyceridemia in conjunction with interacting factors such as diabetes, increased BMI, apo E isoforms, high fat diet etc. Moreover, prevalence of heterozygous LPL gene mutations is increased in familial combined hyperlipidemia or in several forms of

Table 2-7. **Common LPL protein polymorphisms** [581,582]. A summary of effects of variants on LPL activity in vivo and in vitro, and on plasma lipoprotein profile. TC: total cholesterol, TG: triglycerides. Frequencies are given for the general population in Caucasians.

Variant	Asp9Asn	Asn291Ser	Ser447Stop
Frequency	2 - 3 %	2 - 6 %	10 - 20%
LPL activity	↓	↓↓	↑ / ↔
TC	↔	↔	↔
TG	↑	↑↑	↓
LDL Cholesterol	↔	↔	↔
HDL Cholesterol	↓	↓↓	↑

hypertriglyceridemias. LPL protein variants deteriorate the phenotype of heterozygous FH [199]. An interesting link was found between high blood pressure and the LPL gene locus by quantitative linkage mapping and by a careful examination of heterozygous carriers of LPL gene mutations [586,587]. In keeping, Asp9Asn and Asn291Ser mutations have a high prevalence in pre-eclampsia [588]. This may reflect in vivo, functions of LPL at the endothelial-cell surface and in the arterial wall.

Unexpectedly, the common mutation Ser447Stop does not impair LPL catalytic activity, although a functional effect is suspected in vivo. The mutation creates a premature stop codon on the amino acid before the least of LPL primary sequence. However, LPL activity is normal in vitro and mass may be increased as compared with the wild type allele [581,582]. In this case the associated phenotype is opposite: high HDL-low triglyceride. This mutation appears significantly protective against atherosclerosis in populations at large [589,590]. In keeping, with previous observations about apo E alleles and longevity (see below) the Ser447stop mutation is another protective variant, which may determine yet unknown cellular and tissue functions at the LPL gene locus.

Data from molecular genetics of a very rare lipid disorder have demonstrated that the same locus may be involved as a modifier gene in common atherogenic dyslipidemia. They confirm the rate-limiting role of LPL in the regulation of lipoprotein metabolism and its position as an important target in the prevention of atherosclerosis.

2.2.12-   Hepatic   Lipase,   a   Lipase   for   Lipoprotein
            Remodeling

Hepatic lipase (HL) is a triacyl-glycerol hydrolase synthesized and secreted by hepatocytes [591-593]. It is bound to HSPG at the surface of sinusoid cells, the hepatic counterpart of capillary endothelium. Contrary to LPL, it does not require any cofactor for its lipolytic activity. Beside its triacyl-glycerol hydrolase activity it may hydrolyze phospholipids, suggesting its involvement in lipoprotein remodeling. It is specifically active on apo B-rich lipoproteins of smaller size like IDL (Intermediate Density Lipoproteins), accelerating their conversion into LDL. HL also mediates the liver uptake of triglyceride-rich lipoproteins by LRP [594]. Another important function of hepatic lipase is to remove triglycerides from the core of HDL_2 (large HDL) converting them into HDL_3 (*see figure 2-28*). Hepatic lipase also promotes the selective uptake of HDL cholesterol by the liver generating native HDL able to shuttle back to the reverse cholesterol transport pathway [595]. Although steroidogenic organs do not express HL, it is abundantly found there, enhancing selective uptake of cholesteryl esters by SR-BI [73,596,597]. Therefore hepatic lipase plays a significant role in atherogenic particle clearance and in the generation of anti-atherogenic particles.

The gene for hepatic lipase (LIPC) has a structure very similar to that of LPL [598,599]. It is single copy, spanning approximately 35kb. It is composed of 9 exons, localized on the long arm of chromosome 15, in 15q21 [513]. It encodes for 499 amino acids, of which 23 encode for a signal peptide and 476 amino acids encode for a mature protein of 53kD. Exon-intron breakpoints stop on the same nucleotides for both genes and exons 3, 4, 5, 6, and 7 have an identical length (*figure 2-19*). Similarly to LPL, exons 4, 5, and 6 encode for the catalytic site, which includes the triad: Ser145-Asp171-His256. There are also N-glycosylation sites, of which Asn56 is necessary for secretion. Potential sites for heparan sulfate binding are found in exon 6. The noncoding sequence within the exon 9, in 3' of the mRNA is much shorter than in LPL (49 nucleotides). The gene is mainly expressed in the liver [600], consistent with the finding of liver-specific response elements for hepatic nuclear factors (HNF) in the gene promoter. Likewise upstream of the LPL gene there are steroid hormone receptors and cAMP response elements. Estrogens downregulate HL expression. An *Alu* repeat is present

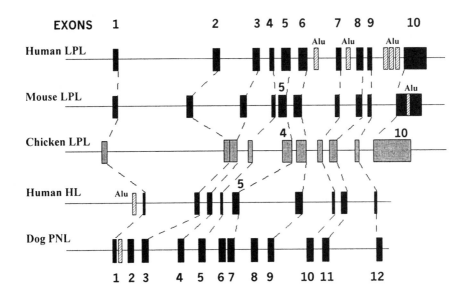

Figure 2-19. **Genomic organization of the lipase superfamily of genes.** Exons are indicated by filled boxes. *Alu* repeats are indicated by hatched boxes. Dotted lines indicate correspondences between lipoprotein lipase (LPL) from different species and between lipoprotein lipase and other lipases: Hepatic Lipase (HL) and Pancreatic Lipase (PNL).

upstream of the promoter. Transgenic experiments have better delineated the role of HL in vivo [508]. Hepatic lipase appears as a major enzyme of HDL remodeling in mice (HL expressors) and in rabbit (a low HL expressor). HL overexpression generates denser LDL and HDL, overall lowering levels of circulating lipoproteins. Interestingly, catalytic activity does not seem necessary for hepatic HDL metabolism [601]. Oppositely HL knock out mice exhibit increased levels of circulating HDL.

Familial hepatic lipase deficiency is a disease seldom reported [593,602], probably because it is difficult to discern from other dyslipidemias. This recessively inherited disease manifests in homozygotes by type IV or type III hyperlipidemia. However, patients with subnormal lipid profiles have been described [603]. Larger buoyant LDL and HDL enriched in triglycerides accumulate in plasma, sometimes with intermediate lipoproteins (IDL). A constant feature seems to be higher levels of larger HDL in plasma. HDL_2 are poor effectors of reverse cholesterol transport, therefore HL deficiency would be atherogenic despite increased concentration of HDL cholesterol in plasma. Plasma hepatic lipase activity measured after heparin injection is significantly reduced. Heterozygotes have usually normal HL activity. Hyperalphalipoproteinemia sometimes associates with moderate combined hyperlipidemia type IIb or type IV. Plasma-clearance of triglyceride-rich lipoproteins is delayed after an oral fat load. Cardiovascular complications after the fifth decade may be present. Naturally occurring mutations have been reported to cause HL deficiency [593,604,605]. Missense mutations occur on conserved amino acids inducing conformational changes disrupting enzymatic activity and protein secretion in-vitro [606,607]. Heterozygous HL mutations are more prevalent in familial combined hyperlipidemia, and associated with higher levels of apo B-rich lipoproteins compared to non-carriers [608,609].

Several gene polymorphisms have been described on the gene for hepatic lipase. More interesting are common gene promoter variants. Sib-pair analysis using neutral HL gene polymorphisms in subjects with normal plasma triglyceride, showed that the hepatic lipase locus could determine 25% plasma levels of HDL cholesterol [610]. Later several polymorphisms of the gene promoter in strong linkage disequilibrium were identified to modulate plasma HDL levels [611]. More particularly a −514C/T polymorphism (or −480C/T, according to different gene numbering) was associated with a significant increase in HDL levels. Despite the fact that in-vitro changes in promoter activity induced by these variants are debated [612,613], changes in HDL metabolism and HL activity were consistently reproduced in several human samples [614-619]. However the association of this variant with coronary disease remains uncertain [615,616,617]. This variant may not be directly involved in HL expression changes in vivo, however it may be linked with other variants, which determine lower HL activity in vivo. In keeping with the example of LPL, partial modifications of HL activity have a clear effect on lipoprotein profile, however with a conditional effect on atherosclerosis. Another promoter variant −250G/A has been associated with changes in HDL metabolism, LDL buoyancy and insulin resistance [620,621].

Therefore, despite an apparently mild phenotype associated with complete HL deficiency, this enzyme may be an important modifier of lipoprotein metabolism through its role on HDL and intermediate lipoproteins.

2.2.13-   Other Lipases

Hormone-sensitive lipase (HSL) hydrolizes triglycerides stored in adipocytes to release free fatty acids, and cholesterol esters in steroid glands, to release cholesterol necessary for steroidogenesis [622]. This esterase is active after reversible phosphorylation by cAMP-dependent kinases that depend upon action of lipolytic hormones (i.e. glucagon, catecholamines, ACTH). Conversely insulin inhibits HSL activity in the adipose tissue. The genomic sequence (LIPE) is unique [623], spanning 11kb. It comprises 9 exons and was localized to the long arm of chromosome 19, 19 cen-q13.3-q13.3 [624]. The gene sequence is highly conserved in mammals having 85% homology with the murine gene [625]. The cDNA encodes for 786 amino acids of a MW 85.5kD mature protein. This lipase does not belong to the LPL superfamily, but rather resembles bacterial lipases, like lipase 2 of *Moraxella* TA144, a bacterial strain from the Antarctic. Exons 6 and 9 encode for the catalytic site containing a consensus pentapeptide for the catalytic site in this family of lipases and a lipid-binding domain [626]. Ser423 seems critical for HSL catalytic activity. Several phosphorylation sites encoded by exon 8 control cAMP-dependent and Ca2+/calmodulin-dependent protein kinase activities [627,628]. There are several mRNA transcripts expressed mainly in the adipose tissue, steroidogenic organs and muscles. Adipose-specific regulatory sequences (AP-2), hormone sensitive (C/EBP), NF-κB and SREBP sensitive elements are found upstream of the gene [625]. An uncoupling mechanism occurs during development between preferential post-transcriptional regulations on triacylglycerol hydrolase activity in the adipose tissue, and transcriptional regulations on cholesterol-ester hydrolase activity in steroidogenic organ [629].

There is no known human disease resulting from hormone-sensitive lipase deficiency. The homology of HSL sequence with a cryoresistant bacterial lipase could mediate a lipolytic activity essential to body temperature homeostasis. In fact, the enzyme is 3 to 5 times more active at 10°C than CEL (see below) or LPL. A balanced regulation is observed in hibernating mammals, which alternate LPL activity and that of hormone-sensitive lipase between summer and hibernation seasons [630]. However, in HSL knock out mice have normal body-temperature adaptation to cold [631]. Moreover, if brown adipose tissue is hypertrophic, as the result of defective intracellular lipolysis and triglyceride accumulation, white adipose tissue (i.e. subcutaneous adipose tissue in mice) was less affected by HSL gene invalidation. Differential regulations are reported between omental and peripheral fat mass, suggesting that HSL is preferably involved in basal energy metabolism homeostasis [632]. However the most striking finding was that male mice were sterile as the result of a defective cholesterol-ester hydrolase activity. Moreover, mice overexpressing HSL in macrophages had a moderate increase in plasma total cholesterol and were susceptible to diet-induced atherosclerosis [633]. This suggests at least in murine species, that HSL may have be more critical for cholesteryl ester tissue mobilization than that for triacyl-glycerols.

Carboxyl-ester lipase (CEL) is the lipase from milk. This multifunctional lipase is synthesized by the pancreas and the mammary gland. It is activated in the digestive tract in presence of bile salts to cleave all ester bonds in lipids (i.e. cholesteryl-esters, triglycerides, retinyl esters and lysophospholipids), facilitating their absorption. It plays a crucial role in the nutrition of newborns. The gene (CEL) spans 9.8kb divided into 11 exons, and was localized to chromosome 9, 9q34-qter [634]. The mature protein is highly glycosylated, with a molecular weight of 100kD. It is not related with other lipases, but rather with serine-esterases (acetyl-cholinesterase, cholinesterase) of the nervous system. The gene is rich in repeated *Alu* repeats of which one is located within the promoter. Moreover, the promoter comprises response elements to glucocorticoids, estrogens, mammary gland-specific and pancreatic-specific transcription factors. The gene is expressed during lactation, confirming physiological data. The protein may also circulate in human plasma and could play a role in lipoprotein remodeling, particularly oxidized lipoprotein remodeling through its lyso-phospholipase activity [635]. A gene, related with CEL is also present on the long arm of chromosome 9, called CELL, for Carboxyl Ester Lipase-Like gene. This gene has 97% homology with the 5' flanking sequences and with 5 exons (1, 8, 9, 10 and 11) of the CEL gene. This gene was suspected to be a pseudogene because of the loss of exons 2, which prematurely interrupts the reading frame. In addition, there is a sequence divergence for exons 10' and 11'. However CELL transcripts are found at low levels in all tissues. Moreover there is an open reading frame for a signal peptide followed by a peptide of 59aa, different from CEL. If this protein were translated, it would lack serine protease activity, but would keep an affinity for lipids. The locus is polymorphic: a RFLP (Eco IH) in the gene CEL and a hypervariable domain rich in GC in the last exon of gene CELL were described [636]. There is no known deficit known for any of these genes, but their regulatory role in the lipid intestinal absorption and in lipoprotein remodeling could influence lipoprotein metabolism.

Pancreatic Lipase (PNLIP) is the digestive lipase, which plays a major role in the absorption of dietary lipids [637]. This lipolytic activity had been identified as early as the 19th century from pancreatic secretions in the dog. The enzyme is active in presence of calcium ions and of a peptidic cofactor, colipase, which stabilizes the enzyme in the presence of bile salts. The primary sequence and the three-dimensional structure of the enzyme were identified by gene cloning and x-ray crystallography [638]. It is a 449 amino acid protein, of molecular weight 49.5kD, which comprises a structure folded around a hydrophobic site. A catalytic triad Ser153-Asp 177-His 264 closed by a mobile «lid» formed by a C-terminal loop represents the active part of the enzyme. The high sequence conservation of lipase genes has allowed predicting a similar secondary structure for other lipases of this family. Gene characterization in dog has revealed a genomic organization divided into 13 exons spanning approximately 35kb, close to that of the putative ancestral gene of the superfamily [512,639]. Genes for pancreatic lipase and colipase were localized on the human genome on chromosome 10 and 6 respectively [640,641]. Pancreatic lipase and colipase deficiencies are characterized by intestinal malabsorption with steatorrhea from childhood [642]. This clinical notion helped to develop inhibitors of pancreatic lipase used in the treatment of obesity [643]. Remarkably, these compounds have a mild effect on lipoprotein profile. To date no natural mutation of pancreatic lipase was reported as a cause of congenital pancreatic lipase insufficiency in humans.

Endothelial lipase (EL) was recently identified as a lipase belonging to the LPL gene family [644]. This lipase, which is primarily a phospholipase is synthesized and secreted by endothelial cells. In addition it is detected in placenta, liver, lung, kidney, thyroid, gonads and macrophages. The cDNA (LIPG) encodes for a 18aa signal peptide followed by a 482aa mature protein, of MW 55kD. It shares strong homology with LPL, being its closest relative cloned so far. The protein is glycosylated and able to bind heparin. It contains the critical catalytic triad Ser-Asp-His of lipases, closed by a lid. EL does not require any cofactor for full catalytic activity. Interestingly, this lipase has low triglyceride-lipase activity despite its capacity to bind lipoproteins. When transiently expressed in the liver in mice, EL markedly decreased circulating HDL and apo A-I, suggesting that this lipase facing the vessel lumen could play a role in lipoprotein remodeling.

Other mechanisms regulating lipolytic activity remain to be identified. The *cld* (Combined Lipase Deficiency) mouse strain has a genetic defect impairing the processing and maturation of LPL and HL out of the ER. In line with LPL KO mice, mice die soon after birth of massive chylomicronemia [645].

2.2.2- Apolipoproteins

2.2.21 - Apolipoprotein C-II, an Activator of Lipoprotein Lipase

Apolipoprotein C-II (apo C-II), is one of the members of small exchangeable apolipoproteins: A-I, A-II, A-IV, C-I, C-II, C-III, E *(figure 2-20)*. This 79 amino acid peptide of molecular weight 8.9kD represents 10% of the protein content of VLDL and 1% of that of HDL [122]. Like in other apolipoproteins, its N-terminal domain contains amphiphathic α-helices, able to bind lipids, and to integrate the apolipoprotein into lipoprotein particles. In the C-terminal domain the LPL activating site is encoded by amino acids 56 to 79 which would be reinforced by a short domain interacting with phospholipids. Apo C-II is a natural activator of lipoprotein lipase [646].

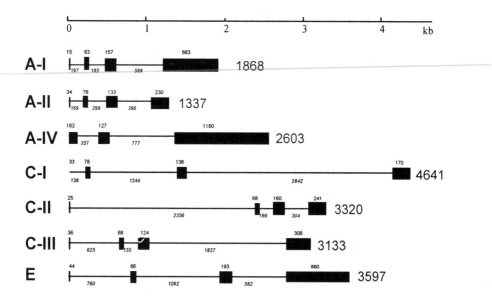

Figure 2-20. **The gene family of small exchangeable apolipoproteins.** Filled bars represent exons. Lines in between represent introns. Lengths in nucleotides are indicated above exons and below introns. The full length of the gene is indicated at the 3' end.

Apo C-II overexpressed in transgenic mice may also inhibit apo E binding to LRP or to the LDL receptor, thereby delaying liver endocytosis of triglyceride-rich particles. It is also an inhibitor of LCAT and of hepatic lipase suggesting that it may be also a modulator of HDL metabolism.

The gene for apo C-II (APOC2) is integrated into a 50kb long cluster, which includes (centromeric towards telomeric *(figure 2-21)* apo E, apo C-I, apo C-II and a recently described apo C-IV [647]. It is a small gene 3320bp long, localized at the long arm of chromosome 19, 19q13.3 [648]. The gene is organized into 4-exons/3-introns, a genomic structure common to the gene family of small apolipoproteins [649]. Exon 1 comprises a noncoding sequence, exon 2 encodes for a signal peptide of 22aa and for the first amino acids of the protein. Exon 3 encodes for a lipid-binding domain, and exon 4 contains the activating sequence for lipoprotein lipase. Three *Alu* repeats are found in intron 1, and a fourth is present in the 3' flanking sequence. The gene is mainly expressed in the liver and to a lesser degree in the intestine.

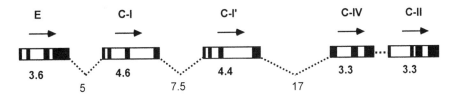

Figure 2-21. **The Apo E-C1-C4-C2 gene cluster.** Filled bars indicate exons. Empty bars indicate introns. Arrows indicate transcription orientation. Dotted lines indicate intergenic regions. Numbers indicate then length in kilobase-pairs of genes and intergenic regions.

The locus is polymorphic. The first intron contains a highly informative microsatellite $(GT)_m(GA)_n$ [650], and intron 3 contains a minisatellite resulting from the repetition of a 37nt motif [651]. There is also a protein polymorphism Lys55→Gln in 10% of normolipidemic subjects originating from Black Africa [652]. The change of charge resulting from this substitution does not disturb LPL activation, though the functional effects of this variant remain unknown.

Familial apo C-II deficiency is the second cause of familial chylomicronemia. It closely resembles the syndrome caused by LPL deficiency, although with small differences [506]. The onset is retarded, from adolescence until beyond the fifth decade. Plasma triglyceride levels are lower, however bouts of acute pancreatitis are frequent. Moreover, in the majority of cases neither eruptive xanthomatosis, nor lipemia retinalis were described. Dietary fat tolerance would be better compared to LPL deficiency. Several defective mutations were described on apo C-II as a cause for this syndrome *(table 2-8).* In many cases nonsense or frameshift

Table 2-8. **Naturally occurring mutations of apolipoprotein C-II.**

Mutation	Location	Type	Phenotype	Reference
A→G : Koln	Promoter	-86	decreased expression	653
A→G : Paris-1	Exon 2	Met^{-22}→Val	nul allele	654
C→T : Paris-2 Barcelona	Exon 2	Arg^{-19}→STOP	nul allele	506,555
G→C : Hamburg Japan	5' intron 2	Splice	nul allele	646,654
ΔC : Japan Venezuela	Exon 3	Gln^2frameshift	nul allele	654
ΔG : Nijmegen	Exon 3	Val^{18}frameshift	nul allele	654
T→C : San Francisco	Exon 3	Lys^{19}→Thr	-	646
T→C : Japan	Exon 3	Ser^{21}→Pro	-	654
T→A : Japan	Exon 3	Trp^{26}→Arg	-	646
C→A : Padova	Exon 3	Tyr^{37}→STOP	truncated protein	654
C→G : Bari	Exon 3	Tyr^{37}→STOP	truncated protein	506
G→A : San Francisco	Exon 3	Glu^{38}→Lys	normal LPL activation	655
ΔT : Toronto	Exon 4	Thr^{68}frameshift	truncated protein	506
Ins 1bp : St Michael	Exon 4	Asp^{69}frameshift	longer protein	506
C→T : Japan	Exon 4	Gln^{70}→STOP	-	654
A→T : Japan	Exon 4	Glu^{79}→Val	-	654

mutations abolish transcription or generate abnormal proteins in their 3' domain. Naturally occurring mutations have confirmed the importance of apo C-II C-terminal domain for LPL activation. The limited number of apo C-II mutations could result from a biased recruitment, selecting only for the most severe cases. Apo C-II mutations are a less frequent cause of familial chylomicronemia, compared to the numerous LPL gene mutations. This may be the consequence of a milder phenotype and from length difference in functionally relevant sequences. Mutation hits having more statistical chances to be observed in longer genomic regions (i.e. LPL gene) than in shorter ones (i.e. APOC2).

Information is still limited on heterozygous apo C-II deficiency. [652,655]. Missense mutations Ile19→Thr and Glu38→Lys were described in heterozygotes but it did not segregate with hyperlipidemia in the family [646,656]. It was assumed that these mutations -like the Lys55→Gln polymorphism did not cause hyperlipidemia directly, but they could intensify the effect of other lipid-increasing factors. Family analysis of a subject carrying the Apo C-II$_{Hamburg}$ mutation showed low levels of apo C-II however plasma triglycerides were inconstantly elevated in heterozygotes [657]. Members of the same family had suffered from coronary events. As in the case of LPL, it remains to be determined whether heterozygous apo C-II deficiency is a factor for atherogenic dyslipidemia.

A previously unrecognized gene encoding for a novel apolipoprotein, **apo C-IV** (APOC4), is located 555bp upstream of apo C-II, in the same transcriptional orientation [658]. The gene comprises 3 exons encoding for a protein of 127aa and for a 25aa signal peptide. There are two amphipathic α-helices in the C-terminal domain. Transcripts are found at low levels in the liver. Overexpression of the protein in transgenic mice induced hypertriglyceridemia as the result of lipolysis inhibition [659].

2.2.22-   Apolipoprotein C-I, A Modulator of Lipoprotein Catabolism.

Apolipoprotein C-I (apo C-I) is the shortest among apolipoproteins, 57aa for a molecular weight of 6.6kD. It represents 10% of proteins in VLDL, and 2% of those in HDL [122]. As for other apolipoproteins it comprises a domain rich in amphipathic α-helices, allowing its integration into lipoproteins. Transgenic experiments in mice shed light on its previously unknown role in humans [646,660]. Transgenic mice accumulate large amounts of VLDL in their plasma as the result of a delayed catabolism. Apo C-I does not impair lipolysis, but prevents binding of VLDL and VLDL remnants to LRP and VLDL receptors. Apo C-I inhibitory effect is likely to result from the displacement of apo E from VLDL and β−VLDL preventing efficient lipoprotein binding with cellular receptors. However, apo C-I does not seem to impair apo B binding to LDL receptors. Apo C-I also interferes with HDL metabolism. Apo C-I is a strong activator of LCAT, whereas it decreases CETP activity. Moreover, apo C-I modulates enzyme accessibility to the surface of lipoprotein particles. Mice strongly overexpressing apo C-I in the skin have reduced sebaceous secretions favoring skin and hair lesions. However, the role of apo C-I in human skin remains unclear. Interestingly, apo C-I knock out mice are normolipidemic on a chow diet and become hyperlipidemic when fed an atherogenic diet. One rationale is the relative enrichment of VLDL particles in apo A-I and apo A-IV, which in turn would impair apo E accessibility to cellular receptors. Therefore apo C-I acts as a modulator of lipoprotein metabolism, by competing with other small exchangeable apolipoproteins in their interactions with circulating enzymes and cellular receptors thereby prolonging triglyceride-rich lipoprotein residence time in plasma.

The apo C-I gene (APOC1) has a 4-exons/3-introns organization common to exchangeable apolipoproteins *(figure 2-20)*, [661]. There are numerous intragenic repeated *Alu* sequences including 2 in intron 2 and 7 in intron 3. It is located within the apolipoprotein gene cluster on chromosome 19, 5.3kb telomeric from the apo E gene, and in the same transcription orientation [647]. While apo E has a diversified tissue expression, apo C-I is mainly expressed in the liver. The physical proximity of both genes (5kb apart) as well as the combined induction of their expression in macrophages has suggested the possibility of concerted regulations. Studies on transgenic animals overexpressing either apo E or apo C-I, or both, showed that a 319bp domain controlling hepatic expression of apo C-I and apo E genes (Hepatic Control region, HCR) was located 5kb downstream from apo C-I, and upstream from apo C-I' pseudogene *(figure 2-22)*. In vivo footprinting have identified complementary liver-specific responsive sequences overlapping an *Alu* sequence adjacent to the HCR [662]. Indeed, transgenic mice overexpressing only human apo C-I, have combined hyperlipidemia whereas strains overexpressing apo E and apo C-I, are normolipemic [660]. The presence of both apolipoproteins in the same lipoprotein particles suggests that concerted regulations of apo C-I and apo E may modulate triglyceride-rich lipoprotein metabolism in the liver.

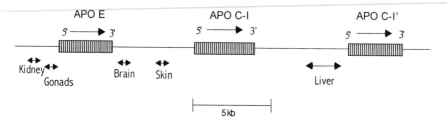

Figure 2-22. **Regulatory sequences of tissue-specific expression of apo C-I and apo E.** Hepatic control regions (HCR) are located far downstream from both genes.

There is a pseudogene apo C-I' 7.5kb downstream from the apo C-I gene. Indeed, a nonsense mutation in the signal peptide Gln-2→STOP interrupts prematurely protein translation. However, this pseudogene is highly similar: 91% sequence homology with the apo C-I gene and repeated *Alu* sequences within introns 2 and 3. *Alu* sequences have been suggested to mediate genomic conversion events favoring exon shuffling observed throughout the evolution of certain gene families like that of the LDL receptor [136]. *Alu* sequences are located within introns 2 and 3 of apo C-I and C-I' genes, between exons encoding for sequence motifs common to small apolipoproteins. Therefore the gene family of small apolipoproteins (*figure 2-23*) may have also evolved through exon shuffling containing ancestral motifs close to those of apo C-I [649]. Comparative analysis of these sequences allowed dating the gene duplication of apo C-I back to approximately 40 million years. Other gene duplications including the HCR region followed later 37 million years ago, contemporary to the separation between Old and New World monkeys [663].

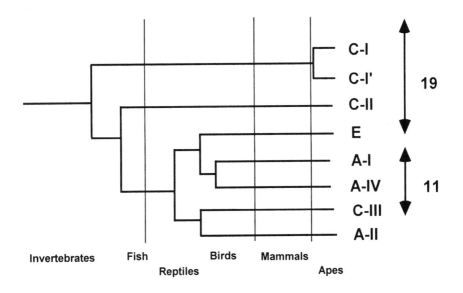

Figure 2-23. **A phylogeny of small exchangeable apolipoproteins across the evolution of animal species.** Genes for apo A-I, C-III, A-IV are clustered on chromosome 11. Genes for apo C-I, C-II and E are clustered on chromosome 19. Apo A-II is localized on chromosome 1.

There are several gene polymorphisms at the apo C-I locus detected by enzymatic restriction (BglI, DraI and HpaI) in strong linkage disequilibrium with apo E polymorphism [664-666]. No human pathology has been linked to defective mutations of apo C-I. However, in line with observations in transgenic mice, post-prandial accumulation of apo C-I rich lipoproteins is exaggerated in normolipidemic CAD patients in comparison with unaffected controls [667]. Moreover, the HpaI polymorphism has been identified as an insertion/deletion polymorphism of a 4-bp sequence: "CGTT" at position -317 from the transcription start site [668]. Interestingly, the common allele corresponding to the 4-bp deletion (H1) was associated with reduced transcriptional activity. On the other hand, the rare allele corresponding to the 4-bp insertion (H2) showed an increased transcriptional activity and a defective binding to a nuclear factor bound to the H1 allele. This suggested that the rare allele increased apo C-I expression through the decreased binding of a negatively acting transcription factor. However modulatory effects of apo E polymorphism interfered with the interpretation of lipoprotein changes in human carriers of this polymorphism. Further studies will clarify this point.

Experimental data have suggested that apo C-I overexpression rather than defective expression would favor the development of atherogenic dyslipidemia in humans. Apo C-I acts as a modulator at the surface of lipoproteins, delaying triglyceride-rich lipoprotein clearance. It remains to be determined if protein redundancy of the family of small apolipoproteins does not compensate in vivo for apo C-I dysregulated expression.

2.2.23-   Apolipoprotein E, the Major Apolipoprotein for Intermediate Lipoproteins.

Apolipoprotein E (Apo E) represents 20% of the protein content in VLDL and 2% in HDL. It is a natural ligand for many cell-surface receptors of the LDL-receptor gene family (LDLR, LRP, megalin, VLDLR, Apo ER2 etc.) and for heparan sulfate proteoglycans (HSPG) [668,669]. It plays a major role in triglyceride-rich lipoprotein metabolism and modulates cellular lipid exchanges with extracellular spaces. This apolipoprotein of 299 amino acids and of molecular weight 34 kD, is expressed mainly in the liver. However, it is ubiquitously expressed in many tissues, particularly those in high lipid demand: acutely for tissue renewal and repair, or in cells of the immune system, or continuously in steroidogenic organs, or in central nervous system. It comprises two domains, individualized by thrombin cleavage. The N-terminal domain (amino acids 1 to 191) contains five amphipathic α-helices of 22aa, characteristic of apolipoproteins, and a domain rich in basic amino acids in domain 140-160 [670]. This domain which interacts with cellular receptors has the form of a four α-helix bundle [671]. Its action is reinforced by residues 171 to 183, which would also interact with heparan sulfate proteoglycans [672]. The N-terminal domain is stable in aqueous medium, which is not the case for the C-terminal domain rich in amphipathic α-helices, involved in lipoprotein core-lipid binding. Moreover apo E is able to form multimers by self-association in lipid-poor lipoproteins.

The crucial role of apo E in lipoprotein metabolism and atherosclerosis was extensively studied from its basic effects in cellular models and transgenic animals up to its clinical impact in human patients and large populations. Transgenic mice have demonstrated that apo E participates in every step of intermediate lipoprotein metabolism [673]. Apo E enhances VLDL assembly and secretion in the liver. In peripheral tissues, it mediates binding with cell-surface HSPG, and enhances lipolytic activity by LPL. In peripheral tissues as well as in the liver, this two-step mechanism (HSPG binding/lipolysis, followed by cell surface receptor binding) facilitates lipoprotein uptake by cellular receptors of the LDL receptor gene family (see page 120) [674]. Moreover, apo E enhances hepatic lipase activity, enhancing apo E redistribution to HDL, selective uptake of cholesteryl esters and receptor mediated lipoprotein endocytosis [675]. Knock out mice or transgenic mice overexpressing defective apo E are massively hyperlipidemic and spontaneously susceptible to severe and premature atherosclerosis [673,676-678]. By reproducing many aspects of human pathology, apo E knock out mice have been widely recognized as an animal model of genetically induced atherosclerosis. These models have established that disorders of triglyceride-rich lipoproteins metabolism are spontaneously and highly atherogenic in vivo in concert with Apo E playing a central role in lipoprotein metabolism.

Apo E has local anti-atherogenic properties in peripheral tissues. A mechanism of local secretion-recapture has been described for apo E [679]. It has been well documented in the liver. Apo E is secreted in the space of Disse and recaptured by hepatocytes after apo E enrichment of lipoproteins. Local apo E secretion from within hepatocytes seems necessary [680]. The same phenomenon is observed in activated macrophages. [681] Therefore, apo E may satisfy cellular lipid demand, by offering new binding sites for membrane receptors at the surface of lipoproteins coming close to the cell surface. As a consequence the catabolism of apo E-containing lipoproteins is accelerated, which could be a protective mechanism against lipoprotein retention in the arterial wall [682]. More surprisingly various cell types (fibroblasts, hepatocytes or macrophages) are able to produce native HDL particles containing only apo E (γ-LpE) which could ensure cellular cholesterol efflux [683]. Together with apo E being a moderate activator of LCAT, the role of apo E in cellular cholesterol efflux could reduce foam-cell formation through the export excessive cholesteryl esters. Apo E is also a natural antioxidant and a modulator of inflammatory response in lymphocytes and macrophages. Apo E may inhibit cytokine production, smooth muscle cell proliferation or platelet aggregation by altering intracellular signalling responses in activated cells [681]. Thus, apo E may be locally anti-atherogenic by accelerating cellular catabolism of atherogenic lipoproteins and by stimulating anti-atherogenic mechanisms.

The local role of apo E within peripheral tissues startled on attractive hypotheses on cellular lipid homeostasis [684]. Contrary to other apolipoproteins apo E has no propeptide and free mature protein is detected in several intercellular compartments. Interestingly, exogenously derived apo E (after endocytosis) is not found in the same compartments as endogenously produced apo E [679]. In the central nervous system Apo E is present at the surface of HDL in the cerebrospinal fluid and can be secreted by vessel astrocytes and macrophages. Following a nervous injury, these cells secrete apo E to mobilize extracellular cholesterol and enhance lipoprotein endocytosis. Schwann cells use these lipid sources to accelerate the myelinating process. Even more surprising, apo E at the surface of lipoproteins acts on neuronal remodeling, enhancing neuronal outgrowth and neurite extension [685]. In addition, apo E promotes cholesterol efflux from astrocytes and neurons. Interactions between apo E at the surface of HDL or β-VLDL and LRP seem necessary to this process [686-688]. Apo E is present within cells from the central nervous system suggesting that its role in neurotrophicity is mediated by yet unknown intracellular mechanisms. A paracrine role of apo E secreted by granulosa cells was also suggested to modulate ovarian follicular maturation by inhibition of androgen secretion [689]. Apo E modulates lymphocyte inflammatory response, suggesting that it could participate to cell protection within damaged tissues. The modulator role of apo E on lymphocytes or in the ovary could involve apo E dimerization [690]. The local role of apo E in peripheral tissues, in response to various stimuli open on fascinating biological questions which largely extend beyond the limits of lipoprotein metabolism and atherosclerosis.

The gene for apo (APOE) is located at the centromeric end of the E-CI-CIV-CII gene cluster of chromosome 19 [647] (*figure 2-21*). It is a small gene 3597bp in length, with a 4-exons/3-introns genomic structure common to exchangeable apolipoproteins [648] (*figure 2-20*). It is localized 5 kb upstream of apo C-I, with which it is subjected to concerted hepatic regulation, mediated by HCR elements, downstream form apo C-I [662]. The first exon is noncoding, the second exon encodes for a signal peptide, and exon 3 encodes for the first 61 amino acids of the mature protein. Exon 4 encodes for most of the mature protein and comprises a hundred noncoding nucleotides in its 3' end. There are several *Alu* repeats one is located approximately 370 nucleotides upstream of the start codon. Two others are head-tail oriented within intron 2 and 150 nucleotides downstream from exon 4. Exon 4 has a particular structure in that it contains 8 repeats of 66 nucleotides encoding for amphipathic α-helices. This domain is very rich in CpG dinucleotides, making of apo E one of the richest human proteins in CpG islands. When CpG islands are methylated upstream of genes, gene expression is abolished. In an unexpected way, this coding domain of the apo E might be methylated in all human tissues including liver, which strongly expresses apo E [691]. This domain is not methylated only in sperm cells. The same phenomenon is observed in baboon and to a lesser degree in rat. Moreover, CpG islands found in similar gene regions of apo A-I and apo A-IV are not methylated. These discrepancies on patterns of apo E fourth exon methylation remain unexplained. However, methylation could have given rise to apo E protein polymorphism in humans, and to deleterious apo E mutations occurring on CpG dinucleotides in this domain (*figure 2-24*).

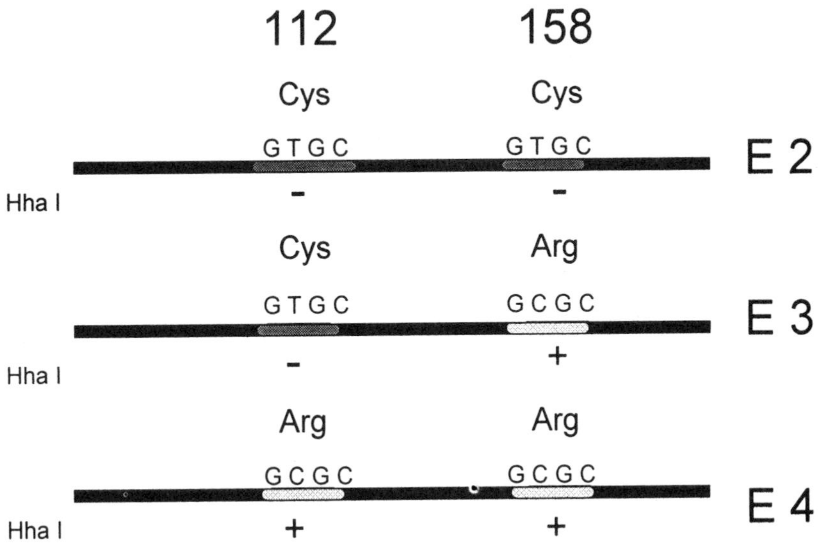

Figure 2-24. **Apolipoprotein E polymorphism.** The genomic sequence encoding for residues 112 and 158 may vary between individuals. Nucleotide changes modify restriction sites for Hha I, allowing direct genotyping.

Apo E expression is ubiquitous. The liver produces 90% of circulating apo E [692]. However, peripheral tissues may locally produce apo E at high rates (in gonads, adrenals or brain). Numerous factors regulate apo E expression and their coordinated actions appear sometimes unclear. At the DNA level apo E is expressed in various tissues under the control of sequence elements located 5' upstream and in the Apo E-apo C-I intergenic domains (*figure 2-22*). Sites sensitive to estrogens and to other steroids, or to *c-fos* and Sp1 sensitive are present in the promoter [693]. Apo E may be glycosylated on Thr194 [694] although it is not necessary for apo E secretion [695]. In hepatocytes fatty acids, cholesterol, lipoprotein endocytosis, a sucrose-rich diet and estrogens stimulate synthesis and secretion of apo E. Conversely, insulin inhibits hepatic synthesis of apo E [696]. Whereas cholesterol and steroid hormones seem to act on transcriptional level, lipoproteins act at a final step of apo E maturation and secretion [697]. In steroidogenic glands apo E synthesis could be induced by cAMP and protein kinases A and C signalling, under the influence of peptide hormones or of intracellular cholesterol influx [698]. Apo E synthesis and secretion are induced by TNFα in activated macrophages [681]. Apo E transcription and synthesis are detectable only in atherosclerotic plaques, whereas apo E gene remains silent in the surrounding healthy arterial wall [699]. The same phenomenon is observed in macrophages involved in nerve or axonal regeneration following an injury [684].

Apolipoprotein E is polymorphic. Three common alleles ε2, ε3 and ε4, are observed worldwide [700]. They result from C→T transitions at residues 112 and 158, that determine 3 isoforms: E2, E3 and E4. In humans, isoform E3 is the most prevalent (2/3 to 3/4 of subjects), isoform E4 comes usually second in frequency, and isoform E2 is generally the least frequent. According to population frequency isoform E3 was supposed the most ancient. However genetic and phylogenic arguments support that it could be isoform E4. Indeed, the E3 allele could have derived from the E4 allele by natural mutagenesis on a CpG dinucleotide *(Figure 2-5)*, whereas the reverse supposes a less probable mutational event. Moreover, apo E4 corresponds to the isoform found in many mammals (murine, guinea pig, dog and primates), whereas isoform E3 is present only in bovine and rabbit. Isoform E2, which could in turn have derived from E3 isoform by mutagenesis on a CpG dinucleotide, is observed exclusively in human [701].

Changes in amino acids at positions 112 and 158 (Cys/Arg) induce isoform specific functional changes [702]. These changes are subtle and may be masked in humans by many other factors *(table 2-9)*. Experiments in

Table 2-9. **Functional and phenotypic changes associated with apo E polymorphism.**

	E2	E3	E4
Physical Properties			
Residue position **112**	Cys (TGC)	Cys (TGC)	Arg (CGC)
Residue position **158**	Cys (TGC)	Arg (CGC)	Arg (CGC)
Charge	-1	0	+1
Dimer Formation	+	+	-
Biological Properties			
Lipoprotein distribution	HDL ↑	VLDL / HDL	VLDL ↑
Receptor Affinity	↓↓	↔	↑
Lipolysis (LPL, HL)	↓	↔	↔
Cholesterol Efflux	↑	↔	↓
Neurite Outgrowth	↑	↔	↓
Associated Phenotype			
Plasma Cholesterol/Apo B	↓	↔	↑
Plasma Triglycerides	↑	↔	↓
Atherosclerosis	↓	↔	↑
Alzheimer's Disease	↓	↔	↑
Longevity	↑	↔	↓

transgenic animals have been recently highly contributory in understanding these mechanisms [673]. Apo E2 has a low affinity for VLDL and is displaced towards HDL *(figure 2-25)*. VLDL lipoproteins containing only E2

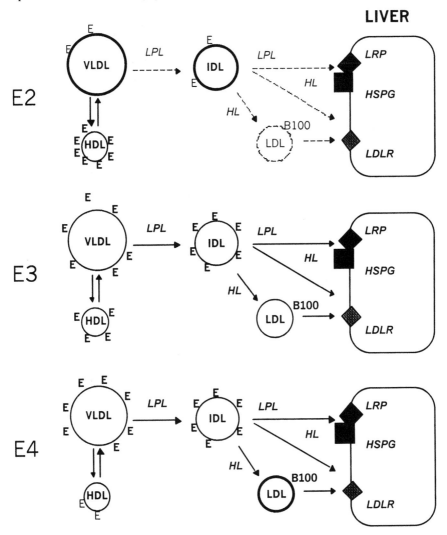

Figure 2-25. **Apo E isoform specific effects on lipoprotein metabolism**. Apo E2 tends to slower remnant lipoprotein catabolism although apo E2 containing HDL are good acceptors in reverse cholesterol transport. Apo E4 has more affinity for VLDL and for cellular receptors. Remnants compete with LDL for LDL-receptor binding and uptake by the liver, resulting in an increased LDL residence time in plasma. Moreover, apo E4 containing HDL are poorer cholesteryl ester acceptors in the reverse cholesterol transport pathway. Apo E3 effects are in-between both the other isoforms.

have a decreased capacity to enhance LPL and HL activities. A decreased rate in lipolysis results in an overall decrease in LDL production. Moreover, the charge modification on residues from the receptor-binding domain impairs apo E interactions with cellular receptors of the LDL receptor gene family (i.e. LDLR and LRP). This results in a tendency to plasma accumulation of remnant-like triglyceride-rich lipoproteins. Besides, apo E2 containing HDL are good acceptors for HDL-mediated cholesterol efflux and competent in reverse cholesterol transport. When triglyceride-rich secretion does not exceed the clearance capacities of the system, subjects are normolipidemic with an efficient HDL metabolism. However, any cause of hypertriglyceridemia (dietary, hormonal, etc.) will be intensified by apo E hypersecretion and VLDL overproduction in the liver. Apo E2 will displace apo C-II in VLDL further reducing lipolysis by LPL and HDL production. VLDL will be poor ligands for cellular receptors enhancing lipoprotein retention in the extracellular matrix and further aggravating plasma accumulation of remnants, as the result of reduced hepatic clearance. Conversely, apo E4 preferentially associates with VLDL. It does not impair LPL or HL activities, and the relative abundance of apo E particles at the lipoprotein surface enhances HSPG binding and remnant clearance by LRP. In addition, remnants will compete with LDL for LDL receptor clearance in the liver. The higher hepatic uptake of cholesterol down-regulates SREBP-responsive genes including that of the LDL receptor. This results in reducing hepatic clearance of LDL and in a tendency to increase levels of plasma LDL. In addition, HDL are poor acceptors for cholesterol efflux from peripheral cells, resulting in a less efficient reverse cholesterol transport. Therefore, any challenge on LDL metabolism will be intensified in apo E4 carriers. When one examines plasma lipid variations in populations as a function of apo E isoforms, genotype E2/E2 is associated with the lowest plasma cholesterol concentrations, and genotype E4/E4 with the highest [22,700]. A dose-effect of apo E alleles from E2 to E4 increases plasma total and LDL cholesterol concentrations. The reverse is observed for plasma triglycerides and this, whatever the population. Biological effects of apo E isoforms are independent of apo E allele frequency. As expected, twin studies have shown that the effect of apo E isoforms on plasma lipid parameters is genetically determined [703]. This effect also modulates the expression of other dyslipoproteinemias, particularly those resulting from disorders of apo E-rich lipoprotein metabolism [198,568,704,705]. Between individual differences in plasma lipid profiles associated with apo E isoforms, although apparently minor in individuals could contribute to 10% of variance of plasma cholesterol levels in populations [111,700]. Thus apo E polymorphism is an example of a functional polymorphism for which, the biological effects are genetically determined by different protein isoforms, directly contributing to the phenotype.

Apo E is the prediposing locus for **familial dysbetalipoproteinemia or type III hyperlipidemia** [669]. A severe combined hyperlipidemia, transient palmar planar xanthoma (see page 5) or tuberous xanthoma characterize this rather rare familial hyperlipidemia (1/104). It complicates with premature coronary and peripheral atherosclerosis. Males are more frequently affected than females, who manifest the disorder more frequently after menopause. Fasting plasma lipoprotein profile is also characterized by the unusual presence of β-VLDL (i.e. lipoprotein remnants). Genetic heterogeneity underlies this syndrome. Type III hyperlipoproteinemia may be dominantly inherited as the results of rare and major heterozygous apo E genetic mutations, which severely impair protein interactions with cellular receptors or gene expression (*table 2-10)*. The other form of inheritance of type III dysbetalipoproteinemia is "conditionally" recessive or pseudo-dominant (*Figure 2-26)*. Segregation analysis of the lipid profile and apolipoprotein E polymorphism has demonstrated this particular type of inheritance in a large

Table 2-10. **Naturally occurring apo E mutations causing hyperlipidemia.**

Mutation	Location	Residue	Protein	Expression	Ref.
G→A	exon 3	Glu^3→Lys	E5	Type IIb, Type IV	706
G→A	exon 3	Glu^{13}→Lys	E5	Type III	707
G→A	exon 3	Trp^{20}→STOP	∅	Type III	708
T→C	exon 3	Leu^{28}→Pro	E4$_{Freiburg}$	Type IIa, IIb,IV,V	709
ΔG	exon 3	Gly^{31}frameshift	∅	Type III	710
A→G	intron 3	3' splice	∅	Type III	711
G→A	exon 4	Ala^{99}→Thr	E3	Type III	691
Dupl. 21nt	exon 4	Dupl.120-126	E3$_{Leiden}$	Type III	712
G→A	exon 4	Gly^{127}→Asp	E1	Type III	708,713 714
G→A	exon 4	Arg^{136}→His	E3	Type III	715
C→A	exon 4	Arg^{136}→Ser	E2$_{Christchurch}$	Type III	716
C→T	exon 4	Arg^{142}→Cys	E3	Type III	717
G→T	exon 4	Arg^{142}→Leu	E2	Type III	718
C→T	exon 4	Arg^{145}→Cys	E2	Type III	707,719
A→C	exon 4	Lys^{146}→Gln	E2	Type III	720
A→G	exon 4	Lys^{146}→Glu	E1$_{Harrisburg}$	Type III	721,722
G→C	exon 4	Ala^{150}→Pro	E3	Type III	691
G→A	exon 4	Trp^{210}→STOP	∅	Type III	723
C→T	exon 4	Arg^{228}→Cys	E2$_{Dunedin}$	Type V	724
G→A G→A	exon 4	Glu244-245 →$Lys^{244-245}$	E7	Type IIb, Type IV	725

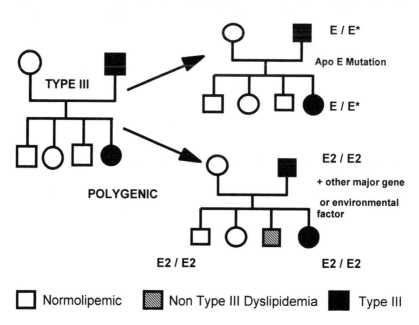

Figure 2-26. **Familial dysbetalipoproteinemia or Type III hyperlipidemia.** The mode of inheritance may vary as a function of apo E alleles. Dominantly inherited type III is caused by heterozygous mutations severely impairing apo E receptor binding or expression. In polygenic type III, only the E2E2 genotype when combined with other genetic or environmental factors results in dysbetalipoproteinemia.

kindred [726]. A homozygous genotype for apo E2 is a necessary condition, although not sufficient. Only subjects who cumulate the E2/E2 genotype and another factor predisposing to hyperlipidemia will manifest with type III hyperlipidemia. This other factor may be genetic, hormonal, or dietary induced. This model accounts for the pseudo-dominant (or conditionally recessive) inheritance of the disease in which, only 25% of subjects express type III, 25% a variable type of hyperlipidemia and 50% are normolipidemic. The second locus may be the LDL receptor as has been reported in certain families, but it is not the only cause [727]. This multifactorial form of the disease represents the large majority of cases with type III as opposed to the monogenic form. Among other factors, hormonal deficiency particularly in estrogens is a critical factor, men with Type III hyperlipidemia being more frequently affected than premenopausal women. Thyroid hormone, renal insufficiency and dietary factors (diet enriched in carbohydrates or in saturated fat) are also factors contributing to type III hyperlipoproteinemia in E2/E2 subjects [22,669,700]. When these factors are missing or when they compensate for the E2 allele-associated defective remnant metabolism, subjects remain normolipidemic or even hypocholesterolemic. Thus type III hyperlipoproteinemia is a real example of a multifactorial dyslipoproteinemia, for which a single locus (apo E) is necessarily involved, in conjunction with other genetic and/or environmental factors, to determine this particular phenotype.

Apo E polymorphism is a modifier of multifactorial diseases. The E4 isoform is an independent risk factor for coronary, peripheral arterial disease and several forms of cerebrovascular diseases, whereas E2 appears protective [700,728]. These associations found in large human cohorts worldwide, could be anticipated from lipoprotein metabolism changes associated with apo E isoforms [729,730]. Moreover the E2 allele is over-represented and its corollary the E4 allele is under-represented in hypocholesterolemia [731] and in longevity [700,732]. This suggests a protective role for isoform E2 with respect to atherosclerosis complications, as opposed to E4. In addition, apo E4 may aggravate the course of primary biliary cirrhosis, as opposed to E2 or E3 isoforms [733]. The primary cause of liver accumulation of bile acids could be aggravated by a higher cholesterol influx within the liver, as the result of higher affinity of E4 enriched lipoproteins for liver receptors.

Even more unexpected, apo E polymorphism has shown to be a major contributor of the genetic susceptibility to Alzheimer's disease [734,735]. As previously mentioned apo E plays an essential role in neuronal protection, repair and remodeling [686-688]. Apo E accumulates in amyloid deposits characteristic of Alzheimer's disease. Apo E binds beta-amyloid to form a protein complex cleared by LRP in cells from the central nervous system. Moreover, apo E4 containing lipoproteins have a lower potential to induce neurite outgrowth and neuronal sprouting, a higher potential to promote beta-amyloid formation as compared with other isoforms [736]. HDL is the predominant form of circulating apo E in the cerebrospinal fluid. HDL containing apo E4 are poor acceptors for cellular cholesterol efflux. Other intracellular factors, regulating apo E expression and localization into cellular compartments remain to be clarified. However, there is a strong relationship between Apo E4 allelic dosage and disease progression, severity and age of onset of Alzheimer disease worldwide. The positive predictive value of Alzheimer's disease in carriers of at least one Apo E4 allele exceeds 95% [737]. However, apo E polymorphism contributes to disease susceptibility and by no means represents a disease causing locus insofar, the genetic heterogeneity and the multifactorial character of the disease being well established [738].

That apo E isoforms operate differently in tissue "repair" and in cell multiplication in response to neuronal loss or injury has been observed in several forms of brain injury or neurodegenerative diseases [684,735]. Apo

E4 retards neuronal repair after brain injury or hemorrhage, and increases the risk of dementia and of peripheral neuropathy in HIV+ patients [739]. However, a surprising observation was made in the case of age-related macular degeneration (AMD) [740,741]. This retinopathy is the most common cause of blindness in the elderly in the United States and in Europe. Apo E and lipids (cholesteryl esters, phospholipids and fatty acids) accumulate in Drusens, the typical lesions of AMD. In this case, apo E4 could be the protective allele, as the result of differential properties, which would facilitate its circulation and maybe that of lipids within Bruch's membrane, an extracellular matrix specific to the retina.

Thus apo E is clearly distinguished from other apolipoproteins by several structural and functional characteristics. These differences were already present during the evolution of animal species. Fish and birds are deprived of apo E but also of apo B editing (see page 56). In mammals the absence of apo B editing in the liver, generates lipoproteins with longer residence time in plasma. Moreover the intestinal production of chylomicrons directs dietary triglycerides to the general blood circulation. This offers an exogenous source of lipids immediately taken up by peripheral organs, like the brain. In fact, it was observed in the rat that central nervous system development uses chylomicrons as the main source of long chain fatty acids. Apo E could provide cells with cholesterol in complement of other lipids, by accelerating endocytosis of triglyceride-rich particles through ligand-specific receptors or heparan sulfate proteoglycans. Thus, the understanding of the fundamental role of apo E in lipoprotein metabolism and in tissue lipid trafficking may offer novel diagnostic and therapeutic solutions to multifactorial diseases.

2.2.3-   Receptors

2.2.31-   LRP, a Multifunctional Receptor for Intermediate Lipoproteins

Early reports had mentioned that hepatocytes defective for the LDL receptor, could persistently take up triglyceride-rich lipoproteins containing apo E, in cell-culture [441,674]. The protein and the cloning of the cDNA from hepatocyte cell lines, uncovered a strong sequence homology with that of the LDL receptor. This protein was named: Low-density lipoprotein receptor-Related Protein (LRP). First suspected as a hepatic receptor for chylomicron remnants, this protein was found identical to the receptor for α 2-macroglobulin [742]. This single copy gene (LRP1) is localized to region q13-q14 of chromosome 12 [743]. It is highly conserved, the protein sequence sharing 83% homology with LRP in chicken [744]. The gene for LRP is 90kb long organized into 89 exons encoding for a 15kb mRNA [745].

Four repeated *Alu* sequences were identified in certain introns one of them being flanked by a microsatellite [746]. The primary structure of the protein is composed of repeated motifs common to the LDL receptor gene family (*figure 2-27*) [747]. These motifs are organized according to the same sequence. First: several cysteine-rich repeats or complement-type repeats, encode for ligand binding sites (class A motifs). LRP contains 31 class A

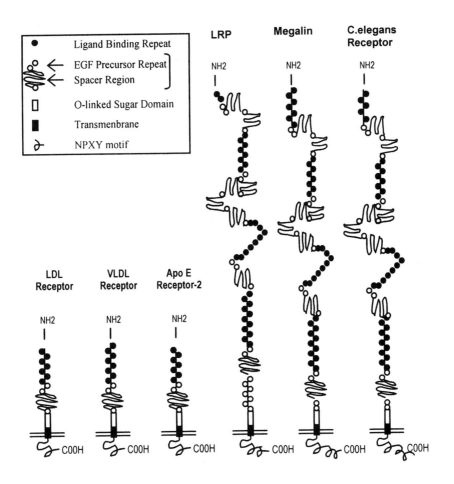

Figure 2-27. **The LDL receptor gene family.**

Each gene is composed of a variable assortment of repeated modules. Ligand-binding type repeats are found in proteins of the complement system. EGF homologous repeats mediate intracellular pH-dependent ligand release and cellular recycling after endocytosis. Spacers stabilize extracellular domains architecture. NPXY motifs trigger coated-pit mediated endocytosis.

cysteine-rich repeats organized into four clusters (I, II, III and IV) containing 2, 8, 10 and 11 repeats each. In LRP, clusters II and IV appear critical for ligand binding [748]. Second: EGF-like modules formed by the succession of two EGF precursor-like motifs (EGF-like, class B2 motifs), YWTD modules and one EGF-like repeat. In the LDL receptor EGF-like repeats mediate pH-dependent ligand dissociation and recycling after endocytosis. LRP contains 8 modules of five motifs containing the YWTD sequence, or spacer, present in the EGF precursor. These motifs group into 6 beta sheets forming a beta-propeller structure, bringing neighboring modules in close proximity [749]. EGF-like modules end with another class B2 EGF-like motif. Third and fourth a single transmembrane domain and a cytoplasmic tail containing NPXY motifs for clustering into clathrin-rich coated pits. In addition, LRP contains 6 motifs homologous to the EGF precursor (class B1 motifs) in its C-terminal part, not found in the LDL receptor. On the other hand, the O-glycosylation domain of the LDL receptor is not found in LRP. As for all members of this superfamily, there is only one transmembrane domain. In LRP, the cytoplasmic C-terminal sequence (NPXY) binding to clathrin and triggering endocytosis is duplicated. This domain also mediates interactions with intracellular adaptors activating protein kinases involved in intracellular signalling [750]. Native LRP is a macromolecule of 600kD cleaved in the Golgi apparatus by the endoprotease furin, into two sub-units of 515kD N-terminal and 85kD C-terminal bound noncovalently to each other.

LRP is a ubiquitous receptor. Similarly with the LDL receptor it is highly expressed in the liver, but also in the brain, lungs, gastro-intestinal epithelium, smooth muscle cells, fibroblasts, gonads and kidney [441,674,751]. LRP gene expression is not regulated by cellular sterols, the promoter lacking SRE (see page 47). It is down regulated in macrophages by lipopolysaccharides, interferon-γ and estradiol, but is stimulated by CSF-1 (colony stimulating Factor 1), suggesting that LRP could participate in foam cell formation in the atherosclerotic plaque. In adipocytes, it is stimulated by insulin, which increases receptor number at the cell surface, favoring lipogenesis by increasing post prandial uptake of TG-rich lipoproteins. In addition, LRP expression is upregulated by components of the extracellular matrix [752]. A significant step regulating LRP expression is post-transcriptional. A 39kD protein copurified with LRP, Receptor Associated Protein (RAP), competitively inhibits ligand binding [753]. RAP binds to all members of the LDL receptor gene family. The genomic sequence was cloned (LRPAP1) and localized to chromosome 4 p16.3 [754,755]. The gene spans approximately 16kb organized into 8 exons, encoding for 323aa [756]. The N-terminal domain (aa 1-86) corresponds to the pathogenic epitope for Heymann nephritis (see page 122). Receptor binding sites are found between residues 85-148 and 178-248 [757]. They contain amphipathic α-helices and a region rich in basic amino acid residues homologous to apo E receptor binding sites. Heparin binding domains are found between residues 261 and 323. RAP binds to LRP ligand-binding clusters II, III and IV inducing a conformational change preventing any ligand interaction [748]. It was shown that apo E or other coexpressed ligands for LRP, could induce its retention in the endoplasmic reticulum and its intracellular degradation when bound to receptors [758]. Therefore, RAP could act as a chaperone protecting newly synthesized receptors of the LDL receptor gene family from interaction with ligands in the ER, allowing proper folding and receptor exposure at the cell surface. Although it is an ER-resident protein, RAP may act in extracellular spaces as a modulator of ligand-induced endocytosis. Its ligand-inhibitory property was used to demonstrate that LRP plays a major role in the clearance of intermediate lipoproteins in vivo. In LDL receptor deficient mice, transient RAP protein expression causes plasma accumulation of chylomicron remnants [759].

LRP is a hepatic receptor specialized in triglyceride-rich lipoprotein remnant clearance [674,760,761]. A two-step process takes place in the space of Disse in the liver. Lipoproteins exposing apo E, LPL or HL at their surface bind to heparan sulfate proteoglycans in a dose-dependent manner. HSPG have a low affinity for lipoproteins, however their abundance represents a slow-high capacity pathway for lipoprotein endocytosis. Moreover, lipoprotein residence in close proximity to the cell surface, allows lipoprotein remodeling by lipolysis and apo E enrichment as the result of local apo E secretion by hepatocytes (see page 107). The higher number of ligand binding sites enhances lipoprotein affinity for cellular receptors such as the LDL receptor and particularly LRP. Moreover, B-100 containing lipoproteins or TG-rich lipoproteins previously enriched in apo E, may be rapidly cleared by receptor-mediated endocytosis.

Beyond its role in triglyceride-rich lipoprotein catabolism, LRP is a multifunctional receptor. It recognizes numerous ligands with high affinity and takes them up into the cytosol by endocytosis. Ligands were shown to be apo E-rich lipoproteins, lipases, vitellogenin, protease-inhibitor complexes (α2-macroglobulin-protease; tPA-PAI-1), antihemophilic factor VIII, tissue factor pathway inhibitor, activated complement component C3, exotoxin A of *Pseudomonas* [441,762]. LRP clears protein complexes that circulate in plasma, in extracellular spaces, or that associate with membranes. α2-macroglobulin is a scavenger of serine-proteases, growth factors or cytokines in plasma, by forming stable complexes allowing their endocytosis by LRP. This role seems critical in the brain since α2-macroglobulin binds to amyloid beta allowing its endocytosis by LRP [763]. LRP also binds and takes up secreted isoforms of amyloid precursor protein. LRP mediates the effect of apo E-rich lipoproteins on neurite outgrowth and on cholesterol efflux (see page 107). Moreover, LRP could trigger signalling pathways involving factors required for neuroanatomical organization of the central nervous system [750]. These properties complementary to those of apo E have suggested that LRP could play a significant role in the pathogenesis of Alzheimer's disease. Associations were reported between Alzheimer's disease and LRP gene polymorphisms. However, the putative defect at the LRP locus remains to be determined [764]. The role of LRP in protein complex clearance was also observed for t-PA (tissue Plasminogen Activator) and its inhibitor PAI-1. t-PA is a strong activator of fibrinolysis and of many processes involving extracellular matrix break-down: tissue remodeling, cellular migration, metastasis dissemination and embryo implantation. Homozygous LRP knock out mice display embryonic failure as the result of impaired tPA-PAI-1/receptor complex clearance preventing continuous tissue breakdown necessary for embryo implantation in the uterus [441]. In addition to its role in embryonic implantation, tPA-PAI endocytosis by LRP or other RAP sensitive receptor could inhibit smooth muscle cell migration induced by the plasminogen activation system, suggesting that LRP could be involved in arterial wall remodeling in atherosclerosis [765].

No naturally occurring mutation was identified on the LRP gene so far as a cause of a human disease. Heterozygous littermates knock out for LRP develop and survive normally indicative of the recessive effect of LRP gene invalidation. Whereas the total absence of LDL receptors is compatible with survival, LRP plays a much more fundamental role in embryonic development and cellular life. A nearly identical gene, with a gene organization resembling that of LRP was identified in the nematode *Caenorhabditis elegans* suggesting a close resemblance with the ancestral gene of the LDL receptor gene family [766]. Together with this evolutionary argument, the multiple functions of LRP in cellular and organ remodeling, and in the clearance of triglyceride-rich lipoproteins points out its importance in lipoprotein metabolism and atherosclerosis.

2.2.32-   Megalin, an Ancient Member of the LDL Receptor Gene Family Expressed in the Kidney.

Megalin, formerly known as glycoprotein gp330, is a large transmembrane 600kD protein belonging to the LDL receptor gene family [747,767]. It is mainly expressed in epithelial cells of the renal proximal tubule (and to a lesser extend in glomerular epithelium), lung, endometrium, epididymis, eye ciliary epithelium, ear labyrinth epithelium, and parathyroid. It is highly expressed during early embryogenesis in the yolk sac epithelium and in the neuroectoderm. Megalin is a multifunctional receptor recognizing several apolipoproteins (apo B-100, apo E, apo H, apo J), enzymes (including lipoprotein lipase, PAI-1 and PAI-1-tPA), vitamin binding proteins (binding proteins for Vit D, Vit A and Vit B12) hormonal and non-hormonal peptides (PTH, insulin, β2-microglobulin, EGF, PRL, lysosyme, cytochrome c), small proteins such as lactoferrin, plasminogen, albumin, thyroglobulin and polybasic drugs (aminoglucosides, aprotinin, polymixin B). Similarly to LRP, megalin binds RAP with high affinity, which inhibits binding to other natural ligands and favors its integration into the plasma membrane.

Megalin is clustered in plasma membrane clathrin-coated pits where it has been proposed to mediate the endocytosis of lipoproteins, of protease:protease-inhibitor complexes, of polybasic drugs thereby contributing to drug-induced kidney and ear toxicity, and of peptides found in the glomerular ultrafiltrate. Megalin was initially identified as the autoantigen (later with RAP) for Heymann nephritis, an experimental model of autoimmune membranous nephropathy [768]. A 46-aa motif in the second cluster of cystein-rich repeats was identified as the causative antigen for Heymann nephritis. Similarly with other members of the superfamily, calcium modulates megalin ligand affinity and binding. In this line megalin has been identified as a calcium-sensor at the surface of parathyroid epithelium. Megalin plays a crucial role in renal reabsorption and secretion of three vitamins: Vitamin D, A and B12, through the endocytosis of vitamin-vitamin carrier complexes. Knock out experiments have shown severe deficiencies in Vitamin A and D leading to multiple tissue malformation and dysfunction. Moreover, in the case of Vit D, megalin-mediated endocytosis of 25-OH-D$_3$ is an obligatory step for its conversion into the biologically active vitamin 1,25-(OH)$_2$-D3 in renal proximal tubules. Megalin has been identified as a receptor mediating endocytosis of lipoproteins (particularly Lp(a) and β-VLDL), and of lipoprotein lipase [418]. Moreover, it may serve as a coreceptor for cubilin to mediate HDL endocytosis in the kidney (see figure 2-28)[74]. Megalin plays a crucial role in early embryonic development. Megalin gene invalidation in mice leads to defective forebrain development and severe malformations as a result of defective uptake of lipoprotein-derived cholesterol and other vital nutriments [769]. Megalin was suggested to play a role in Alzheimer's disease pathogenesis. Through its interaction with apo J-amyloid beta complexes, it could prevent beta-amyloid accumulation in neurons and promote their cellular degradation [770].

The cDNA for megalin is 15.4kb long, encoding for a mature protein of 4660 amino acids. The gene (LRP2) has been localized to chromosome 2 q24-q31, and exhibits a fair degree of sequence homology with other members of the LDL receptor gene family (see figure 2-27) and with its homologs in other species [771,772]. The N-terminal end begins with a 25-aa signal peptide followed by a 4400-aa long extracellular domain. The single transmembrane domain is 22-aa long followed by a 213-aa cytoplasmic tail. The extracellular domain contains 36 class A cysteine-rich repeats organized in 4 clusters (7, 8, 10 and 11 repeats). The second cluster includes binding sites for several ligands including RAP [773]. Furthermore there are 16 EGF-like repeats separated by 8 spacer regions containing the YWTD motif, necessary for acid-dependent dissociation of ligands in endosomal compartments. Contrary to the LDL receptor Megalin does not contain any O-glycosylation domain. Contrary to LRP, Megalin does not contain any recognition signal for furin cleavage in the trans-Golgi, indicating that the mature protein is a single amino-acid chain. The cytoplasmic domain contains 3 NPXY motifs necessary for coated pit clustering and endocytosis. This domain also contains Src-homology binding regions, casein kinase II sites, and protein kinase phoaphorylation sites suggesting other roles in signal transduction. A nearly identical gene, with a similar gene organization was identified in the nematode *Caenorhabditis elegans* [766]. These observations suggest that an ancestral gene resembling megalin encoded for a multifunctional protein able to collect substances from external sources essential for tissue remodeling and cellular homeostasis. This high molecular weight receptor has diversified into specialized large multifunctional receptors in the brain, the liver (LRP) or the kidney (Megalin) or into smaller specialized receptors like the LDL receptor, the VLDL receptor or Apo ER2. No natural mutants have been identified so far on megalin as a cause of human dyslipidemia. However, its role as a multifunctional receptor together with its role of coreceptor with cubilin for HDL endocytosis in the kidney (see page 158) suggest that it might regulate important steps of lipoprotein metabolism.

2.2.33-   VLDL Receptor (LR8), a Multifunctional Receptor with Species–Dependent Requirements

Another member of the LDL receptor gene family recognizes apo E-rich lipoproteins [774,775]. This receptor is able to mediate the endocytosis of VLDL, of chylomicron remnants and of IDL in tissues in a high demand in fatty acids (muscles, heart and adipose tissue). However, the VLDL receptor is a multifunctional receptor able to bind ligands recognized by LRP: PAI-1:uPA complexes, thrombin:proteinase nexin 1, LPL, rhinoviruses of the minor serotype group and RAP. The gene (VLDLR) was first identified in rabbit then in human [776]. It is single copy and localized to chromosome 9, 9p23-pter. The gene is strongly conserved between species and shares high sequence homology with its closest relative the LDL receptor (*Figure 2-27*). It spans approximately 40kb and contains 19 exons ending at similar breakpoints with those of the LDL receptor gene *(Figure 2-3)*. An additional exon encodes for the first cysteine-rich motif upstream of the seven repeats found in the ligand-binding domain of the LDL receptor gene. A ligand binding domain for RAP was identified in the second cystein-rich repeat, encoded by exon 4 [777].

The VLDL receptor is predominantly expressed in peripheral tissues, particularly in heart, muscle and adipose tissue. A weaker expression is found in kidney, murine brain and in the ovary [776,778]. The VLDLR is not expressed in the liver or in the intestine. A differential splicing generates two mRNA transcripts (3.9 and 5.2kb) differing by the inclusion or the exclusion of the O-glycosylation domain. Promoter sequences comprise motifs recognized by Sp1, heat shock protein sensitive sites, an octamer binding transcription factor (OTF-1). Two imperfectly conserved SRE 1 motifs are present, however sterols do not suppress VLDLR expression in THP-1 cells. This insensitivity may result from several nucleotidic substitutions in the binding sites for SREBP1. There are also partial sites of recognition for steroid hormone receptors accounting for its upregulation by fibrates in adipose tissue and muscle. Insulin and GM-CSF (Granulocyte-macrophage colony stimulating factor) also induce VLDLR expression. A hypervariable domain resulting from triplet (CGG) repeats is observed 19nt upstream of the transcription start site. This domain does not seem to be unstable in hyperlipidemic subjects. This polymorphism has been inconsistently reported in association with Alzheimer's disease [775].

The role of this receptor remains unclear. No natural mutant causing dyslipidemia or any other human disease was described so far on the VLDLR gene. Although this receptor strongly resembles the LDL receptor, and has shown to serve as a back-up receptor when engineered to express in the liver [779,780], its ligand binding properties and tissue distribution are completely different. Like the LDL receptor, it is made of a series of building blocks typical of the LDLR superfamily. A homologous gene (VLDLR/VTGR) was cloned in chicken [781]. This gene encodes for a 95kD membrane receptor able to bind VLDL and vitellogenin (VTG). It is expressed primarily in the ovary in hen, under the influence of estrogens and plays a predominant role in the accumulation of yolk lipids in the ovocyte. A major locus for ovogenesis was localized on the female sexual chromosome (chromosome Z) in a strain of chicken with a recessive phenotype: the restricted ovulator phenotype (R/O). This disorder is characterized in hen by a failure to lay eggs and by a severe hyperlipidemia complicated with atherosclerosis. A defective Cys682→Ser mutation, inducing a protein conformational change, was identified on the VLDLR/VTGR gene as the cause for the R/O phenotype [782]. The identification of the crucial role of this gene in a species deprived of apo E, suggest that functional and evolutionary links might exist with the VLDL receptor in mammals. In birds, this receptor recognizes vitellogenin, the chicken homologue of apo B and apo E (see page 17), in lipoproteins. This receptor is mildly expressed in heart and muscles in chicken, whereas it is mildly expressed in the ovary in the mouse and rabbit. In mammals, more particularly in humans, the loss of function in ovogenesis vital in birds, and the autosomal location of the gene, could determine other specializations. For example the function of delivery of lipids in tissues in high demand by lipoproteins containing ligands homologous to vitellogenin appears to be conserved. VLDL receptors are highly expressed in cells of the arterial intima and media during atherosclerotic lesion development [783]. However, in mammals the primary function of the VLDL receptor does not seem to confine to lipoprotein metabolism or reproduction. VLDL receptor knockout mice develop apparently harmoniously although with a modest decrease in adipose body mass [784]. They have normal plasma lipoprotein profile challenged the same way by dietary induced hyperlipidemia in mutant and in control mice.

However, VLDLR knockout mice helped to demonstrate a previously unsuspected role of VLDL receptors in cellular signalling during brain development [785]. Other brain-specific members of the LDL receptor gene family have been identified. Apo E receptor 2 (ApoER2,) also called LR7/8B and LR11 (or SORLA1, Sortilin related Receptor 1) were named after the number of class A ligand-repeats found in their primary sequence. They both bind RAP and apo E [786,787]. Apo ER2 is the closest relative of the VLDL receptor, sharing a short linker sequence between cysteine-rich repeats 5 and 6, whereas the LDLR linker sequence is between repeats 4 and 5. Double VLDLR and Apo ER2 knockout mice develop severe cerebellar dysplasia as the result of the disruption of a signalling pathway (Reelin-VLDLR/ApoER2-Disabled) involved in neuronal positioning during brain development. Extracellular Reelin binding to the VLDLR or to ApoER2 induces intracellular Disabled-1 phosphorylation, which in turn activates tyrosine kinase pathways. Moreover Apo ER2 and LR11 could be involved in axonal growth and LR11 is upregulated in atherosclerotic lesions [788,789]. Thus, smaller multifunctional receptors of the LDLR superfamily could act predominantly in tissues in high lipid demand like adipose tissue, muscle or in signalling pathways of the nervous system. This contrasts with the LDL receptor, which has low impact on development and on the CNS, but a high influence on lipoprotein metabolism in mammals. Other proteins containing ligand-binding type repeats named LRP3 and LR6 were identified [747]. A liver receptor, lipolysis-stimulated receptor (LSR), inhibited by RAP and binding LDL in the presence of fatty acids has been described. Despite its capacity to take-up lipoprotein remnants, it does not share sequence homology with members of the LDL receptor gene family [790].

Enzymes, apolipoproteins and receptors involved in the metabolism of triglyceride-rich lipoproteins, play a major role in the handling of exogenous or endogenous sources of lipids. These genes have shown to interact with other pathways of lipoprotein metabolism and to play biological roles far beyond the limits of atherosclerosis. It may be predicted from their fundamental biological functions that they may significantly contribute to the most frequent forms of dietary sensitive atherogenic dyslipidemia and to other late-onset multifactorial diseases.

2.3- GENES OF THE REVERSE PATHWAY OF LIPOPROTEIN METABOLISM

2.3.1-   Apolipoproteins

2.3.11-   Apolipoprotein A-I, a Major Component of HDL

Apolipoprotein A-I (apo A-I) is the major constituent (70%) of the protein mass of HDL [122]. These lipoproteins represent a heterogeneous group, whose metabolism is not fully clarified. However, several decades of epidemiological and clinical evidence of an inverse relationship between plasma HDL cholesterol concentration and cardiovascular disease suggest the existence of important underlying biological links. [791]. Apo A-I constitutes the backbone of HDL and interacts with cell surface components to enhance cholesterol efflux [792]. It is a protein 243 amino acid long, with a MW 28kD molecular weight, synthesized in liver and intestine. It is secreted as a propeptide becoming mature after cleavage of its 6 N-terminal amino acids. It may be secreted directly in the form of native HDL, or integrated into chylomicrons or VLDL *(figure 2-28)*. After hydrolysis of triglyceride-rich lipoproteins by lipoprotein lipase, surface components are detached to form native HDL of discoidal form. Surface components (mainly phospholipids) detachment and HDL formation is facilitated by Phospholipid Transfer Protein (PLTP). Native HDL avidly take-up unesterified cholesterol from membranes at the cell surface. In membrane microdomains named caveolae, ABC1 (or CERP) enhances cellular efflux of unesterified cholesterol towards HDL [70]. Apo A-I is the primary activator of LCAT (Lecithin-Cholesterol-Acyl-Transferase), in native HDL. LCAT esterifies cholesterol so that hydrophobic cholesteryl esters reach the core of the particle. Another receptor mediated mechanism of cholesterol

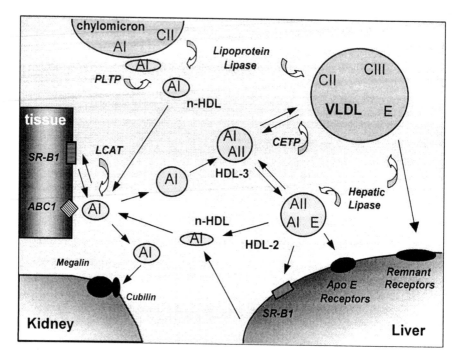

***Figure 2-28*. HDL metabolism.**

• Native HDL (n-HDL) are discoidal particles mainly containing apo A-I, produced by the liver or the intestine. They may also detach with other surface components from triglyceride-rich lipoproteins after lipolysis by LPL (lipoprotein lipase), and circulate in plasma through the action of phospholipid transfer protein (PLTP). Native HDL avidly take up free cholesterol at peripheral cell surfaces through an ATP-dependent cholesterol transfer mediated by ABC1. LCAT esterifies cholesterol allowing the loading of cholesteryl esters into the HDL core. Cholesteryl ester efflux may alternately be enhanced by SR-BI, towards lipid poor HDL.

• Mature HDL circulate and exchange lipids and small apolipoproteins with VLDL or chylomicrons remnants through the action of CETP. Cholesterol enriched remnants will be taken up by liver receptors. Triglyceride enriched HDL will be remodeled in the liver by hepatic lipase. HDL unload their cholesterol content towards SR-BI, a receptor mediating cholesteryl ester selective uptake and their use in liver bile acid synthesis. Smaller HDL may cycle back to the general circulation for another round in reverse cholesterol transport.

• HDL may be cleared by endocytosis in the liver through apo E recognition by multifunctional receptors or in the kidney through interactions with cubilin and megalin as a coreceptor.

efflux involves SR-BI, which enhances cholesteryl esters efflux towards the core of native HDL [73]. Cholesteryl esters and apolipoproteins become exchangeable with triglycerides from LDL and VLDL via cholesteryl ester transfer protein (CETP). In the liver, HDL apo A-I interacts with SR-BI, a membrane receptor, which mediates selective uptake of cholesteryl esters [73]. Here, contrary to the situation in peripheral tissues where SR-BI mediates cholesteryl ester efflux, HDL are rich in cholesteryl esters favoring their influx into hepatocytes and their secretion into bile. This mechanism is an important step of reverse cholesterol transport. Alternately, HDL enriched in apo E and/or bound to hepatic lipase may be taken up by multifunctional receptors. In addition, hepatic lipase may remove triglycerides from the core of HDL, which in addition to cholesteryl esters unloading via SR-BI, allows HDL to shuttle back into the circulation [592]. HDL are eventually cleared from the circulation in the liver by lipoprotein endocytosis through apo E recognition by multifunctional receptors or in the kidney through interactions with cubilin and megalin as a coreceptor [74]. An additional role of HDL is to deliver cholesteryl esters to steroidogenic organs, a mechanism facilitated by hepatic lipase and SR-BI. Thus HDL play a role of shuttle, which promotes cholesterol efflux from cells and interacts with other lipoprotein species to ensure lipid exchange and the reverse transport of cholesterol. The protective role of apo A-I as an essential component of reverse cholesterol transport and of antiatherogenic protection has been demonstrated in transgenic animals (mice and rabbits) [793]. However, relative composition changes in apolipoprotein content of circulating lipoproteins seems critical since apo A-I knock out mice despite very low plasma HDL levels do not seem susceptible to atherosclerosis. This suggests that other mechanisms remain to be clarified to explain discrepancies in the association between low plasma HDL cholesterol and atherosclerosis in humans.

The gene for apo A-I (APOA1) is short, spanning 1868 bp, integrated into the A1-C3-A4 gene cluster on the long arm of chromosome 11 [794,795] (*figure 2-29*). Its organization is common to that of other apolipoproteins: 4 exons/3 introns (*figure 2-20*). The 3rd and 4th exons encode for the mature protein and a 3' noncoding domain. The protein is organized into a globular amino-terminal domain (amino acids 1 to 43) and a lipid-binding carboxyl-terminal domain (amino acids 44 to 243) [796,797]. Exon 4 comprises a CpG island within which 10 repeated motifs encode for amphipathic α-helixes. Atomic map resolution studies have suggested that in discoidal HDL two antiparallel apo A-I molecules pair by their C-terminal domains to form a double belt structure stabilizing a micellar phospholipid bilayer. In larger particles another pair of molecules joins the structure allowing more lipid loading in the spherical particle. The α-helix domain contains regions interacting with LCAT and transfer proteins. Although apo A-I interacts with SR-BI and possibly with ABC1, the functional domains precisely involved remain to be identified. Amphipathic α-helices confer antioxidative and anti-infectious properties to apo A-I [798]. For example it has been shown that apo A-I inhibits HIV infectivity and virus induced syncytium formation by

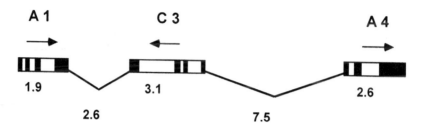

Figure 2-29. **The Apo A1-C3-A4 gene cluster.** Genetic distances are indicated in kilobases below genes and intergenic sequences. Arrows indicate transcription orientation.

interfering with proteins of the viral envelope [799,800]. Native apolipoprotein A-I is synthesized in the form of a propeptide, whose role would be to direct certain protein maturation steps in endoplasmic reticulum and Golgi apparatus [801]. Phosphorylation on Ser201 might also contribute to apo A-I maturation [802].

The gene for apolipoprotein A-I belongs to a 15kb **gene cluster** including genes for apo A-I, C-III, and A-IV, from centromeric to telomeric on the long arm of chromosome 11, 11q23 [794]. The gene for apo C-III is located 2.6kb downstream from APOA1 in opposite orientation. The gene for apo A-IV is located 6.2kb downstream from APOC3 and 12kb downstream from APOA1, in the same orientation with APOA1. Several *Alu* repeats are present: one is 1kb 3' flanking of apo A-I, the second is in intron 3 of APOC3, and the third is 5' flanking of APOA4. Gene cluster organization and length are conserved between mammals [649,803] and has a similar organization in chicken, although the genomic region is only 6kb long [804]. There is also a high conservation of flanking and intergenic sequences. That the clustering of apolipoprotein genes occurred prior to the divergence between birds and mammals, 280 million years ago, and that the structure of the A1-C3-A4 gene cluster was maintained across evolution, suggest that a strong pressure has maintain concerted regulations of these genes.

The promoter of the apo A-I gene comprises four sites necessary for expression in liver and intestine [805]. ARP-1 (Apo A-I Regulatory Protein-1) modulates apo A-I liver expression [217]. This protein, which belongs to the family of nuclear hormone receptors, forms a homodimer binding by its C-terminal domain to recognition sequences also found in many other genes including apo B, apo C-III or insulin. This 47kD protein may inhibit apo A-I gene transcription, however neighboring sequences within the promoter may modulate its effect. PPARα have also a variable action on apo A-I expression depending on the combination of factors interacting with the promoter [806]. Retinoic acid is an activator of apo A-I gene transcription [807]. Steroid hormones, insulin and dietary sources of lipids up regulate apo A-I expression [808,809]. Moreover, apo A-I expression is species specific varying sometimes in opposite directions under the same stimuli from one species to the other.

Data in transgenic animals have also better specified the mechanisms controlling apo A-I gene expression. Transgenic mice were generated to overexpress a human genomic region including 5kb upstream sequences, apo A-I and 3.5kb downstream sequences corresponding to the intergenic region between A-I and C-III [810,811]. As expected, mice had higher plasma levels of small sized HDL, of apo A-I, than control littermates. Plasma apo A-I levels were strongly inducible by a saturated fat diet, confirming the role of dietary fatty acids in the regulation of HDL metabolism [812]. However in these mice, human apo A-I expression was restricted only to the liver and absent in the intestine. It was shown after transfection of longer genomic fragments, that regulatory sequences for intestinal expression of apo A-I (ICR, intestinal control region) were located downstream from apo C-III, in the CIII-AIV intergenic region [813]. In addition tissue specific apo A-I gene expression could be modulated by DNA methylation in 5' regulatory regions [814]. There is complete methylation of the apo A-I gene in sperm and complete demethylation in the ovum, until the 8-cell embryonic stage. Chromosome 11 is subject to genomic imprinting. Therefore Apo A-I methylation raises the question of its imprinting at an early stage of embryonic development [338].

The apolipoprotein AI-CIII-AIV gene cluster is polymorphic; several SNPs were reported soon after its identification. These polymorphisms are neutral, located in noncoding regions with low sequence conservation during evolution. They proved to be useful to show the existence of linkage disequilibriums within the locus in various populations [815,816]. These data strengthen the hypothesis of a particular physical organization of the gene cluster, which would have been maintained across evolution. Like in the case of chromosome 19 gene cluster (see page 104), intestinal control regions of apo A-I gene expression are located in a genomic region distant from the coding sequence [813]. Motifs located in the apo CIII-AIV intergenic region would also control intestinal expression of apo C-III and apo A-IV. All three apolipoproteins are constitutive of chylomicrons, VLDL and HDL, which production is highly dependent on diet. It is thus possible that intergenic sequences within the A1-C3-A4 gene cluster could coordinate the expression of the three genes. These physiological arguments reinforce the idea that linkage disequilibriums between neutral polymorphic markers are indicative that selective pressure has maintained a particular organization of this genomic region on a short physical distance to ensure concerted gene expression.

The influence of genetic factors on apo A-I expression was shown in humans by biometric studies. Different models were applied to analyze the relative influence of genetic factors and of environment on plasma HDL cholesterol and apo A-I levels. A major locus determining apo A-I in plasma was found in a series of families selected from a proband with CHD [817], and in pedigrees of baboon [121]. In this second study, two major loci were responsive to dietary factors particularly to the percentage of saturated fat in the diet. Sib-pair analyses were also used to identify a dose effect of allelic combinations of the AI-CIII-AIV gene cluster, on the resemblance between pairs for plasma HDL-cholesterol in normolipidemic individuals [610]. Reminiscent of what has been observed for apo B, biometric studies suggest that major genes control plasma levels of apo A-I, but their exact nature remains to be clarified. Soon after the cloning of apolipoproteins A-I, C-III and A-IV genes, genomic probes were used to seek for morbid associations with neutral polymorphisms localized in the gene cluster. Numerous association (case-control) studies were carried out similarly to those performed on apo B, on groups of subjects with frequent and/or complex quantitative traits [818,819]. Associations have been contradictory from one study to the other. However, several studies have been consistent in showing that genetic variations at this locus contribute to individual responsiveness to diet and dietary interventions.

In this line, genetic polymorphisms have been suggested to be directly functional [820]. A mutation G→A at nucleotide -75 in the apo A-I promoter is neighboring a motif recognized by nuclear factors, creating an additional repeat of a "CAGGGC" motif [805]. The rare allele (frequency = 17.6%) was associated a reduced apo A-I production rate in vivo. The activity of both alleles was analyzed after their transfection in liver derived cell lines. A reduced transcriptional activity was associated with the rare allele. Later, numerous studies were conducted to assess the effect of this polymorphism on plasma lipoprotein profile [821]. Despite several discrepancies from one study to the other, apo A-I levels appear to be moderately influenced by this variant. In addition, five gene variants in strong linkage were identified in the intestinal control region between nucleotide -770 and -194 upstream of the apo C-III gene. A combination of alleles, or a haplotype (frequency 0.44), has shown to decrease apo A-I synthesis in intestinally derived cell lines and apo A-I plasma level [822]. Furthermore, this haplotype was reported to increase resistance to dietary induced hypertriglyceridemia, as the result of down regulation of apo C-III gene expression. These data underscore the role of intergenic and regulatory genomic regions, in which genetic variations may modulate the expression of dyslipidemia.

Several natural mutations of apo A-I were reported (*table 2-11*).

Table 2-11. **Naturally occurring mutations of human apolipoprotein A-I.** hypoα: hypoalphalipoproteinemia, CO:corneal opacity, CAD:coronary artery disease. nr: not reported

Mutation	Exon	Residue	Protein	Hypoα	CO	CAD	Other	Ref.
C→T Toronto	3	$Gln^{-2}STOP$	null allele	+	-	+	Retinopathy	823
C→G	3	Pro^3Arg	charge +1	-	-	-	ProapoA-I +	797
Ins C Turkey	3	Pro^5STOP	null allele	+	+	nr	Xanthoma	824
G→T Baltimore	3	$Arg^{10}Leu$	charge -1	+/-	-	+	-	797
G→T Yame	3	$Asp^{13}Tyr$	charge +1	+	-	-	-	797
G→C Iowa	3	$Gly^{26}Arg$	charge +1	+	-	-	Amyloidosis	797
T→C	4	$Trp^{50}Arg$	charge +1	nr	nr	nr	Amyloidosis	797
T→G	4	$Leu^{60}Arg$	charge +1	nr	nr	nr	Amyloidosis	797
Δ 35/ Ins 5 nt	4	$Δ(60-71)$	charge +1	+	-	-	Amyloidosis	825
Δ 9 nt	4	$Δ(70-72)$	α-helix	nr	nr	nr	Amyloidosis	797
C→T	4	$Pro^{90}Leu$	α-helix	nr	nr	nr	Amyloidosis	797
Δ 3 nt Marburg	4	$ΔLYS^{107}$	α-helix	+	+	-	LCAT ↓	826
T→G Pisa	4	$Leu^{141}Arg$	charge +1	+	-	+	LCAT ↓	797
C→G Giessen	4	$Pro^{143}Arg$	charge +1	+	-	?	LCAT ↓	827
Δ 45 nt Seattle	4	$ΔGlu^{146}?$ Arg^{160}	α-helix	+	+	-	TG ↑	797
C→T Paris	4	$Arg^{151}Cys$	charge -1, α−helix	+	-	-	LCAT ↓ TG ↑	797
T→A Oita	4	$Val^{156}Glu$	charge -1	+	-	-	LCAT ↓	797
C→A	4	$Ala^{158}Glu$	charge -1	+	-	-	-	797
T→G Fin	4	$Leu^{159}Arg$	charge +1	+	-	-	LCAT ↓	797
G→T Oslo	4	$Arg^{160}Leu$	charge -1	+	-	-	-	797
Ins 23 nt Sasebo	4	$Fs\ Thr^{161}STOP$	α−helix	+	+	-	Xanthoma LCAT ↓	828
C→G Munster	4	$Pro^{165}Arg$	charge +1	+	-	-	LCAT ↓	797
C→T Milano	4	$Arg^{173}Cys$	charge -1, α−helix	+	+	-	LCAT ↓ TG ↑	797
G→C	4	$Arg^{173}Pro$	charge -1	+	-	-	Amyloidosis	797
Δ G Munster	4	Thr^{202} $STOP^{230}$	α−helix	+	+	-	LCAT ↓	829
Δ 3 nt Nichina	4	$ΔGlu^{235}$	α−helix	+	+	-	LCAT ↔ CE efflux ↓	830
G→T	4	$Glu^{237}STOP$	null allele	+	-	-	-	797

Although dominantly inherited hypoalphalipoproteinemia is relatively constant, the phenotype may vary from one family to another. Atherosclerosis complicated with premature cardiovascular disease has been reported in homozygous carriers of null alleles, in whom apo A-I expression is abolished [831]. Corneal lipid deposits resembling those of "fish eye disease" (see page 144) may be reported. Abnormal apo A-I is found in LpA-I and in LpAI-AII particles, or with an abnormal lipoprotein species, LpA-II [832]. Sometimes hypertriglyceridemia or on the contrary low plasma triglycerides are observed. LCAT activity may be lowered to normal. In carriers of mutations of the N-terminal globular domain, despite an apparent normal lipoprotein profile and the absence of atherosclerosis, disseminated or localized amyloidosis has been reported. Dimerized isoforms of apo A-I resulting from mutations introducing an additional cysteine residue in the α-helix domain (Apo A-I Milano, Apo A-I Paris) were associated with a low HDL/high triglyceride trait and the absence of cardiovascular disease. These variants may increase HDL capacity to mediate cholesterol efflux. Hypoalphalipoproteinemia, even when dominantly inherited does not always result from a mutation at the AI-CIII-AIV locus [833]. On the other hand, other apo A-I mutations were identified, with no specific changes in the lipid profile or in the predisposition to any disease [834]. Studies of natural apo A-I mutations have nevertheless contributed to better determine the functional domains of apo A-I.

It remains a paradoxical fact: although many subjects with apo A-I mutations have very low HDL cholesterol and apo A-I in plasma, they generally do not suffer from cardiovascular disease. These observations agree with the moderate atherosclerosis observed in apo A-I knock out mice, even under an atherogenic diet [835]. In vivo, other exchangeable apolipoproteins, in particular apo E, may take over several of apo A-I functions in its absence. However, these observations confirm that the link between low plasma HDL and/or apo AI, and risk of atherosclerosis is not univocal. It suggests that in the general population, other mechanisms determine this common trait, offering a good example of the absence of direct biological relationship between two clinical facts, despite their strong and universal association.

2.3.12- Apolipoprotein C-III, a Modulator for Triglyceride-rich Lipoprotein Catabolism

Apolipoprotein C-III accounts for 50% of the protein content of VLDL and 2% of that of HDL [122,646]. It is the most abundant of apo Cs in fasting plasma (12 mg/dL). It is a small 8.8kD apolipoprotein, 79 amino acids long, expressed mainly in the liver and to a less degree in the intestine. It is exchangeable between lipoprotein species, although 1/3 to 2/3 may remain non-exchangeable in VLDL and HDL. During fasting it is mainly found in HDL, whereas post-prandially it is found in large amounts associated with large triglyceride-rich lipoproteins (VLDL and chylomicrons). Apo C-III decreases the catabolism of triglyceride-rich lipoproteins. Biological effects of apo C-III seem to depend highly on its concentration. It is a specific inhibitor of LPL and a noncompetitive inhibitor of hepatic lipase at high concentrations in vitro [836,837]. In vivo, apo C-III overexpression in transgenic mice inhibits the cellular uptake of triglyceride-rich lipoproteins by LDL receptors or LRP, inducing significant hyperlipidemia [646,838]. VLDL are larger, enriched in triglycerides and apo C-III, and very poor in apo E. Overexpression of apo E, for example by cross-breeding with apo E transgenic mice may correct this lipoprotein phenotype [839]. In contrast, apo C-III deficient mice are hypolipidemic and resistant to dietary induced hyperlipidemia. In addition to its role on hepatic lipase apo C-III modulates reverse cholesterol transport. Apo C-III is a moderate inhibitor of LCAT and of CETP activities by displacing activating apolipoproteins (e.g. apo A-I, or apo C-I). Levels of plasma apo C-III are increased in various pro-atherogenic conditions: combined hyperlipidemia, subgroups of patients with type IV hyperlipidemia and overt cardiovascular disease [840]. Therefore excessive expression of apo C-III could contribute to certain forms of atherogenic dyslipidemia resulting from disturbances of triglyceride-rich lipoprotein metabolism in human.

The gene for apo C-III (APOC3) is 2.6kb telomeric from APOA1, on the A1-C3-A4 gene cluster of chromosome 11 [841], (*figure 2-29*). It is 3.1kb long, and comprises 4 exons encoding for a 519nt cDNA (*figure 2-20*). The first exon includes a 5'nontranslated region; the second exon encodes for a prepeptide of 20aa. The third exon encodes for amphipathic α-helices having a weak affinity for phospholipids, preferentially involved in protein-protein interactions. Conversely, exon 4 encodes for class A amphipathic α-helices [842], including an helix formed by residues 50 to 69 with strong affinity for phospholipids. It ends in a 3' noncoding region. As mentioned above (see page 128), there is an *Alu* repeat flanked by a microsatellite $(TTTC)_n$ within intron 3 [843]. In addition to this VNTR, single nucleotide polymorphisms in the apo C-III gene are in linkage disequilibrium with other gene markers along the gene cluster. The rare allele of a SstI RFLP described soon after the gene cloning in the 3' noncoding region of exon 4, has been consistently reported in association with hypertriglyceridemia [844]. Molecular mechanisms underlying this association have not been elucidated yet.

The promoter of the gene for apo C-III includes motifs found in other apolipoprotein (A-I, A-IV and B) gene promoters [845]. Two enhancer regions (distal: -821 to -685 and proximal: -110 to -68) seem to cooperate to control maximum hepatic expression. The distal activating motif is homologous to glucocorticoid-receptor sensitive binding elements. It may be inhibited under the action of insulin and phorbol-esters. A region from nt -461 to -453 contains an insulin response element, down regulating apo C-III gene expression. A repressor element, a counterpart with IRE (Interferon Regulatory Element) located between these motifs (-210 to -110), could bind to nuclear factor $NF_{-\kappa} B$ that mediates intracellular cytokine signalling. Nuclear hormone factors, either activators like HNF4 (Hepatic Nuclear Factor 4) or inhibitors like ARP-1 (Apolipoprotein Response Element or Ear3/COUP-TF (Chicken Ovalbumin Upstream Promoter Transcription Factor) and PPARs interact specifically with the proximal promoter region in hepatocytes [846]. PPAR responsive elements upstream of apo C-III were shown to contribute to the plasma triglyceride-lowering response induced by fibrates [847]. Competitive interactions with the same motifs could result in opposite levels of expression depending on relative concentration of transcription factors within the nucleus. Thus apo C-III expression appears to respond to anabolic steroids, insulin or cytokines thereby contributing to hypertriglyceridemic states observed in hypercortisolemia, insulin resistance or inflammation. Five polymorphisms in regulatory sequences of APOC3 (-641C/A, -630G/A, -625ΔT, -482C/T, -455T/C) are in strong linkage disequilibrium with the neutral Sst I RFLP [848]. Allelic combinations of these markers were underrepresented in patients with severe hypertriglyceridemia (plasma triglycerides higher than 1000 mg/dL), in comparison with normolipemic subjects. Promoter activity of these allelic combinations has been tested in vitro [849], exhibiting decreased transcriptional activity for the rare allele. Interestingly, the rare allele reported as overrepresented in hypertriglyceridemic patients diminishes the repressor action of insulin, thereby increasing apo C-III liver expression and possibly plasma triglyceride-rich lipoproteins. Moreover, the same allele was shown to lower apo A-I expression in the intestine, which could contribute to the low-HDL/high triglyceride phenotype, frequently reported in patients with hypertriglyceridemia. This confirms that intergenic domains within the apolipoprotein AI-CIII-AIV gene cluster contribute to concerted expression of apolipoproteins in plasma, and could contribute to the genetic predisposition to several phenotypes of hypertriglyceridemia. However the role of these variants may be modulated by other factors, association of these promoter polymorphisms being inconsistently reported with hypertriglyceridemia [844].

Apo C-III is O-glycosylated. According to the degree of sialylation of Thr74 different patterns of protein migration are detected on isoelectric focusing. Three major isoforms C-III$_0$, C-III$_1$ and C-III$_2$ are found whether they have none, one or two molecules of sialic acid [122]. Apo C-III would be secreted in C-III$_2$ isoform then desialylated in plasma so that the main circulatory isoform is C-III$_1$. The role of sialylation is unknown: an artificial mutant of apo C-III deprived of Thr74 although non-glycosylated, was normally secreted by hepatocytes and integrates normally into VLDL and HDL [850]. Moreover subjects expressing either non-sialylated or sialylated apo C-III isoforms, do not seem to have an abnormal lipoprotein profile [851].

Very Few molecular defects were reported to date on the apo C-III gene. A genic inversion within the AI-CIII-AIV cluster was described in two homozygotes [852]. Patients had a significant hypoalphalipoproteinemia; they suffered from premature coronary disease, presented corneal opacifications and xanthomas. Remarquably, their plasma triglycerides were normal and VLDL catabolism was accelerated [853]. This complex gene rearrangement was generated by an unequal recombination between two repeated *Alu* sequences: one flanking apo A-I and the other in intron 3 of apo C-III. There is a "*Chi*" sequence -372bp upstream of apo C-III, shown to intervene in mechanisms of homologous recombination in the λ phage, in immunoglobulins, and in a genic conversion of the β-globin cluster [848]. It is not known whether this sequence was involved in this recombinational event.

Point mutations in the apo C-III gene were detected in heterozygotes [854-856] by abnormally migrating apo C-III bands on systematic screening by isoelectric focusing of human plasma. One was an A→C nucleotide substitution in exon 3 causing a Gln38→Lys mutation. A moderate elevation in plasma triglycerides and apo C-III levels was reported in heterozygous carriers within a family. It was supposed that the additive positive charge could further interfere with triglyceride-rich lipoprotein clearance. Another mutation was an A→G transition in exon 4, causing a Lys58→Glu substitution in a highly conserved amphipathic α-helix. This mutation segregated with hyperalphalipoproteinemia (HDL cholesterol = 112 mg/dL), and low apo C-III concentrations in VLDL and HDL. Hyperalphalipoproteinemia seemed to result from an accumulation of abnormally large HDL, enriched in apo E. Another mutation was a heterozygous G→A transition in exon 4, causing an Asp45→Asn substitution. In this case, the lipid profile was normal although the propositus had premature coronary heart disease. Moreover in heterozygotes VLDL contained twice more apo C-III that those of non-carrier subjects in the family. In line with observations in knock out mice, which exhibit lowered plasma cholesterol and triglycerides [646], the metabolic changes induced by these mutations suggest that it is more overexpression than defective apo C-III, which may induce an atherogenic lipoprotein profile. They confirm the role of apo C-III as a natural inhibitor of triglyceride-rich lipoprotein clearance, in which the apo E/C-III ratio seems particularly critical.

2.3.13- Apolipoprotein A-IV, a Component of Intestinal Lipoproteins.

Apolipoprotein A-IV is a small exchangeable apolipoprotein present primarily in HDL. It is produced in the intestine and secreted into chylomicrons. It may also circulate in free dimeric form in interstitial spaces, in lymph and in the cerebrospinal fluid [122,857,858]. It is catabolized in the liver and the kidney [859]. This 4.5kD protein, 376 amino-acid long, is the least lipophilic of small apolipoproteins [41], so that it may easily move away from the lipoprotein surface. Its biological role is not fully clarified. Like other small exchangeable apolipoproteins, it may activate LCAT, however with lower affinity than apo A-I. It also modulates LPL activation by apo C-II [860]. It has been suggested that apo A-IV could play a role in reverse cholesterol transport as a component of HDL. It is detected in high amounts in inflammatory tissues and in regenerating nervous fibers where it may mediate cholesterol efflux from mortified tissues, and its redistribution to regenerative tissue [861]. Apo A-IV has also been proposed to be a potent endogenous antioxidant protecting dietary derived lipids from oxidative modification [862]. Antioxidative protection of lipids from lipoproteins could also be mediated by enhancement of paraoxonase activity [863]. In line with anti-atherogenic properties of HDL, overexpression of apo A-IV in transgenic mice reduces the extent of lesions in dietary induced atherosclerosis and in apo E knock out mice [863,864].

More unusual seems the role of apo A-IV in the regulation of dietary lipid metabolism. It could act as a regulator of food intake in the central nervous system [857]. Intestinal secretion of apo A-IV is maximal post-prandially. Contrary to apo A-I, its concentration increases in the cerebrospinal fluid where it circulates in free form, inducing a reduction of food intake in the rat. In human newborns, plasma concentrations of apo A-IV increase up to the levels found in adult, soon after the beginning of enteral feeding [865]. Dietary modulations may be also qualitative. In the obese and hyperlipidemic Zucker rat, a diet enriched in sucrose stimulates apo A-IV expression, whereas a diet enriched in fish oil will repress it [866]. In addition it was shown that circulating rates of apo A-IV were correlated with plasma triglycerides. The rate of apo A-IV catabolism is reduced in severely hypertriglyceridemic subjects (type V), a lipid disorder known to be sensitive to dietary fat intake [867].

The gene for apo A-IV (APOA4) is integrated into the A1-C3-A4 gene cluster, 6.2kb telomeric of APOC3, and approximately 12kb downstream of APOA1 and in the same orientation *(figure 2-29)*. It is 2.6kb in length and comprises only 3 exons [122]. Contrary to other exchangeable apolipoproteins, the 5' noncoding region is not individualized within a single exon *(figure 2-20)*. The first exon encodes for the signal peptide. The first amphipathic α-helix is encoded by the second exon [41]. The third exon encodes for a series of amphipathic α-helixes more or less lipophilic, and comprises a 3' noncoding sequence. There is an Alu *repeat* upstream of the transcription start site. The promoter of the apo A-IV gene comprises two responsive elements (-293 to -233 and -127 to -60) interacting with nuclear proteins of intestinal specificity, for its maximal activity [868].

Apolipoprotein A-IV is polymorphic. Common protein isoforms were described in various populations by isoelectric focusing according to their charge, from the most acidic to the most basic: A-IV_0, A-IV_1, A-IV_2 and A-IV_3. These isoforms result from genetic polymorphisms *(table 2-12)*. The

Table 2-12. **Apo A-IV gene variants.** Rare allele frequencies are given in Caucasian populations. NC: non coding.

Isoform	Mutation	Residue	Frequency	RFLP	Ref.
-	G→A	Val^8→Met	0.007	-	869
-	G→A	Asn^{127}→Ser	0.22	Hinc II	870
A-$IV_{seattle3}$	G→T	Ala^{141}→Ser	0.03	-	871
A-$IV_{seattle1}$	C→T	Ser^{158}→Leu	rare	-	871
A-IV_3	G→A	Glu^{165}→Lys	0.002	Sst I	872
A-IV_0	A→G	Lys^{167}→Glu	rare	Ava I	872
A-IV_3	G→A	Glu^{230}→Lys	0.001	Sty I	872
A-$IV_{seattle2}$	G→A	Arg^{244}→Gln	rare	Pst I	871
A-IV_1	A→T	Thr^{347}→Ser	0.016	Hinf I	872
A-IV_2	G→T	Gln^{360}→His	0.07	Fnu 4HI Sfa NI	872
A-IV_5	12 nt insertion	Ins Glu^{361} «Glu-Gln-Gln-Gln»	0.003	-	872
-	«CTGT» ins/del.	NC	0.39	-	869

most common isoforms are isoform AIV_1 (Gln360) and AIV_2 (His 360). Allelic frequencies of apo A-IV isoforms may vary between populations. A North→South gradient of increasing frequencies for A-IV_2 was observed in Europe [873]. The Thr347Ser variation, which does not induce any change in protein mobility (A-IV_1), is also common in Caucasians [869,874].

The effect of apo A-IV isoforms on plasma lipoprotein profile was studied in several populations. Apart from rare variants found in subjects with familial combined hyperlipidemia [871], contradictory reports have been made on the effect of apo A-IV polymorphism on baseline lipoprotein levels [819]. These discrepancies could result from a dietary-dependent effect. A relationship was reported between apo A-IV isoforms, body mass index, and overweight and oral glucose tolerance [874]. Interestingly, plasma LDL-cholesterol lowering under a low fat diet is less pronounced in subjects carrying apo A-IV$_2$ (His360) compared with carriers of apo A-IV$_1$ (Gln360) [875]. Moreover, a diet voluntarily enriched in saturated fat and cholesterol induces a less pronounced increase in LDL-cholesterol in subjects carrying the apo A-IV$_2$ allele compared with carriers of the apo A-IV$_1$ allele [876]. In addition, the Ser347 allele is associated with a more pronounced decrease in post-prandial levels of triglyceride-rich lipoproteins than the Thr347 common allele [877]. These clinical observations suggest that apo A-IV isoforms could modulate certain steps of dietary lipid absorption and triglyceride-rich lipoproteins metabolism. Two common isoforms of apo A-IV were reported in baboon, resulting from a Lys76→Glu substitution [878]. In these animals, apo A-IV polymorphism modulates plasma levels of HDL cholesterol under saturated fat diet, whereas this effect is undetectable under chow diet. These observations underscore the fact that polymorphisms in candidate gene of lipoprotein metabolism may unmask their effect in dynamic conditions (e.g. dietary changes) rather than in basal, fasting conditions.

Another interesting observation was made on apo A-IV polymorphism. The 3' end of exon 3, encodes for 4aa repeated motifs (Glu-Gln-X-Gln) between residues 354 and 369 of the mature protein. These motifs which take part in the formation of amphipathic α-helices are highly conserved, but their number may vary between species. A rare polymorphism inserts or deletes one of these 4aa repeats, with variable frequencies in different human populations [872,879]. In keeping, this domain may vary by the number of repeated motifs in cynomolgus monkey or in baboon [879,880]. This polymorphism, which modifies the charge and the helicoidal structure of the protein, is also found in various strains of mice, independently from their phylogenic difference [881]. These observations remain unexplained. They suggest however that evolution of this amphipathic α-helix rich domain was permissive on the variability of apo A-IV structure from one species or even from one strain to the other. It could have served metabolic functions adapted to the nature of food available in the natural environment encountered by different human populations or wild-type animal strains.

2.3.14-   **Apolipoprotein   A-II,   a   Modulator   of   HDL**
                  Metabolism.

Apolipoprotein A-II belongs to the family of small exchangeable apolipoproteins [122]. It is a 17.4kD protein comprising 77 amino acids, with several amphipathic α-helixes having strong affinity for lipids [41]. Human apo A-II is present in plasma in homodimeric form stabilized by a disulfide bridge on Cys6. Apo A-II may also heterodimerize with apo E [882]. Apo A-II dimerization has been observed only in humans so far, however, lack of dimerization does not seem to induce major lipoprotein changes [883,884]. Apo A-II is encoded by a single gene (APOA2) located on the long arm of chromosome 1, bands q21-q23 [885,886]. It has a structure in four exons, similar to that of other exchangeable apolipoproteins (*Figure 2-20*). Its sequence is poorly conserved between species (60% primary sequence homology between mouse and human). An intragenic microsatellite $(CA)_n$ [887] and a 3' flanking Msp I RFLP were used for linkage analyses [888]. Apo A-II is mainly expressed in the liver and the intestine [889]. The promoter sequence contains several responsive elements for transcription factors (HNF, EAR3/COUP, ARP1 and PPAR) also controlling the expression of other apolipoproteins. In line with, Apo A-I or apo B, certain transcriptional factors seem to act in synergy in the control of hepatic and intestinal expression of apo A-II [890].

Although this apolipoprotein is the second major protein component of human HDL (20% of protein content), its physiological role had remained unclear. More recently apo A-II has been identified as a modulator of HDL metabolism by studies in genetically modified animals [793]. HDL are heterogeneous particles: some contain only apo A-I (LpAI), others contain apo A-I and apo A-II (LpAI-AII). Compared to LpAI, LpAI-AII would be less efficient to ensure cholesterol efflux and reverse cholesterol transport in part through LCAT and hepatic lipase inhibition [891,892]. Apo A-II overexpression also depresses plasma LPL activity when associated with VLDL in the fed state [893]. Its absence in knock out mice results in increased remnant clearance, lower plasma concentrations of free fatty acids and insulin hypersensitivity [894]. Moreover, Apo A-II would not confer anti-oxidative properties to HDL as opposed to apo A-I. This mechanism has been associated with a decreased HDL content in paraoxonase [895]. In addition, apo A-II could act as a proinflammatory protein when expressed at high levels. Altogether, apo A-II modifications of lipoprotein structure and metabolism could contribute to its atherogenicity when overexpressed in vivo [896]. Apo A-II concentrations have been shown be abnormally low in the cerebrospinal fluid of patients with AIDS [897]. Although the mechanisms were unclear, this could result from down-regulated expression, or from the "consumption" of apo A-II amphipathic α-helices as antiviral components [799]. Another example of extracellular regulation of circulating apo A-II was given in familial amyloidotic neuropathy resulting from mutations in transthyretin (a plasma transporter of thyroxin and retinol). Plasma levels of apo A-II were shown to decrease along with the progression of the disease [898]. Other plasma apolipoprotein levels and liver production of apo A-II were normal. It has been suggested that apo A-II could be displaced from HDL accelerating its catabolism.

One natural mutant a guanine substitution on a consensus splice site abolishing apo A-II expression was described in a Japanese family. There was no apparent effect on plasma HDL cholesterol concentration or on atherosclerosis [899]. However, a Pro5→Glu mutation of apo A-II and apo A-II accumulation was identified in a mouse strain susceptible to early senescence and to a particular form of amyloidosis [900]. Here, as in the case of apo A-I a conformational change of apo A-II and its tissue accumulation could contribute to disease development.

Animal experiments have elucidated the pathophysiological role of **apo A-II in atherogenesis.** Crossbreeding studies in mice (C57BL/6J, an atherosclerosis susceptible strain, and *Mus Spretus,* wild type) showed that variations at the apo A-II locus may determine circulating levels of apo A-II and of free fatty acids [901]. The same effect was observed on sib-pairs in humans, for plasma apo A-I and apo A-II concentrations in families of myocardial infarction survivors [902]. Moreover, transgenic mice overexpressing murine apo A-II have HDL with a lowered content in apo E, which seems displaced by excessive levels of apo A-II [903]. They have higher plasma levels of free fatty acids and are spontaneously likely to develop atherosclerosis [896], despite higher fasting concentrations of HDL than normal littermates. This effect is species specific. Transgenic mice overexpressing human apo A-II have contrasting differences in lipoprotein metabolism [793]. These data suggest that the composition of HDL and the nature of their interactions with other components of lipoprotein metabolism (e.g. remodeling enzymes or proteins) and of the arterial wall could generate subtle responses, which could prevent or accelerate atherosclerosis. They confirm that it would be more an excess of apo A-II, than its absence that would favor atherogenic dyslipidemia.

2.3.2-   Circulating Enzymes and Transfer Proteins for Lipoprotein Remodeling.

2.3.21-   Lecithin:Cholesterol Acyl Transferase (LCAT), a Rate Limiting Enzyme for HDL metabolism

Lecithin:Cholesterol Acyl Transferase (LCAT) plays a major role in extracellular cholesterol esterification and its circulation within the core of lipoproteins [904-906]. LCAT is bound at the surface of HDL and produces esterified cholesterol and lysolecithin by translocation of a fatty acid chain (i.e. acyl group) from phosphatidyl-choline to the 3β-hydroxyl group of cholesterol. This enzyme uses preferably lipid components of native HDL as substrates. Although the majority of small exchangeable apolipoproteins are potential activators, apolipoprotein A-I is its major natural cofactor. Lysolecithin produced by the reaction binds to albumin or is taken up by tissues. Native HDL are good acceptors of free cholesterol transferred at the cell surface by ABC1. The action of LCAT allows newly formed highly hydrophobic cholesteryl esters to reach rapidly the core of the HDL particle. This results in HDL enlargement and remodeling from a discoid shape to a mature spheroid shape, suitable for transport and exchanges of neutral lipids and apolipoproteins. Thus, because LCAT allows the maturation of HDL this enzyme has been coined as a major enzyme of reverse cholesterol transport. LCAT deficiency is a human disease associated with impaired lipoprotein transport of cholesteryl esters and very low plasma HDL [907]. However, LCAT deficiency is inconstantly associated with atherosclerosis. Genetically modified animals have confirmed the role of LCAT in lipoprotein metabolism and atherosclerosis. Besides its role on HDL metabolism, LCAT lowers levels of circulating apo B-rich lipoproteins, particularly LDL, by mechanisms that require the presence of LDL receptors [904]. LCAT has antioxidant properties particularly on phospholipids at the surface of lipoproteins [905,908]. LCAT knockout mice have reduced paraoxonase and platelet-activating factor acetylhydrolase activities, both major components of antioxidative protection of circulating phospholipids and lipoproteins [909]. Together with impaired lipoprotein metabolism, defective antioxidative protection could contribute to glomerulosclerosis, a major complication of LCAT deficiency in human [910]. LCAT overexpression is protective against atherosclerosis in mice, provided that components of human lipoprotein metabolism (i.e. circulating transfer proteins and mainly apo A-I) are present. On the other hand, LCAT deficiency in knock out mice reproduces the human disease in that it does not result in increased susceptibility to atherosclerosis, as the result of combined reductions of HDL (antiatherogenic) and of apo B-rich (atherogenic) lipoproteins.

LCAT is a monomeric glycoprotein of MW 67kD mainly synthesized by the liver, and has been shown to express in neuroglial cells [911]. It is encoded by a single gene (LCAT) localized on band q22.1 of human chromosome 16, in a genomic region also containing the CETP locus [912]. In addition, mouse LCAT and CETP are located on the same chromosomal region as LPL, another regulatory protein of HDL metabolism [913]. The 1.55kb cDNA contains two short 5'- and 3' non-coding sequences (28 and 23 bp respectively). It encodes for a 24-residue signal peptide and for the 416aa mature protein [914]. There is high sequence conservation of LCAT between animal species. The gene is 4.2kb long, comprising six exons (*figure 2-30),* and several *Alu* repeats in its 5th largest intron and in its 3' flanking region.

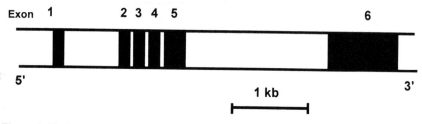

Figure 2-30. Genomic organization of the LCAT gene.

LCAT shares strong sequence homologies with members of the LPL gene family in C-terminal domains involved in its catalytic activity [915]. This sequence similarity has helped together with site directed mutagenesis experiments, to reconstruct a model of LCAT three-dimensional structure and of its functional domains [905]. Exon 1 encodes for the signal peptide and for the first 27aa of the mature protein. N-glycosylation sites on Asn20 in exon 1 and on Asn84 in exon 3 have been identified. Other glycosylation sites are encoded by exon 6: Asn272, Asn384, Thr407, Ser409. Similarly to the family of lipases a triad formed by Ser181 (exon 5), Asp315 and His 377 (exon 6) is crucial for catalytic activity. The protein folding divides the molecule into two major domains: a highly conserved seven beta-strand structure separated by loops and four amphipathic α-helices involved in lipid binding and catalytic activity, and a long excursion domain between residues 211 and 323 sharing little species conservation.

That significant functional domains are distributed along the gene sequence is confirmed by the wide dispersion of naturally occurring mutations causing familial **LCAT deficiency** (FLD) (*table 2-13).* This recessively inherited dyslipoproteinemia described worldwide is generally discovered in adulthood [907]. Corneal cholesterol deposition known as «fish eye» characterizes the disease, after the whitish-opaque aspect it gives to the cornea. Corneal opacification is usually the main patient's complaint and

Table 2-13. **Naturally occurring mutations of the LCAT gene.** Class 1 defect: null allele. Class 2 defect: missense allele causing complete LCAT deficiency. Class 3: partial LCAT deficiency. Class 4: alpha-LCAT deficiency.

Mutation	Location	Protein	Class	Population	Ref
Familal LCAT Deficiency (FLD)					
Ins 30 nt	Exon 1	Dupl. aa$^{-6} \to$aa^{+4}	2	Germany	915
A→T	Exon 1	Asn5→Ile	2	Japan	916
Ins C	Exon 1	Fs Pro9→Stop16	1	Japan, Finland	916
C→T	Exon 1	Thr13→Met	2	USA	917
G→A	Exon 2	Gly30→Ser	2	Japan	915
T→C	Exon 2	Leu32→Pro	2	USA	915
G→C	Exon 2	Gly33→Arg	2	Germany	915
C→A	Exon 3	Tyr83→STOP	1	Italy, USA	916
G→A	Exon 3	Ala93→Thr	2	Denmark	916
C→T	Exon 4	Arg135→Trp	2	Canada	916
G→A	Exon 4	Arg135→Gln	2	Holland	915
Δ TG	Exon 4	Fs Thr138→Stop144	1	France	918
G→A	Exon 4	Arg140→His	2	Austria	915
Ins G	Exon 4	Fs Ala141→Stop	1	Japan	916
A→G	Exon 4	Tyr144→Cys	2	Australia	916
C→T	Exon 4	Arg147→Trp	2	Italy, France	916, 919
T-22→C	Intron 4	lariat branchpoint	1	Holland	916
T→A	Exon 5	Tyr156→Asn	2	USA	916
T→G	Exon 5	Tyr171→STOP	1	France	919
G→A	Exon 5	Gly183→His	2	Canada	915
T→C	Exon 5	Leu209→Pro	2	France	916
C→A	Exon 6	Asn228→Lys	2	Japan	916
G→C	Exon 6	Gly230→Arg	2	Finland	920
T→A	Exon 6	Met252→Lys	2	Norway	915
Δ G	Exon 6	Fs Val264→Stop	1	Japan	916
C→T	Exon 6	Met293→Ile	2	Japan	916
C→T	Exon 6	Pro307→Ser	2	USA	917
C→T	Exon 6	Thr321→Met	2	Italy	916
G→A	Exon 6	Gly344→Ser	2	Japan	915
C→T	Exon 6	Thr347→Met	2	Germany	916
Ins A	Exon 6	Fs Gln376→Stop	1	Canada	916
C→T	Exon 6	Arg399→Cys	2	Finland	916
Fish Eye Disease (FED)					
C→A	Exon 1	Pro10→Gln	4	Holland	915
C→T	Exon 1	Pro10→Leu	4	Sweden	915
C→T	Exon 3	Arg99→Cys	3	Spain	915
C→T	Exon 4	Thr123→Ile	4	Caucasian	915
A→G	Exon 4	Asn131→Asp	3	Holland	915
C→T	Exon 5	Arg158→Cys	4	Denmark	915
Δ CTC	Exon 6	Δ Leu300	3	France	915
A→G	Exon 6	Asn391→Ser	3	Canada	915

may require corneal replacement by grafting. Patients have very low plasma HDL cholesterol, down to approximately 10% of normal, moderate hypertriglyceridemia and normochromic anemia. Plasma LCAT activity is abolished although LCAT mass may be present. Circulating lipoproteins are abnormal in their shape and composition. Their core is deprived of cholesteryl esters and enriched in phospholipids, with higher susceptibility to oxidation. Small discoidal HDL and lipoprotein LpX (an abnormal lipoprotein containing apo B, phospholipids and albumin) are found in plasma. Although cholesterol deposits are observed, atherosclerosis and xanthomas are inconstant. HDL and LDL catabolism are accelerated accounting for their low circulating concentration. The outcome depends on renal insufficiency. Renal disease characterized by a glomerular accumulation of cholesterol and phospholipids has been correlated with plasma accumulation of LpX and oxidized lipoproteins [910].

Another recessive disease has been linked to the LCAT gene: **"Fish Eye Disease"** (FED), which distinguishes from the former by isolated corneal deposits and partial enzymatic deficiency. In some cases, only cholesterol esterification of HDL (α-LCAT activity) is defective, whereas apo B containing lipoproteins have a normal content in cholesteryl esters (β-LCAT activity). A thorough structure-function analysis of naturally occurring mutations causing familial LCAT deficiency has shown that mutations causing FLD are found mainly on highly conserved residues involved in ligand binding and catalytic domains of the protein [915]. A classification of LCAT mutations causing FLD and FED has been proposed [916]. Class 1 defects correspond to null alleles caused by nonsense mutations. Class 2 defects correspond to defective alleles resulting from missense mutations. In the latter case, the protein may be secreted however completely catalytically defective. Class 3 defects correspond to partial LCAT deficiency and class 4 defects are confined to α-LCAT deficiency only. An unusual case of intronic mutation was described in intron 4, on a residue involved in lariat formation for proper mRNA splicing, as a cause of class 1 mutation [921]. Cases of intronic mutations are rarely reported as cause of human diseases. Another case of cryptic mutation was reported on the LDL receptor gene [922]. In contrast, mutations on the LCAT gene causing FED are confined to less conserved domains on the hydrophilic face of certain helices at the surface of the protein. They indicated that particular residues of LCAT bind specifically to HDL, probably through protein-protein interaction with apo A-I. This hypothesis is reinforced by the fact that certain mutations of apo A-I can result in FED as a result of defective α-LCAT activity [829]. In addition, β-LCAT activity was used to show that Lp(a) assembly is dependent on the integrity of LDL. Lp(a) is absent in plasma from subjects with LCAT deficiency, however after incubation of their LDL with apo (a) normal LCAT restores the assembly of Lp(a) in vitro [923].

Some points remain to be clarified about this enzyme. Contrary to other animal species, human LCAT expression is not highly dependent on diet, but its activity clearly depends on the shape and content of lipoproteins. In LCAT deficiency, even if the abnormal lipoprotein profile is correlated with the severity of the disease, the molecular pathogenesis of renal insufficiency and of corneal deposits remains unclear. LCAT was initially proposed, because of its role of major component of HDL metabolism and reverse cholesterol transport [65], to be an important antiatherogenic target. However, its overexpression is not necessarily protective against atherosclerosis, depending on the physiological context and of other plasma components of lipoprotein remodeling. LCAT mediates one of the multiple steps involved in reverse cholesterol transport. LCAT activity should be finely tuned with other upstream or downstream factors along the chain of reactions remodeling HDL. For example, apo A-I concentration as an upstream cofactor, and CETP activity downstream of LCAT action will represent limiting factors in the face of highly expressed LCAT, despite increased HDL concentration [904]. Moreover, LCAT deficiency does not appear to be proatherogenic in knock out mice. Despite very low HDL cholesterol concentrations in plasma form LCAT deficient humans, atherosclerosis is not a prominent feature of the disease. This underscores that other mechanisms ensure reverse cholesterol transport. Aside from "passive" cholesterol efflux, cellular receptors like SRB-I and ABC1 have shown to mediate cellular efflux. Moreover liver uptake of cholesterol involves several pathways (SR-BI, CETP, HDL uptake etc), independent of LCAT activity. LCAT has been found, exclusive of CETP, to be a modifier locus on chromosome 16 for familial combined hyperlipidemia [924]. It remains unknown if LCAT acts directly or if this linkage reflects the effect of another closely related locus. Recently, a novel enzyme (LRO1) sharing sequence homology with LCAT was identified in yeast to mediate the final step of triglyceride biosynthesis (diacylglycerol esterification), suggesting that other members of this gene family may provide other functional links between lipid and HDL metabolism [925].

2.3.22-   Cholesteryl Ester Transfer Protein (CETP), a "Points man" of Lipoprotein Metabolism.

Cholesteryl Ester Transfer Protein (CETP) promotes the net transfer of esterified cholesterol from HDL, in exchange for triglycerides coming from apo B-containing lipoproteins (*figure 2-28*). It may also promote the transfer of phospholipids and free cholesterol from triglyceride-rich lipoproteins or cells, towards HDL [926,927]. Through its action on surface and core lipid components of lipoproteins, CETP appears as an important factor of lipoprotein remodeling in plasma, at the crossroads of major lipoprotein pathways. In addition, CETP favors liver detoxification of oxidized lipids, by transferring oxidized cholesteryl esters from oxidized LDL towards HDL for rapid liver uptake and degradation. Its activity may be modulated by the apolipoprotein content of HDL and triglyceride-containing lipoproteins [646,928]. In humans, it daily ensures 50% of the phospholipid transfer between various types of lipoproteins and 50% to 70% of cholesterol transfer in the opposite way [913]. There is a universal negative correlation between plasma CETP activity and plasma HDL concentration. CETP acts as an enhancing factor of reverse cholesterol transport by redirecting cholesterol towards high capacity loading vehicles (i.e. apo B-rich lipoproteins), which may be taken up by multiple liver remnant receptors (LRP, HSPG, LDLR etc). Moreover by forming smaller HDL more efficient in their capacity to deliver cholesterol to the liver CETP accelerates HDL turnover and reverse cholesterol transport. Therefore CETP appears to act as a "points man", which reorients lipid transport and singularly cholesterol to different lipoprotein pathways.

CETP was given a contrasted role in atherosclerosis. In line with observations made on LCAT, overexpressed or inactive CETP do not necessarily result in opposite effects on atherosclerosis. In many animal species, spontaneous levels of hepatic CETP expression correlate positively with the relative part taken by apo B-containing or HDL lipoproteins in lipid transport, as well as with natural susceptibility to atherosclerosis [927]. Rodent species express low levels of CETP; mouse strains the most susceptible to atherosclerosis expressing higher plasma CETP levels, than atherosclerosis-resistant strains. In transgenic mice, CETP overexpression may enhance susceptibility to atherosclerosis particularly in models with altered remnant lipoprotein catabolism (apo E knock out, high fat diet). The lowering of HDL is accompanied by a down-regulation of hepatic LDL receptors, as a consequence of increased intrahepatic influx of cholesteryl esters and of free fatty acids [929]. However, on the metabolic context of apo C-III overexpression (an activator of CETP activity) [646], CETP appears protective against atherosclerosis by enhancing HDL turnover and reverse cholesterol transport, despite very low plasma levels of HDL and high levels of triglyceride [930]. Thus, by its action on lipoprotein remodeling and trafficking, CETP may be protective in metabolic contexts where either reverse cholesterol transport or remnant clearance are functional. However, in any condition impairing either of these pathways or both, unadjusted levels of CETP activity may favor the development of atherosclerosis, a situation found in human CETP deficiency (see below).

CETP a highly hydrophobic glycoprotein of MW 74kD, and 476 amino acids long [931]. It is encoded by a single gene (CETP) 25kb in length, containing 16 exons, localized on chromosome 16, band q22 [932,933]. The cDNA is about 1783bp long, with short 5' and 3' non-coding sequences. CETP belongs to a family of lipid transfer/lipopolysaccharide binding proteins (BPI gene family) including PLTP (Phospho Lipid Transfer Protein), LBP (Lipopolysaccharide Binding Protein) and BPI (Bactericidal Permeability Increasing protein), which modulates bactericidal activity of neutrophils [926]. They all bind phospholipids and lipopolysaccharides. They are all secreted; CETP, PLTP and LBP are bound to HDL, while BPI is bound to the surface of neutrophil secretory granules. The relative functional and primary sequence conservation of this family of proteins allowed deducing a common structure from the crystal structure of BPI. Lipid transfer proteins are "boomerang" shaped molecules, with two similar domains shaped like barrels separated by a central beta sheet domain. The barrel-shaped domains are involved in lipid binding. One domain is involved in neutral lipid binding domain, while the other domain is involved in phospholipid binding. In CETP a short C-terminal amphipathic α-helix determines neutral lipid (i.e. cholesteryl esters and triglycerides) transfer activity. Amino acids Lys233 and Arg259 are highly conserved in the family and are necessary for cholesteryl ester transfer and HDL binding in CETP.

Several silent RFLPs were described at the CETP locus: TaqIa, TaqIb, BamHI, StuI [934] and Eco NI [935]. The TaqIb polymorphism located in intron 1, has been associated with baseline HDL, lipid transfer activity and CETP mass in plasma in different populations, the strength of association depending on study design, metabolic context and genetic background [936]. Moreover, it was shown in the REGRESS cohort of Dutch male patients with CAD, that the B1 allele (frequency =0.59) was associated with higher CETP mass, lower HDL and higher angiographic progression of CAD over 2 years [937]. Similarly to the case of the apo C-III SstI RFLP, linkage disequilibrium with some functional sequence within or nearby the CETP locus could account for this association. However, this functional sequence would act depending on the metabolic context, since patients who had the B2 allele had lower CETP and higher HDL levels and did not benefit from lipid lowering therapy contrary to B1 carriers. Although these observations need confirmation in other populations, this study had demonstrated that it may be more dynamic phenotypes (i.e. arterial disease progression, dietary responses etc) than static phenotypes (baseline, fasting plasma lipid levels, etc.) that may reflect the effect of candidate gene polymorphisms of lipoprotein metabolism. In addition, this underlines that CETP is a modifier protein, which may be beneficial at some critical level of activity finely tuned with the metabolic context, whereas it could act as a proatherogenic protein outside of this window.

CETP is expressed primarily in the liver, spleen and adipose tissue, and at lower levels in intestine, adrenals, skeletal muscle, heart and kidney [931]. CETP may be expressed in white-blood cells lineages and is also found in interstitial fluids including cerebrospinal fluid and in sperm [926]. In the majority of tissues only one mRNA transcript is detectable. However an alternate splicing can generate mRNAs deprived of exon 9, which encodes for 66aa in the central part of protein. This truncated isoform is not secreted, and would even inhibit secretion of the full-length protein by dimer formation, blocking the complex within the endoplasmic reticulum. It was suggested that this alternate splicing could modulate CETP secretion and activity within tissues. CETP expression is strongly dependent on diet. Saturated fatty acids and cholesterol enriched diets strongly upregulate CETP liver expression and plasma activity. Monounsaturated and omega 3 fatty acids (fish oil) diminish this response. CETP is a cholesterol sensitive gene. SREBP responsive elements are found upstream of the transcription start site and SREBP-1a was shown to transactivate CETP basal expression in the liver [926]. However, LXR (Liver X receptor) a nuclear hormone receptor activated by oxysterols mediates dietary cholesterol induction of CETP expression, by heterodimerization with RXR (Retinoid X receptor) [938]. CCAAT/enhancer binding protein alpha (C/EBPα), ARP-1 (apolipoprotein A-I regulatory protein 1) and HNF (hepatic nuclear factor) are also found upstream from CETP. Depending on the presence of other tissue specific transcription factor, these elements may be enhancing or inhibiting CETP expression in a tissue-specific manner. In addition, CETP expression is strongly dependent on hormones and inflammatory signalling. CETP activity is strongly inhibited by lipopolysaccharide (LPS) through a corticosteroid dependent mechanism [939]. TNFα and Interleukin-1 also down-regulate CETP suggesting that it may have other functions modulating inflammation in peripheral tissue and/extracellular spaces. Insulin down regulates CETP expression in adipose tissue and muscle. Plasma CETP mass and activity have been correlated with the degree of insulin resistance in humans.

CETP deficiency is a dominant disorder caused by mutations on the CETP gene [926,927]. Mutations (mainly nonsense mutations) have been described in various populations, however most were found in Asia, some being particularly prevalent in Japan *(Table 2-14)*. CETP deficiency is characterized by a dominantly inherited hyperalphalipoproteinemia sometimes associated with moderately lowered LDL cholesterol and moderately increased triglycerides in plasma. CETP mass and activity are profoundly altered. In Japan, where mean plasma HDL cholesterol is higher than in western populations, the Asp442Gly and intron 14 splice mutations account for about 10% of the variability of HDL cholesterol. Although

Table 2-14. **Naturally occurring mutations causing CETP deficiency.**

Mutation	Location	Type	Effect	Population	Ref.
ΔC	Exon 1	Stop^{38} (frameshift)	Null	Germany	927
T→G	Exon 2	Tyr^{57} →Stop	Null	China	940
G→T	Exon 6	Gly^{181} →Stop	Null	Japan	927
C→T	Exon 9	Arg^{268} →Stop	Null	Nova Scotia	941
C→T	Exon 10	Arg^{309} →Stop	Null	Japan	926
T→G	Intron 10	T^{+2} →G (splice)	Null	Japan	927
G→C	Exon 11	Ala^{373} →Pro	Defective	Germany	927
A→G	Exon 12	Ile^{405} →Val	Conservative	Caucasians	927
G→A	Intron 14	G^{+1} →A (splice)	Null	Japan	927
Ins T	Intron 14	Ins T^{+3} (splice)	Null	Japan	927
A→G	Exon 15	Asp^{442} →Gly	Defective	Japan	927
G→A	Exon 15	Arg^{451} →Gln	Defective	Germany, Finland	942

CETP deficiency was first suggested to be associated with a longevity syndrome, studies of large cohorts of heterozygous carriers have shown that in fact, CETP deficiency is a risk factor for cardiovascular disease [934,943]. Despite high plasma HDL levels (>1.5g/L) and a low TC/HDL ratio, HDL are dysfunctional in reverse cholesterol transport, particularly in the liver. Dense and oxidized LDL tend to accumulate further increasing atherogenic risk. Interestingly, the risk is even more pronounced in patients who combine hepatic lipase deficiency, despite very high plasma HDL concentrations [944]. Because HL is also a modulator of reverse cholesterol transport and of remnant clearance, both disorders act synergistically upon increasing plasma HDL and the risk of cardiovascular disease. On the other hand, partial CETP deficiency no longer represents a risk factor in conditions where plasma HDL is very high, and reverse cholesterol transport very efficient. A common Ile405Val CETP variant (frequency ≈0.30) has been studied in several populations and was uniformly associated with increased plasma HDL cholesterol [945,946]. Although, this variant was identified as a conservative mutation upon gene cloning, it has been generally reported in subjects with moderately altered plasma CETP mass or activity. Moreover, the Val405 allele has been associated in a dose-effect manner with increased risk of coronary heart disease (relative risk ≈1.4 to 2.1) in premenopausal women or in hypertriglyceridemic men. It is not known if this biological effect is a direct consequence of other non-measurable functional changes on the CETP protein (independent of its neutral lipid transfer activity), or of the effect of other linked variants on the CETP locus.

Recently, a promoter variant -629A/C has shown to display a variable pattern of association with nuclear factors of the Spl family and mildly decreased promoter activity [947]. The -629A allele was also associated with higher plasma HDL cholesterol and lower CETP mass in Caucasians.

One important conclusion was drawn from studies of common variants together with those from CETP deficient patients. Partial CETP deficiency is an independent risk factor for atherosclerosis in subjects with moderately increased plasma HDL cholesterol, a condition supposedly protective against atherosclerosis in the general population. The detection of partial CETP deficiency could be a way to avoid misleading cardiovascular risk estimation in the face of a low TC/HDL ratio or moderately high HDL cholesterol. However, in conditions where reverse cholesterol transport is enhanced (e.g. estrogen treatment in post-menopausal women) or if apo B-rich lipoprotein metabolism is accelerated (statin treatment), the effects of partial CETP deficiency may be no longer significant. It has been proposed that partial CETP deficiency could represent a "thrifty" way to slower lipid trafficking by modulating lipoprotein metabolism and reverse cholesterol transport. Therefore, it may be predicted that CETP will be pro- or anti-atherogenic depending on the metabolic status.

CETP may have other physiological roles. It has been shown to enhance sperm capacitation in follicular fluid. In the majority of mammals where it mainly expresses in liver, LCAT is less expressed and reciprocally [913]. In species where it is rather highly expressed like primate, human and rabbit, it would compensate for lower hepatic synthesis of LCAT. Contrary to LCAT, CETP expression is induced by food intake, reorienting cholesterol to high capacity lipoprotein pathways. Moreover, by deriving most of cholesterol transport towards the apo B-containing lipoprotein pathway, it would enhance the load taken up by the liver through the remnant receptor pathway. It appears as a modulator protein, enhancing lipoprotein capacities in their functions of lipid transport. It is not known what evolutionary advantage CETP expression could have conferred by redirecting cholesterol transport from a system depending mainly on HDL (like in murine species), to a transport system mainly relying on larger apo B-containing lipoproteins like in primates or humans. In addition, human CETP deficiency has given another paradoxical example of an atherogenic genetic condition associated with high plasma HDL cholesterol.

2.3.23-   Phospholipid   Transfer   Protein,   a   Major
Component of HDL Formation in Plasma

Phospholipid Transfer Protein, or PLTP is the other lipid transfer protein
bound to HDL [926,948]. PLTP transfers phospholipids from the VLDL or
chylomicron surface towards HDL thereby regulating an important process
of lipoprotein remodeling; the redistribution of surface components. Upon
lipolysis of triglyceride-rich particles by LPL, the redistribution of surface
phospholipids and free cholesterol towards native HDL by PLTP is a crucial
step in HDL lipoprotein formation in plasma (*figure 2-28*). HDL become
mature enriched in other apolipoproteins, able to cycle in the reverse
cholesterol transport pathway. Moroever, apo B-rich lipoproteins may be
further processed down the remnant or LDL pathways. Besides
phospholipids and lipopolysaccharide, PLTP has broader binding capacities
towards diacylglycerols, sphingolipids and alpha-tocopherol (or vitamin E).
Genetically modified animals have confirmed the physiological role of
PLTP. PLTP knockout mice have decreased plasma HDL and apo A-I. They
tend to accumulate triglyceride-rich lipoproteins under a high fat diet. Mice
overexpressing PLTP have an increased capacity to produce HDL despite
lower plasma HDL concentration. Moreover, HDL have an increased
capacity for cellular cholesterol efflux from macrophages, suggesting that
they could prevent foam-cell formation.

The cloning of the PLTP gene (PLTP) came late after that of CETP and of
other members of the BPI gene family [949]. Although PLTP primary
structure is closer to LBP (lipopolysaccharide binding protein) than to
CETP, the gene has a similar organization (16 exons/15 introns) with highly
conserved exon-intron boundaries [950]. It has been localized to human
chromosome 20q12-q13.1 [951]. Moreover, the crystal structure obtained for
BPI combined with site directed mutagenesis has allowed a better
understanding of PLTP structure-function relationships [952]. The general
boomerang shape structure common to the family is organized in two lipid
binding barrel-shaped domains. Highly conserved Cys146 and Cys185 form
a disulfide bridge necessary for protein activity.

PLTP is expressed in a variety of tissues particularly in placenta, pancreas,
adipocytes and lung. Lower expression levels are found in liver, muscle
kidney and heart. Interestingly, PLTP expression seems to mirror CETP
expression in animal species, to a degree correlated with natural resistance to
atherosclerosis [953]. In mouse, rat, dog or cat species known to be naturally
resistant to atherosclerosis, plasma PLTP activity levels were the highest
whereas CETP were the lowest to absent. Oppositely in human, pig or rabbit,
PLTP activity was lower while CETP was highly expressed. Like for CETP,
the type of dietary fatty acid ingested, modulate PLTP expression [954]. Bile

acids transactivate PLTP liver expression through FXR (Farnesoid X activated Receptor) activation, a family of nuclear receptors that heterodimerize with RXR (Retinoid X Receptor) [955]. Conversely fibrates down-regulate PLTP expression through promoter responsive elements binding PPAR, another family of nuclear receptors acting as heterodimer with RXR [956]. In humans, PLTP activity correlates with body mass index, plasma triglycerides and inversely with apo A-I in HDL (LpA-I) [957]. It has also shown to correlate with plasma glucose but not with insulin levels in type II diabetics [958]. Although several neutral gene variations have been described, no natural mutations have been reported in human [957]. From the data of mice studies it could be anticipated that PLTP deficiency could result in impaired reverse cholesterol transport and plasma triglyceride-rich lipoprotein metabolism. Together with the loss of PLTP capacity to transfer protective alpha-tocopherol (a natural antioxidant) to apo B-rich lipoproteins [948], it could be anticipated that PLTP deficiency could represent a cause of atherogenic hypoalphalipoproteinemia in human.

2.3.3-   Receptors

2.3.31-   ABC-1, or Cholesterol Efflux Regulatory Protein, a Gene for Tangier Disease and Familial Hypoalphalipoproteinemia

Tangier disease was first described in 1961 in children born from a consanguineous union, who originated from a small, secluded community living on Tangier Island (Virginia) [959]. Large yellow-orange tonsils, as well as orange lipid depositions in adenoids and reticulo-endothelial cells (directly visible on the rectal mucosa on endoscopy) characterize the disease [960]. Hepato-splenomegaly, peripheral neuropathy characterized by cholesteryl ester accumulation in Schwann cells and disseminated atherosclerosis generally manifesting by coronary events are also hallmarks of the syndrome [961]. The disseminated tissue accumulation of cholesteryl esters contrasts with low plasma total cholesterol and the absence of HDL and apo A-I. The abnormal lipoprotein profile readily detectable in childhood is characterized not only by quantitative lipoprotein disorders like moderate hypertriglyceridemia and decreased LDL cholesterol but also in qualitative abnormalities of lipoprotein shape and composition. Cellular cholesterol efflux is profoundly impaired on primary cultures of fibroblasts or of monocyte-derived macrophages [962,963]. Moreover, phospholipid efflux is also decreased; HDL interactions at the cell surface and intracellular lipid trafficking are profoundly impaired. HDL catabolism is accelerated despite normal apo A-I, A-II or LCAT gene sequences. The disorder is recessively inherited, however heterozygotes generally exhibit low HDL, decreased cellular cholesterol efflux, sometimes complicated with cardiovascular disease during or after the fourth decade [964]. Although the disease is very rare it had drawn attention because it appeared related with crucial mechanisms of reverse cholesterol transport, lipoprotein metabolism and atherosclerosis.

Reverse genetics approaches in parallel with functional molecular and biochemical analyses allowed identify the Tangier disease-causing gene [114-116,965]. A candidate locus was assigned to chromosomal region 9q31 by a linkage exclusion method [966]. In the meantime, a membrane lipid transporter belonging to the family of ABC transporters (characterized by an ATP Binding Cassette) ABC1, and assigned to a broader region, was found to have a sterol-dependent regulation in macrophages [967,968]. These observations were confirmed by gene expression array analysis in fibroblasts from normal or Tangier diseased patients [965]. Naturally occurring mutations of ABC transporters were known to cause several human diseases: cystic fibrosis, X-linked adrenoleucodystrophy, Zellweger peroxisomal disease, liver cholestasis or Stargardt macular degeneration [969]. ABC transporters are transmembrane molecules, which use ATP to transport ions, amino acids, phospholipids, sugars, vitamins, bile acids or toxic products across the plasma membrane. Altogether, these data made ABC1 a good candidate to analyze by positional cloning and gene sequencing in families with Tangier disease [114-116]. Soon after several groups consistently reported the identification of ABC1 gene mutations as a cause of Tangier disease in families originating from different parts of the world, including the original Tangier disease family [70,970,971].

ABC1, or ABCA1 (ABC sub-family A, member 1), or CERP (Cholesterol Efflux Regulatory Protein), is encoded by a 149kb long gene (ABC1), organized in 50 exons, localized to chromosome region 9q31 [972]. The first exon is noncoding as well as the 5' part of the second exon, which encodes the first 21 amino acids of the mature protein. The first intron is very long (>24kb), and up to exon 8, intronic sequences extend over more than 10kb. Except for exon 50, which is more than 3.4kb long, most exons are short in length (33 to 245 bp). The mature protein in encoded by exons 3 to 49. The first two introns contain numerous *Alu* repeats, out of the 62 *Alu* repeats found along the locus. In addition a repeated sequence belonging to the LINE family (Human Endogenous Retrovirus element) is found within intron 5. Therefore, interspersed repeated sequences at the ABC1 locus could represent as many potential unequal crossover breakpoints leading to large gene rearrangements. The gene structure and sequence are highly conserved in mouse and has a similar organization with ABCR, another member of the gene family of ABC transporters causing Stargardt retinopathy [972]. Knock-out mice were generated, closely reproducing the human disease [71,973]. Interestingly there was also a gene dosage effect in heterozygotes and an increased absorption of dietary cholesterol under a western diet, suggesting novel functions of ABC1 in dietary modulation of lipoprotein metabolism. Unexpectedly, mice displayed lung foamy lesions and extracellular cholesterol deposition. Moreover it was shown that caveolins failed to reach the cell membrane, remaining in the *trans*-Golgi compartment. This demonstrated that ABC1 is involved in cellular trafficking of unesterified cholesterol and phospholipids between the Golgi apparatus and the plasma membrane. Therefore ABC1 would regulate important pathways of intracellular trafficking of unesterified cholesterol and particularly its export across the plasma membrane towards the core of HDL. In this process, competency of the acceptor lipoprotein (i.e. HDL) seems important [70].

The ABC1 mRNA transcript is 6783nt long is widely expressed, however mainly in liver, small intestine, lung, placenta, adrenal glands and brain [968,972]. ABC1 expression is induced in macrophages by cholesterol loading, acetylated LDL or cAMP treatment [965,967,972]. Conversely, HDL apo A-I binding or interferon γ down-regulate ABC1 expression in monocyte derived macrophages [974]. The promoter region contains several response elements for ubiquitously expressed transcription factors: Sp1, NF-κB, AP-1 (and other related members) and E-boxes. As expected, binding sites for HNF-3β, a liver specific nuclear factor and for macrophage specific nuclear factors like STAT c-myb or GATA are also found. All these nuclear response elements have been described in promoter regions such as those for the LDL receptor or scavenger receptors (SRA, CD36 or SR-B1). Moreover, E-boxes are recognition sites for the β-HLH family of transcription factors to which belong SREBP. However, the sterol dependent transactivation of ABC1 is mediated by LXR-RXR heterodimerization [975]. LXR is a liver specific nuclear receptor for oxysterols, which up regulates CYP7α or CETP gene expression. RXR is a ubiquitous nuclear receptor for 9 cis-retinoic acid, suggesting that concerted regulations operate as well over reverse cholesterol transport. It has been shown that LXR plays a global role on ABC1 expression and reverse cholesterol transport [976]. Particularly LXR by upregulating oxysterol-dependent ABC1 expression in macrophage stimulates cholesterol efflux from peripheral cells [977]. In the liver, LXR stimulates bile acids secretion through CYP7α activation. However in the intestine, cholesterol reabsorption is limited by ABC1 up-regulation resulting in cholesterol excretion in the lumen. Consistent with this hypothesis is the increased cholesterol absorption in the intestine of ABC1 knock out mice and cholesterol deposition in the intestinal mucosa of humans with Tangier disease.

ABC1 is a 2261 amino acid long protein of MW 220-240kD (*figure 2-31*). It contains 2 symmetrical domains containing each typical motifs of the ABC family: 6 membrane spanning domains and a nucleotide-binding fold (ATP binding) containing Walker A and B motifs and an active transport signature, N-terminal to the B motif [70,116,971]. A highly hydrophobic linker is found between the symmetrical regions. Together with the transmembrane domains a vesicle like aqueous chamber may form a pore within the membrane, open to the outer side of the membrane and into the lipid phase (closed at the inner face of the membrane) favoring lipid export.

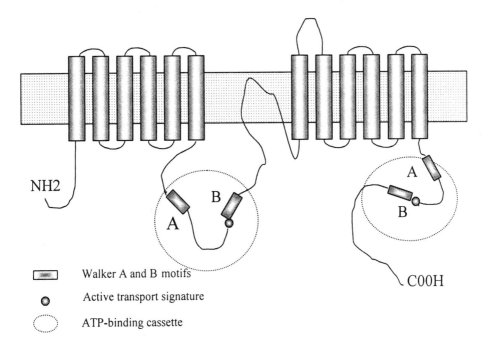

NH2

A   B

A

A   B

Walker A and B motifs

Active transport signature

ATP-binding cassette

COOH

Figure 2-31. **Predicted structure of the ABC1 protein** [70,971]. The protein is divided in two symmetric parts containing each 6 transmembrane domains and one ATP-binding cassette, separated by a large hydrophobic region.

Disease causing mutations were found to disrupt highly conserved residues in functional domains of the ABC gene family. Interestingly, ABC1 mutations extend beyond the restricted and rare phenotype of Tangier disease (TD). As mentioned above, Tangier heterozygotes may exhibit lowered plasma HDL cholesterol and atherosclerosis. Heterozygous mutations have been described in familial hypoalphalipoproteinemia (FHA), a dominantly inherited disorder in which plasma HDL cholesterol is below the 5th percentile [114,978]. These families represent a subset of high-risk families with defective cellular cholesterol efflux and consequently with premature cardiovascular disease. It is not clear at present if nonsense or missense ABC1 mutations induce qualitative or quantitative differences in lipoprotein profile or in cardiovascular disease risk. Moreover, some TD patients do not suffer from neurological or heart disease and some heterozygotes remain normolipidemic. These cases indicate incomplete penetrance and suggest that other factors may modulate the expression of ABC1 mutations. Four common variants have been identified along the ABC1 protein: Arg159Lys, Val765Ile, Ile823Met and Arg1527Lys [979]. The Ile823Met mutation located upstream of Walker A motif of the first ABC, has been associated with higher plasma HDL cholesterol in Inuits. A significant proportion of patients with premature coronary heart disease exhibit low HDL levels, despite normal total cholesterol. The recent discovery of ABC1 as the disease-causing gene for certain forms of inherited hypoalphalipoproteinemia was a significant breakthrough in the understanding of molecular mechanisms underlying defective reverse cholesterol transport in humans. Moreover, its discovery has reconciled with the low-HDL/high cardiovascular disease risk hypothesis.

2.3.32-   Scavenger Receptor Class B Type 1 (SR-B1), A Multifunctional Receptor for the Selective Uptake of Cholesterol

Studies of HDL turnover had shown that the cholesteryl ester content of HDL had a much shorter plasma half-life than the protein content of HDL, suggesting a selective lipid removal from the core of lipoproteins [979,980]. Proteins involved in HDL remodeling such as LCAT, PLTP, CETP or hepatic lipase, were known to regulate reverse transport of cholesterol to the liver. However, they could not account for the selective uptake of cholesteryl esters observed in the liver or in steroidogenic glands. Moreover, there were reports that steroidogenesis was disturbed as a consequence of deficient cholesterol uptake in apo A-I deficiency in mice despite normal cholesterol delivery through the LDL receptor pathway [981]. This suggested that alternate pathways of cholesterol delivery mediated by HDL would exist at the cell surface. Scavenger Receptor class B type 1 (SR-B1), also called CLA1 (for CD36 and LIMPII Analogous 1), was first identified as another member of the gene family of scavenger receptors class B, which includes CD36 (see page 81) and the multifunctional protein LIMPII (Lysosomal Integral Membrane Protein II) [982]. SR-B1 was found to bind HDL and modified LDL (acetylated, oxidized or maleylated LDL) and more unexpectedly non-HDL lipoproteins and native LDL [983]. It was identified as a receptor involved in HDL-mediated selective uptake of cholesteryl esters in the liver and steroidogenic glands [984,985].

The human gene (CD36L1) encoding SR-B1, spans approximately 70kb, comprises 13 exons, localized to chromosome 12q24 [482,986]. The primary sequence encodes for a 509amino acid hairpin shaped, monomeric protein (see *figure 2-15*, page 82). The sequence and gene structure are very similar to that of CD36 [480]. The N-terminal and C-terminal ends are cytosolic, there are two transmembrane domains and a long extracellular domain. SR-B1 differs from other members of the gene family by the extracellular domain, which has been shown to enhance cholesteryl ester uptake from HDL [479]. The extracellular domain contains cysteine-rich regions and N-glycosylation sites. Moreover it is palmitoylated which could favor its clustering in caveolae [987]. The C-terminal domain longer than this of CD36 was shown to bind an intracellular protein CLAMP (C-terminal Linking And Modulating Protein), which could mediate the intracellular transport of cholesterol taken up from HDL by SR-B1 [988]. Although the exact mechanisms remain to be determined, SR-B1 could bind HDL and might form a hydrophobic channel through which cholesteryl esters could move from the HDL lipid core down to the plasma membrane or associate with membrane compounds, which could mediate cholesterol transfer [989]. Cholesterol is then irreversibly internalized through a non-lysosomal non-endocytic pathway.

SR-B1 in highly expressed in the liver and in steroidogenic glands [985]. It is highly sensitive to steroid hormone regulation through SF-1 (Steroidogenic Factor 1), C/EBP responsive elements. SR-B1 knockout female mice are infertile, as the result of cytosolic cholesterol depletion in the oocyte. SR-B1 expression is regulated by cellular cholesterol. There are several SRE upstream of the coding sequence. SREBP1a bound to these elements synergistically upregulate SR-B1 expression with SF-1 in the ovary [990]. In the liver SR-B1 mediated cholesterol uptake seems to be primarily directed to biliary cholesterol excretion, giving further support to its role in reverse cholesterol transport [991,992]. Dietary polyunsaturated fatty acids increase SR-B1 expression in the liver. SR-B1 gene expression may be induced in thickened intima of atherosclerotic arteries. In macrophage foam cells PPARs were shown to induce SR-B1 expression [993]. Interestingly SR-B1 may selectively take up cholesteryl esters from LDL (without mediating their endocytosis) and enhance peripheral HDL mediated free cholesterol efflux [994,995]. In peripheral tissues SR-B1 facilitates the uptake of free cholesterol by native HDL through interactions with their phospholipid moiety (see *figure 2-28*). Here, free cholesterol is exported along an increasing concentration gradient towards native HDL, which avidly take up cholesterol. Transgenic mice overexpressing SR-B1 have low HDL levels, increased biliary cholesterol production and are resistant to dietary or genetically induced atherosclerosis [985]. Conversely, SR-B1 knockout mice exhibit high plasma HDL cholesterol, decreased biliary cholesterol and increased susceptibility to atherosclerosis. Several SNPs have been identified, however most of them are silent base pair changes [986]. A Val135Ile mutation was identified in exon 3, however no specific phenotype has been associated with this gene variant. Further studies will determine if SR-B1 gene variants may modify lipoprotein metabolism and susceptibility to atherosclerosis in humans.

2.3.33-Cubilin, A Receptor for HDL in the Kidney

If the issue of selective uptake of HDL cholesterol was solved by the identification of SR-B1, the problem of the clearance of the protein moiety of HDL remained elusive. Kidney had long been recognized as playing an important role in lipoprotein metabolism. Many kidney diseases result is lipoprotein disorders, which in turn may accelerate atherogenic processes, which further disrupt renal functions. Moreover, aside from the liver, kidney had been recognized as a major site for HDL catabolism in vivo [996]. HDL were shown to be taken-up by an unknown receptor-mediated endocytosis mechanism. The answer came from a protein known to mediate intestinal absorption of Intrinsic Factor/vitamin B12 complexes: Cubilin [74].

Cubilin (or gp280), was identified as a large 460kD protein belonging to a family of extracellular proteins containing particular "CUB" modules named after: C1r/s (Complement subcomponents C1r/C1s), Uegf (sea Urchin EGF like domain), and BMP1 (Bone Morphogenic Protein-1) [997]. The protein is expressed at high levels on the apical face of epithelia of the ileum, the yolk sac and renal proximal tubules in the kidney. Cubilin was first identified to mediate the uptake of Intrisic Factor (IF) bound to vitamin B12 also named cobalamin (Cbl), at the apical side of the ileal epithelium [998]. Because Cubilin has no transmembrane domain, it uses megalin (see page 120) as a coreceptor for coated-pit receptor mediated endocytosis [999]. After endocytosis, the complex is directed towards lysosomes where IF is degraded. Cobalamin is further processed to the basal side where it is secreted with transcobalamin into the blood circulation. Cubilin has been identified as a multiligand receptor, binding RAP, IgG-light chains, and more recently albumin in addition to IF-Vitamin B12 complexes [1000].

However the most unexpected finding was that cubilin binds Apo A-I in HDL and mediates the uptake of the whole particle in the kidney [1001,1002]. It uses Megalin as a coreceptor, in renal proximal tubules, through a mechanism very similar to that observed for IF-VitB12 in the ileum (see *figure 2-28*). After endocytosis, lipoprotein components are directed towards lysosomes where they are degraded. It is not known if the protein or other HDL components taken up by the cubilin-megalin pathway could also undergo transcytosis in the kidney. Because Cubilin faces the urinary face of the tubular epithelium, HDL is supposed to reach this compartment after glomerular filtration. Cubilin binds Apo A-I with high affinity, whereas Apo A-II, Apo C-III or Apo E are poor ligands. The charge, lipid-content and size of HDL were recognized as important for HDL clearance in the kidney [1003]. Therefore the small, lipid-poor and positively charged HDL have greater chances to be ultrafiltrated than larger HDL. This observation is in line with the fact that many conditions associated with low or abnormal plasma HDL or Apo A-I, are also associated with increased catabolic rates of smaller particles.

A 11.4kb long cDNA for Cubilin (CUBN) was cloned from a human kidney cDNA library, and has been localized to human chromosome 10p12.33-13 [1004]. The primary sequence has 69% identity with rat cubilin. The 10.9kb long mRNA transcript encodes for a 3597 amino acid propeptide. After cleavage of a 24aa signal peptide, the protein undergoes further cleavage after recognition of a "Arg^7-Glu^8-Lys^9-Arg^{10}" site for trans-Golgi proteinase furin, shortening the mature protein by 10 more amino acids. The N-terminal part contains 8 EGF-like repeats followed by 27 CUB domains. Interestingly, the EGF-like domain contains an amphipathic α-helix, which could mediate hydrophobic interactions. The CUB repeats domain contains ligand affinity domains: CUB domains 5-8 bind IF-vitamin B12; CUB domains 13-14 bind RAP. Gene mutations in cubilin were shown to cause a recessively inherited form of megaloblastic anemia due to vitamin B12 malabsorption, the Imerslund-Gräsbeck syndrome [1005]. A Pro1297Leu mutation was identified as a frequent cause of the disorder in Finns [1006]. This severe and juvenile form of vitamin B12 malabsorption may be corrected by vitamin B12 intravenous supplementation. Interestingly, vitamin B12 supplementation does not correct proteinuria in these patients. Moreover, in a patient carrying a splice site mutation, leading to premature truncation of the protein, abnormal apo A-I urinary excretion was neither prevented, suggesting that cubilin mediates specific uptake of apo A-I [1001]. For unclear reasons at present, the patient exhibited high plasma HDL concentration, suggesting that the cubilin-megalin pathway could modulate levels of circulating HDL. Further studies in animal models or humans will clarify the role of this novel pathway of lipoprotein metabolism.

The recent identification of receptors involved in HDL metabolism has further demonstrated the multiple metabolic connections that HDL establish with peripheral tissues and the main regulatory pathways of lipoprotein metabolism. In a broader perspective, the identification of candidate genes regulating HDL metabolism has confirmed the ancestral role of this class of lipoproteins of shuttle between tissues tightly connected with other pathways of lipoprotein metabolism, singularly those regulating triglyceride-rich lipoproteins. Moreover, they have given a molecular basis to reverse cholesterol transport and it's role in atherosclerosis, emphasizing that it is more the functionality of HDL than it's plasma concentration that contribute to cardiovascular protection.

2.4- OTHER REGULATORY PATHWAYS OF LIPOPROTEIN METABOLISM

2.4.1-   The Traffic of Intracellular Lipids

Lipid homeostasis depends primarily on pools and flows of lipids between cellular compartments and their membranes, however mechanisms are poorly understood. Indeed the control of cellular lipid trafficking is crucial at least for cellular membrane maintenance and energy homeostasis. The identification of regulatory cascades induced by LDL receptor endocytosis on intracellular and extracellular homeostasis of cholesterol, was early evidence that molecular specific regulations of intracellular lipid homeostasis could resemble those of hormone regulated cellular processes. Moreover, studies on lipoprotein assembly have shown that lipid pools and their flow were regulated by proteins which could strongly determine lipoprotein secretion. In addition, tissue specificity may determine specialized mechanisms: muscle and adipose tissue will handle fatty acids in the opposite way (energy expenditure/storage), just as liver and steroid glands will not handle cholesterol in the same metabolic pathways for obvious physiological reasons. Moreover, the adaptability of lipid metabolism to physiological (e.g. food intake, growth and development, muscle activity etc.) or pathological conditions (inflammation, tissue repair, infections etc.) is determined by intracellular lipid homeostasis. Intracellular compartments (endoplasmic reticulum, mitochondria, lysosomes, peroxisomes and membranes) contain proteins essential for intracellular lipid homeostasis and trafficking. Moreover, specialized nuclear proteins have been uncovered as essential controllers of lipid trafficking within cells and in the organism.

2.4.11-   Lipid Trafficking in Cellular Compartments

- The intracellular traffic of cholesterol after lipoprotein endocytosis.

Following receptor-mediated endocytosis in clathrin-coated pits of the cellular membrane, lipoprotein-receptor complexes are found in vesicles named early sorting endosomes (*figure 2-32*). Although some of the lipoprotein content may return to the cell surface, endosomes acquire

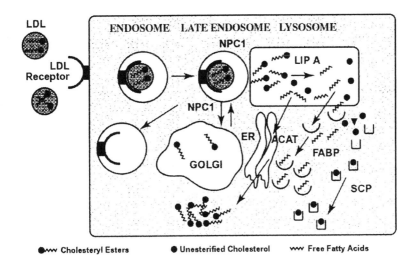

Figure 2-32. **Intracellular processing of cholesterol after LDL endocytosis.**

•   Lipoproteins detach from membrane receptors in the low pH lumen of late endosomes. Lipids are directed towards lysosomes or Golgi apparatus through a sterol sensing mechanism mediated by NPC1. Neutral lipids are hydrolyzed by lysosomal acid lipase (LIPA).

•   Unesterified cholesterol or free fatty acids may be directed to sites of metabolic needs or transported into the cytosol by SCP (sterol carrier protein) or FABP (fatty acid binding protein) respectively.

•   Alternately, both may be used by ACAT (Acyl:Cholesterol Acyl Transferase) in the endoplasmic reticulum (ER) to form cholesteryl esters which will be stored into cytosolic lipid droplets.

membrane proteins including acid hydrolases and phospholipids exchanged with the trans-Golgi network [1007]. This results in changes in membrane composition (microdomain or rafts formation), and shape forming intralumenal whorls, and in a decrease in the intralumenal pH of what has become late endosomes. This allows the recycling of membrane receptors back to the trans-Golgi network and to the cell membrane. These organelles may further fuse with lysosomes where complete degradation of lipoproteins will be achieved.

Niemann-Pick Disease type C was identified as a lipid storage disease resulting from disorders of intracellular cholesterol trafficking [1008]. The onset (usually in infancy or childhood) and severity of the disease may be variable. It is characterized by progressive neurological manifestations (ataxial gait, dystonia, dysarthria etc) and hepatosplenomegaly. NPC is characterized by vertical supranuclear gaze palsy. Neither major lipoprotein disorders nor atherosclerosis were described. The outcome is severe the disease progressively complicates with seizures, dementia, liver failure and mortality before adolescence [1009]. Impaired cellular LDL cholesterol processing, late endosomal and lysosomal accumulation of unesterified cholesterol and of phospholipids characterize the disease. In brain phospholipids particularly gangliosides tend to accumulate, with a composition different between the central nervous system and the liver. A similar clinical picture describes Niemann-Pick Diseases type A or B, which result from primary sphingomyelinase deficiency.

The NPC locus (NPC1) was first localized to chromosome 18q11 [1010], followed several years later by the identification of the causal gene by positional cloning and gene complementation in mutant CHO (Chinese Ovary) cell lines [1011,1012]. The strong homology with genes identified in mammals, in the *Caenorabditis Elegans* nematode and in the *Saccharomices Cerevisiae* Yeast, confirms its fundamental role in cell biology. A frameshift mutation resulting from an insertion of retrotransposon-like sequences was identified in a mouse model for the disease. The human gene spans 47kb and contains 25 exons [1013]. There is a CpG island 5' upstream and within intron 1. Several SNPs were identified along the gene sequence. NPC1 is a glycosylated integral membrane protein with an estimated MW of 142kD encoded by 1278 amino acids. The N-terminal region includes a signal peptide and a highly conserved region, the NPC1 domain. This domain contains leucine-zipper motifs involved in protein dimerization and essential for lysosomal cholesterol unloading [1014]. Most important, NPC1 contains 13 to 16 putative transmembrane domains, and a so-called "sterol sensing domain", highly conserved in other proteins involved in cholesterol trafficking: HMGCoA reductase, SREBP cleavage-activating protein (SCAP), and Patched, the membrane receptor for the *Sonic Hedgehog* morphogen in *Drosophila* [37]. All these key regulatory proteins induce crucial cellular responses after sterol binding. In HMGCoA reductase, sterol-sensing domains trigger protein degradation when sterol level rises in the endoplasmic reticulum. In keeping, SCAP activation by low sterol level within the endoplasmic reticulum is a rate limiting step of SREBP activation, which in turn controls multiple pathways of intracellular energy and cholesterol homeostasis (see below). After covalent binding of cholesterol to the *Sonic Hedgehog* morphogen, the binding to the sterol-sensing domain of Patched triggers preferential clustering of *Sonic Hedgehog* to the basolateral side of the ectoderm, inducing proper embryonic patterning [38]. NPC1 would thus play a crucial role in the intracellular redistribution of recently internalized cholesterol to the transGolgi network and further to the endoplasmic reticulum, where important signalling responses in lipid homeostasis take place. As expected a great majority of NPC patients described worldwide have mutations within the highly conserved domains of the NPC1 protein [1008]. Despite great allelic heterogeneity, several mutations may be more prevalent in several human subgroups (Acadians from Canada, Mexican Americans and Europeans) as the result of a founder effect [1015]. Phenotype-genotype relationship studies are underway to identify if specific mutation could predict some disease characteristics. Genetic heterogeneity has been found for NPC, suggesting that other regulatory proteins regulate early endosomal lipid trafficking [1016]. However, genetics of NPC1 have established the crucial role of this protein for the intracellular trafficking of LDL-derived cholesterol and other membrane derived lipids. At present, it is not known if there are physiological or pathological situations where NPC1could be overactive.

Lysosomal Acid Lipase (LAL) ensures one of the first degradation steps of cholesterol delivered from LDL endocytosis into lysosomes *(figure 2-32)*. Cholesteryl esters contained in the lipid core of the particle are hydrolyzed by lysosomal acid lipase (which also degrades triglycerides) releasing unesterified cholesterol and free fatty acids. **Wolman disease** (WD) and **Cholesteryl Ester Storage disease** (CESD) were associated with primary defects in lysosomal acid lipase [1017]. Both diseases are allelic: WD is more severe with early onset soon after birth; CESD has later onset generally after adolescence. A disseminated cholesterol overload is observed in liver, spleen, lymph nodes bone marrow, intestinal mucosa etc. Liver steatosis evolving into liver cirrhosis determines the main clinical manifestations. Lipid storage is characterized by cellular overload in esterified cholesterol and triglycerides within lysosomal droplets. Adrenal calcifications may be observed in Wolman disease. In cholesteryl ester storage disease, hypercholesterolemia, hypertriglyceridemia and hypoalphalipoproteinemia are associated with premature atherosclerosis. However, skin xanthomas are not typical of the disease. Both diseases are recessively inherited.

The gene (LIPA) was cloned and localized to chromosome 10 (10 q24-25), [1018]. The gene is approximately 36kb long and comprises 10 exons [1019]. Primary sequence analysis has shown that LIPA belongs to the family digestive acid lipases (gastric lipase, salivary etc.). The promoter comprises motifs for type SP1 and AP2 with no particular tissue specificity. However, mRNAs are preferably found in organs corresponding to the pathological involvement of the disease: liver, spleen and lymphoid organs, adrenals, intestine, pancreas and kidney [1020]. The first exon is noncoding; the second exon encodes for signal peptide. Exon 5 comprises a specific pentapeptide typical of the catalytic site in this family of lipases. Exons 8 and 9 are almost identical and seem to result from genomic segment duplication. A microsatellite comprising at least 10 alleles was localized in the first intron. Studies of LIPA mutations and of their functional consequences have identified the molecular basis for phenotypic heterogeneity of the locus [1021]. Severe mutations (frameshift, stop codon) that completely abolish acid lipase activity are found in Wolman disease. However, if one of the defective alleles preserves enzyme activity at least partially, patients will manifest with the milder phenotype of CESD. In line with these clinical findings, knock out mice for the LIPA gene, are a good model of human disease, heterozygotes having a normal phenotype despite a 50% decrease in enzyme activity [1022].

Acyl-CoA: Cholesterol Acyl Transferases (ACAT) are microsomal rate limiting enzymes for intracellular cholesterol esterification, a natural protection mechanism for neutralizing sterols by rendering them hydrophobic [1023]. When intracellular pools of free cholesterol increase, ACAT esterifies cholesterol or oxysterols using acyl-CoA as substrates. The acyl group may be any long chain fatty acid, although endogenously produced fatty acids (oleate or palmitoleate) through stearoyl-coA desaturase (see below) are predominantly found. Dietary enrichment in polyunsaturated fatty acids results in enrichment of polyunsaturated acyl chains in cholesteryl esters. ACAT are cellular sensors of pools of free cholesterol, which augment cellular pools of cholesteryl esters. Esterified cholesterol may be stored in cytosolic droplets (pathologically excessively numerous in foam cells), or may be added to the lipid core of lipoproteins during their assembly in the liver or the intestine. Drugs inhibiting ACAT activity decrease VLDL synthesis and increase the pool of HDL in several animal models [1024]. At least two enzymes have been identified which display ACAT activity, in human, mouse and yeast [1025]. ACAT-1 is a 50kD integral membrane protein of the endoplasmic reticulum. The protein contains 550 amino acids, including a hydrophobic region containing 7 transmembrane domains [1026]. The gene (ACAT-1) is about 80kb long, containing 16 exons. A highly informative microsatellite $(GT)_n$ was identified in intron 12. At least four mRNA species are generated. Strikingly, if 3 mRNA species are generated from a genomic region where most of the coding sequence is located on chromosome 1q15, an additional 4.3kb transcript is generated by an unusual transplicing mechanism with another mRNA encoded by a genomic region located on human chromosome 7 [1027]. Although ACAT-1 is ubiquitous, it is expressed predominantly in the liver, steroidogenic glands and macrophages [1028]. ACAT-1 accounts for 80% of ACAT activity in the liver [1029]. Naturally ACAT-1 deficient mice i.e. *"ald"* mice, develop adrenocortical lipid depletion mainly at the expense of cholesteryl esters appearing otherwise healthy [1025,1030]. However, ACAT-1 knock out mice develop severe xanthomatosis complicating in alopecia and diffuse tissue cholesterol deposition when crossed on a hyperlipidemic genetic background (apo E or LDL knock out) [1031,1032]. Interestingly, atherosclerosis develops despite lower plasma cholesterol and a reduced number of foam cells in lesions. Genetic redundancy and therefore the presence of ACAT2 were suspected by the observation that mice with spontaneously or genetically induced ACAT-1 deficiency had preserved ACAT activity [1033]. ACAT-2 shares homology with ACAT-1 only for its hydrophobic C-terminal domain, whereas the N-terminal domain is different. ACAT-2 has more restricted expression to the intestine apical villi in human. No human pathology was associated with spontaneous defects in ACAT-1 or ACAT-2 so far. However, the significant role played by this family of enzymes makes them interesting therapeutic targets for the correction of lipid disorders and in the prevention of atherosclerosis.

Cholesterol intracellular trafficking is still largely unknown. However it does not restrict to post endocytic pathways. Free cholesterol may derive from cholesteryl ester hydrolysis, although a main source of cellular cholesterol comes from the endogenous biosynthesis of cholesterol by the mevalonate pathway (see page 11). Newly synthesized cholesterol may be further processed to other metabolic pathways such as steroid hormone synthesis, bile acid synthesis (see below), or lipoprotein assembly. It may be used as a modifier molecule like in the example of the *Sonic Hedgehog* morphogen, or esterified for innocuous storage. It is processed however in most cells towards membranes, and particularly to the cell membrane. It may be oriented to specialized membrane microdomains (caveolae) and it may be driven out of the cell by passive or active cholesterol efflux.

The intracellular traffic of fatty acids is obviously very different from one organ to the other (i.e. brain, adipose tissue, heart etc). However, several factors in lipoprotein producing organs or produced in the adipocyte may have a relevant influence on lipoprotein metabolism. Fatty acids released by triglyceride-rich lipoprotein lipolysis or albumin may enter the cell either passively once un-ionized or to a lesser extent actively with several transporters [1034]. **Acylation Stimulating Protein** (ASP) or complement C3a-desArg regulates fatty acids entry in adipocytes, in part by up regulating diacylglycerol acyltransferase (DGAT), an enzyme involved in the last step of triglyceride synthesis [82]. ASP also regulates adipocyte fatty acid release in an independent and complementary way to that of insulin.

Once within cells, fatty acids bind to soluble carriers, particularly fatty acid binding proteins (FABP) to easily shuttle between cellular compartments. **Fatty Acid Binding Proteins** represent a diversified family with little conservation [1035,1036]. They are involved in the transport of fatty acids, their Coenzyme A derivatives, bilirubin and other small molecules, between different cellular organelles and the plasma membrane. FABP1 is mainly found in liver, whereas FABP2 in mainly expressed in intestine. Other isoforms have been shown to express in muscle, skin or brain. FABP1 gene encodes for a 162aa protein and is located on chromosome 2p11 [1037].

FABP2 gene is 3.3kb long, organized in 4 exons, encoding for 240aa and is located on chromosome 4q28-q31. A common Ala54Thr FABP2 mutation has been reported in several populations of Native Americans to be associated with insulin resistance or dyslipidemia [1038]. It has been suggested that the Thr54 variant could enhance dietary uptake of fatty acids. However, inconsistent findings in other human subgroups suggest that it may be a mild modifier allele of other more significant metabolic disorders such as diabetes, familial combined hyperlipidemia or atherosclerosis.

Lipid biosynthetic pathways could also represent important cellular regulatory mechanisms of lipoprotein metabolism. **Fatty acid synthase** (FAS) is the rate-limiting enzyme of long chain fatty acid synthesis in the cytosol [1039]. It acts as a homodimer of a multifunctional enzyme (554kD) elongating the fatty acid chain by sequentially adding two carbons from malonyl CoA. Monomers of 2504aa are organized in a head to tail fashion for full catalytic activity. The enzyme is ubiquitous however highly expressed in organs in high lipid demand (i.e. liver, adipose, brain, muscle, etc.). The structural gene (FASN) is located to human chromosome 17q25 [1040,1041]. Its expression is highly sensitive to hormones, particularly insulin by a mechanism mediated by SREBPs in the liver [1042,1043]. Mirroring the fundamental role of HMGCoA reductase in cholesterol biosynthesis, the absence of naturally occurring FAS mutation reflects its fundamental role in lipid biogenesis. However, it may represent an interesting therapeutic target; FAS inhibitors reduce food intake and body weight in genetically obese *"ob/ob"* mice [1044]. Another important issue of lipoprotein metabolism is the biosynthesis of triglycerides (i.e. the condensation of a glycerol molecule with 3 fatty acids). Different pathways may achieve this synthesis: the intestine monoacylglycerol pathway and the ubiquitous glycerol-phosphate pathway. In the first pathway fatty acids are added sequentially to monoacylglycerol to form triacylglycerol (i.e; triglyceride). In the second pathway, two fatty acids are condensed on a glycerol-3-phosphate molecule to form phosphatidate. **Diacylglycerol acyl transferase** (DGAT) catalyzes the last step of triglyceride biosynthesis by removing the phosphate group and covalently joining a fatty acid group [1045]. The recent cloning of the gene has shown that it belongs to the ACAT family of genes: ACAT uses sterols whereas DGAT uses diacylglycerol as acyl group acceptors. DGAT contains seven putative transmembrane domains and a putative diacylglycerol-binding domain in its C-terminal part. DGAT could contribute to enrich the cytosolic pool of triglycerides for storage (e.g. in the adipocyte) or the lipid core of lipoproteins for secretion. Similarly to ACAT, knock out experiments have shown functional redundancy of the DGAT pathway, since DGAT deficient mice are healthy and fertile. Transacylase mechanisms mediating the transfer of an acyl group from one molecule of diacylglycerol to another one may form triglyceride and monoacylglycerol. Moreover, a recently identified LCAT-like gene in yeast (LRO1) with transacylase activity between diacylglycerol and phospholipids could again produce triglycerides and lysophospholipids, like LCAT produces esterified cholesterol and lysolecithin [925].

Fatty acids may be modified in their length and degree of unsaturation. An interesting group of lipid modifying enzymes is fatty acid desaturases [1046]. Lipid unsaturation plays a major role in modifying physicochemical properties of lipids. Thereby it has a significant biological influence on processes such as membrane fluidity, cellular signalling, intermediate metabolism, or lipoprotein assembly and secretion. The most abundant fatty acids found in mammals are mono-unsaturated palmitoleate (C16:1) and oleate (C18:1). However, cellular de novo lipogenesis produces palmitate (C16:0), a saturated fatty acid. Moreover, stearate (C18:0) is insoluble in natural biological mediums. **Stearoyl CoA desaturase**, SCD (or delta 9 desaturase) introduces a double bond in the Δ9 position of the acyl chain of palmitate or stearate. One human SCD gene has been identified [1047]. It is 24kb long and contains 6 exons, localized to chromosome 10. It encodes for 359 amino acids, containing highly conserved domains in mice, rat and yeast, involved in lipid binding and catalytic activity. Two mRNA species of 3.9kb and 5.2 kb are expressed in brain and liver, although they are found ubiquitously. SCD1 and SCD2 proteins have been identified in mice, with SCD1 predominantly expressed in liver. The expression of SCD has been extensively studied in murine species. Diets enriched in carbohydrates, cholesterol, or saturated fatty acids strongly induce its liver expression whereas it is down regulated by unsaturated fatty acids. Responsive elements for SREBP, C/EBP, NFY and more specific repressor elements sensitive to PUFA are found upstream of SCD1 and SCD2 genes, accounting for dietary and hormonal sensitivity of their expression. Moreover a natural mutant strain, the "*asebia*" mouse has been identified with an extensive gene deletion abolishing SCD1 expression in homozygotes [1048]. Hypoplastic sebaceous glands, its corollary sparse and abnormal hair, corneal opacities and hypoplastic meibomian glands characterized the phenotype. Interestingly, liver content in cholesteryl esters and triglycerides was very low, only partially restored by a diet enriched in triolein and tripalmitolein, resulting in low plasma levels of VLDL and LDL triglycerides [1049]. Observations from this mouse model suggest that SCD may be an important regulatory gene in the control of dietary-derived lipoproteins.

Aside from qualitative modifications an important issue in intracellular fatty acids metabolism is their degradation into the β-oxidation pathway in mitochondria, providing with a major source of energy [31]. An important regulatory mechanism is fatty acids transfer into the mitochondrion by the carnitine pathway. Systemic carnitine deficiency is a recessively inherited disease, which may results from defective transport and cellular uptake of carnitine [1050]. The disease causes myopathy, cardiomyopathy, hypoketotic hypoglycemia, and hyperammonemia, which responds in liver to carnitine supplementation. Liver steatosis and abnormal fatty acid metabolism were also reported. Several mutations for a **sodium ion-dependent carnitine transporter, OCTN2**, a high affinity transporter of carnitine, have been identified as a cause of systemic carnitine deficiency. The human gene for OCTN2 (named after its sequence homology with organic cation transporter OCTN1) has been localized to chromosome 5q31.1-32 [1051,1052]. It is 30kb long organized in 10 exons encoding for 557aa. The protein contains 12 membrane-spanning domains, glucose transporter specific motifs, a nucleotide binding folding motif. It is mainly expressed in kidney, muscle and placenta. A spontaneously mutant mouse model of the disease, the "jvs" mouse (for juvenile visceral steatosis) was identified to carry a homozygous missense mutation of the mouse gene for OCTN2 [1053]. A genetically engineered mouse strain deleted for a large genomic region 450kb long, encompassing OCTN2, was studied for its lipid metabolism [1054]. Mice accumulated fatty acids and triglycerides in their heart, liver and plasma. Triglycerides were found in VLDL, which production rate was increased, whereas lipase activities and VLDL hepatic uptake were normal. Liver lipid abnormalities were not corrected by carnitine supplementation, suggesting that the disease was more severe in the null mutant compared with the defective mutant "*jvs*" mice. Although the disease pattern observed in mice and humans appears slightly different, the identification of a gene involved in liver overproduction of VLDL as the result of impaired intracellular fatty acid degradation, underlines the role of available pools of fatty acids for lipoprotein assembly and secretion. Moreover, these data show that dietary compounds like carnitine have molecular targets that enhance liver fatty acid degradation thereby modulating lipoprotein circulation.

In addition to these examples, there are many more lipid-carriers, lipid-modifier proteins, located in different cellular organelles (lysosomes, peroxisomes or mitochondria, endoplasmic reticulum, Golgi apparatus, membranes etc.), which may result in other metabolic and cellular lipid storage diseases, which fall beyond the scope of this chapter [117].

2.4.12-   **Nuclear   Coordinators   of   Intracellular   Energy Resources**

As was mentioned along this chapter, regulatory sequence elements, upstream of genes for apolipoproteins, enzymes, transfer proteins or receptors, are sensitive to specific nuclear factors coordinating intracellular lipid metabolism. **Sterol Responsive Element Binding Proteins** or SREBPs were recognized as major regulators of cholesterol homeostasis [84]. SREBPs bind to Sterol Responsive Elements (SRE) upstream of target genes to regulate their transcription in conjunction with other nuclear factors. Surprisingly, SREBPs were found to be also integral membrane proteins of the endoplasmic reticulum (ER), a pivotal compartment for the regulation of intracellular pools of cholesterol. SREBPs form a family of microsomal proteins about 1150aa long [1055]. Three members have been identified: SREBP-1a, SREBP-1c generated from a single gene (chromosome 17p11.2) by alternative splicing and SREBP-2 (chromosome 22q13) [1056]. They have 2 transmembrane domains exposing the N-terminal and C-terminal ends towards the cytosol, whereas the loop separating transmembrane domains is turned towards the ER lumen. Their N-terminal part contains a beta helix-loop-helix-leucine-zipper (b-HLH-ZIP) DNA binding motif. The 31aa central loop contains a "RSVL" motif recognized by a protease (site 1 cleavage). The N-terminal transmembrane domain contains another cleavage site for a second protease (site 2 cleavage). The C-terminal cytosol domain contains a regulatory domain binding SCAP (SREBP Cleavage Activating Protein).

When cellular levels of cholesterol drop below a certain level, SCAP senses this change in the ER cholesterol concentration and activates SREBP cleavage. Cholesterol sensing in SCAP is mediated by a 5 transmembrane "sterol-sensing domain" common to other proteins like HMGCoA reductase, the morphogen Patched or NPC1 [37]. SREBP binds its C-terminal regulatory domain to SCAP, which translocates the complex to the Golgi apparatus for cleavage at site 1 [1057]. The first protease recognizing the SCAP-SREBP complex cleaves the intralumenal SREBP loop domain. The N-terminal part of SREBP is then released by the second protease that recognizes site 2. The activated proteolytic fragment of SREBP containing the b-HLH ZIP motif detaches from the ER membrane and moves into the nucleus to induce gene transcription. When intracellular cholesterol (and 25-hydroxycholesterol) concentration increases, cholesterol prevents SCAP from escorting SREBP to the Golgi, thereby preventing SREBP cleavage by protease 1. SREBP remains in the ER membrane, and in turn SRE remain unbound in the nucleus.

Negative feed back regulation by sterols has major physiological and pathological consequences. SRE sequence elements are found upstream of many genes controlling intracellular energy homeostasis [84,1056]. As a consequence SREBPs have shown to operate concerted regulations on fundamental biological pathways. SREBP gene targets include proteins involved in cholesterol homeostasis: HMGCoA reductase, HMGCoA synthase, farnesyl diphosphate synthase, squalene synthase (endogenous synthesis), the LDL receptor (cholesterol entry), ABC-1, SR-B1, caveolin 1 (cholesterol in membranes and cholesterol efflux) [1056,1058]. SREBP may also down regulate MTP, decreasing lipoprotein assembly and secretion in the liver [1059], an effect which could contribute to plasma lipoprotein lowering under strong HMGCoA reductase inhibition in human. Negative feed back regulation by sterols explains the dose effect of defective alleles of the LDL receptor gene in familial hypercholesterolemia. In this disease, appropriate levels of intracellular cholesterol are maintained by concerted regulation on pathways, alternate to cholesterol entry by LDL receptors at the expense of circulating LDL. Negative regulation by sterols is the target mechanism accounting for the strong apo B-rich lipoprotein lowering effects as well as the overall anti-atherogenic effects of HMGCoA reductase inhibitors (statins). The transient suppression of intracellular cholesterol *de novo* synthesis activates SREBP signalling, increasing the number of LDL receptors at the cell surface. Moreover, the mevalonate pathway generates essential components modulating inflammatory, proliferative and cell-activating processes taking place in the atherosclerotic plaque. Beyond their central role in cholesterol homeostasis, SREBPs monitor other genes of energy homeostasis, by regulating ATP citrate lyase (acetyl CoA availability for fatty acids and cholesterol synthesis), fatty acid synthesis (Fatty acid synthase, acetyl CoA carboxylase, stearoyl CoA desaturase), triglyceride synthesis (glycerol-3 phosphate acyl transferase) or lipid uptake (LPL, HSL, CETP) [1056,1060].

Genetically modified mice have better delineated the role of SREBPs in vivo. As expected, SREBPs play important functions during development since SREBP-2 knock out mice are 100% embryonic lethal and that SREBP-1 knock out mice have a 85% mortality rate in utero, sometimes rescued by SREBP-2 up regulation [1061]. SREBP-1 knock out mice, which survived appeared normal and healthy, as the result of SREBP-2 overexpression. Here some compensatory mechanisms upregulate the expression of SREBPs to compensate for the defective gene. SREBP-1a and SREBP-2 have higher transcriptional activity over cholesterol metabolism related genes, than SREBP-1c previously named ADD1 (Adipocyte Determination and Differentiation factor 1) after its cloning from adipocyte cell-lines, which preferentially determines fatty acid synthesis [1062]. However, SREBP-1c is the most abundant species expressed in human liver or adipose tissue. Moreover, if liver overexpression of SREBP-1c had modest effects on liver fatty acids synthesis, it induced a dramatic phenotype of generalized lipodystrophy with insulin resistance and hyperlipidemia when expressed in mice adipocytes [1063]. In this line, SREBP-1a overexpression in the liver induced massive liver steatosis. Interestingly mice remained normolipidemic, a phenotype similar to that of mice overexpressing mutant SCAP resistant to cholesterol down-regulation [1056,1064]. From these experimental data, it could be anticipated that a human disease resulting from SREBP deficiency has hardly any chance to be observed in comparison with situations where SREBPs could be upregulated. Moreover, they have confirmed that SREBPs are crucial coordinators of energy and lipid metabolism, and significant targets in atherosclerosis prevention.

Another group of major factors regulating lipid metabolism is the **Peroxisome Proliferator Activated Receptors** (PPAR) family of transcription factors [85,86,1065]. PPARs were named after the property of several chemical compounds (plasticizers, herbicides or fibrates) named peroxisome proliferators, and a high-fat diet to induce liver peroxisomes proliferation in murine species [1066]. Peroxisomes are small vesicular organelles clustered in the cytosol, which contain membrane and lumen enzymes involved in oxidative, energetic and lipid metabolism [1067]. In particular they contain sets of enzymes necessary for lipid degradation: the β-oxidation pathway. Peroxisome proliferators were shown to strongly induce the synthesis of peroxisomal enzymes, through a transcriptional mechanism mediated by PPARs, resulting in peroxisome proliferation and in accelerated β-oxidation of lipids. PPARs belong to the superfamily of nuclear hormone receptors. Nuclear hormone receptors are a family of ligand-induced transcription factors including steroid, thyroid hormone, vitamin D, retinoid receptors containing a Zinc-Finger domain involved in DNA binding and a C-terminal domain involved in ligand-binding. PPARs belong to the subfamily of receptors requiring heterodimerization with RXR (retinoid X receptor or 9-cis Retinoic Acid Receptor) for full transactivation of target genes. PPARs are intracellular sensors of long chain fatty acids metabolism. Three members have been identified: PPARα, PPARγ, PPARβ/δ (also named NUC-1/FAAR). Although they have a broad pattern of expression and ligand binding affinities, they differ by some characteristics.

PPARα are closely involved in lipid and lipoprotein metabolism. Although ubiquitous, they are highly expressed in the liver and kidney and represent specific targets for fibrates and leucotriene B4. That PPARα coordinates the upregulation of apo A-I, apo A-II, LPL and β-oxidation, and the down regulation of apo C-III has explained the classical pleiotropic lipid-lowering effect of fibrates in humans [1068]. PPARα mediate the adaptive response of lipid metabolism to fasting/feeding states [1069]. Moreover they may modulate activating and inflammatory responses in the arterial wall [1070]. PPARα knock out mice were generated and surprisingly they did not display a grossly defective phenotype, apart from lower plasma HDL and prolonged inflammatory response to leucotriene B4 [1071]. However, when homozygous mice were allowed to grow in age over 6 months they developed delayed and progressive combined hyperlipidemia, female obesity and male steatosis under a chow diet. The human PPARα gene (PPARA) spans 83.7kb. It is composed of 8 exons and is located on chromosome 22q12-q13.1 [1072]. Interestingly, a common Leu262Val mutation (allele frequency =6-10%) with decreased transactivation activity was found associated with mild hyperapobetalipoproteinemia in middle-aged adults of both sex at the heterozygous state [1073]. Further studies will confirm the impact of PPARα mutations on common forms of hyperlipidemia or atherosclerosis in humans.

PPARγ have been also extensively studied because they represent a specific target in energy balance and insulin resistance [86]. PPARγ have also a broad spectrum of ligands including glitazones (insulin-sensitizing drugs). However, their natural activators belong to eicosanoids (15-deoxy-12,14-prostaglandin J2) and to oxidized derivatives of linoleic acid (9-HODE, 13-HODE). PPARγ are ubiquitous, although they are predominantly expressed in adipose tissue. PPARγ expression may be induced by oxidized LDL and have shown to upregulate CD36 in macrophages, suggesting a potential role in foam cell formation and atherosclerosis [472]. In addition PPARγ have shown to stimulate cell differentiation, to be involved in apoptosis and to be overexpressed in tumors [1069]. The human gene for PPARγ (PPARG) spans 100kb and has been localized to chromosome 3p25 [1075]. It is composed of 9 exons, and generates two isoforms PPARγ1 and PPARγ2 by alternate splicing [1076]. PPARγ gene invalidation is embryonic lethal in mice, displaying severe developmental abnormalities. However, heterozygous knockout mice display a mild phenotype of dietary induced obesity and insulin sensitivity [1077]. Interestingly, dominant negative PPARγ germ-line mutations were reported to cause severe insulin resistance and hypertension in humans [1078]. Common PPARγ mutations Pro12Ala and Pro115Gln have been reported in different human populations and associated with body mass index, insulin resistance and obesity. The most consistent association has been a moderate effect although frequent (25% attributable risk), predisposing to type II diabetes in carriers of the common Pro12 allele, after an analysis of 3000 individuals [1079]. In line with observations with PPARα, it seems that contrary to the situation of gene invalidation in mouse, dominant mutations of expressed receptors may result in metabolic phenotypes in humans. Somatic mutations of PPARγ gene were reported in human colon cancer [1080]. Altogether, these observations confirm PPARγ as important regulators of energetic balance and of cellular differentiation.

Through their basic functions and cell-specific expression, PPARs could sense energetic resources related with fatty acids and operate adaptive responses essential for cell survival, suggesting that they have the potential to participate to still unknown functions [1069]. PPARβ/δ are less characterized than other members of the PPAR family although they also bind eicosanoids and have shown to regulate cellular differentiation [1069]. The relative abundance of other transcription factors, the phosphorylation of PPARs and the availability of RXR agonists in the nucleus may modulate PPAR action. RXR agonists may stimulate PPAR related functions [1081]. Moreover, RXR may modulate through heterodimerization, the regulatory activities of other related members of this family of previously named orphan nuclear hormone receptors, and more particularly LXR (Liver X receptor) and FXR (Farnesoid X Receptor) [1082]. **LXR** is mainly expressed in the liver [1083]. Through a feed-forward regulation LXR has shown to be a key regulator of cholesterol excretion into bile in mice. LXR binds to oxysterols (22- and 24-hydroxycholesterol) and is a strong and specific inducer of 7α-hydroxylase (CYP7A1), a rate-limiting enzyme of bile acids synthesis (see below). Therefore, any liver influx of cholesterol may stimulate bile acid synthesis through this mechanism. LXR also stimulates CETP and ABC1 expression increasing reverse cholesterol transport and elimination in the intestine [938,975,976]. However, LXR cannot explain the strong up-regulation of bile acid synthesis induced by bile acid sequestrants in humans (e.g. colestyramine). This was suggested to be mediated by **FXR** (Farnesoid X Receptor), another receptor expressed in liver, intestine and kidney with high affinity for bile acids, particularly chenodesoxycholic acid [1084,1085]. In this case, ligand binding to FXR down regulates 7α-hydroxylase. Thus decreased amounts of recycled bile acids taken up by the liver will result in the suppression of a negative FXR feed back regulation. FXR also upregulates PLTP and a protein involved in bile acid absorption in the intestine, suggesting that FXR could regulate pathways of lipoprotein metabolism complementary to LXR [955]. Other regulatory pathways controlling intracellular lipid metabolism will be probably identified uncovering novel targets for future lipid-lowering and anti-atherogenic treatments.

A striking observation on another category of nuclear compounds came from reverse genetics. **Lamins** are prenylated proteins involved in the formation of a mesh-like structure covering the inner membrane of the nucleus. Mutations of the gene encoding for lamins A/C (LMN A/C) were identified as a cause of dominantly inherited Emery Dreifuss muscular dystrophy and of dilated cardiomyopathy with conduction-system disease [1086,1087]. More unusual, it was found a cause of dominantly inherited lipodystrophy (with partial lipoatrophy) associated with insulin resistance and atherogenic hypertriglyceridemia: the Dunningan syndrome [1088]. In Emery-Dreifuss myopathy, it has been postulated that protein-protein interactions between emerin (the X-linked encoded gene) and lamins were disrupted within the nucleus [1089,1090]. Moreover, lamin mutations causing Dunningan lipodystrophy seemed to cluster in gene domains differing from those involved in myopathy or cardiomyopathy [1091]. This suggests that lipodystrophy related domains of lamins could be involved in other protein-protein interactions within the nucleus. This could be a protein expressed in adipose tissue involved in insulin resistance. It has been shown for example that glucokinase (GK) a major liver glucose-sensing enzyme is mutated in maturity onset diabetes of the young (MODY). A mechanism of nuclear sequestration of GK by a nuclear regulatory protein is involved in glucose sensing regulations [1092]. Therefore, apart from transcriptional regulations, the nucleus could take part to several post-transcriptional regulations of energy metabolism and atherosclerosis. Moreover, the lamin A/C locus is another example of a single locus with multifunctional roles giving rise to independent morbid consequences.

2.4.2-   The Biosynthesis of Bile Acids

Bile acids are an efficient way to remove cholesterol from the organism [1093]. In addition, bile acids combine with dietary lipids forming water-soluble emulsions promoting their hydrolysis and intestinal absorption. Bile acids also enhance the absorption of lipid-soluble molecules (e.g. vitamins A, E or K). They are also an important source of endogenous cholesterol. It has been estimated that 95% of biliary salts are reabsorbed in the distal ileum. Enterocytes use the newly absorbed cholesterol for chylomicron assembly. Cholesterol may be delivered to peripheral tissues or taken up by the liver, which spares the energy required for its de novo biosynthesis. Bile contains phospholipids, which undergo entero-hepatic recycling as well. The synthesis of biliary acids, by controlling the entry-exit movements of liver cholesterol, is a major pathway of lipid homeostasis. This major regulatory pathway has long been used as a target for cholesterol lowering therapy, through the use of bile-acid sequestrants.

More than forty enzymes are involved in the synthesis of major biliary acids. Intermediate metabolites follow pathways of maturation (generally oxidative) shuttling between different cellular compartments (endoplasmic reticulum, mitochondrion, peroxisomes or cytosol) before being excreted in the bile duct, through ATP-dependent transporters [1094]. Two enzymes were shown to be rate limiting in bile acid biosynthesis: 7α-hydroxylase and 27-hydroxylase. The first enzyme directs cholesterol towards the biosynthetic pathway of biliary acids, the second allows the maturation of final products: cholic acid and chenodesoxycholic acid (*figure 2-33*). This metabolic pathway is closely connected with other pathways of cholesterol homeostasis. Concerted regulations operate on intrahepatic cholesterol homeostasis in response to biliary acid turnover, reciprocally biliary synthesis and secretion adapts to endogenous cholesterol homeostasis.

Several genes regulating bile acid synthesis were identified. There are at least 3 microsomal enzymes of the cytochrome P450 with **7α-hydroxylase** activities with different substrate specificity, tissue distribution and regulatory mechanisms [1095]. The classical pathway 7α-hydroxylase (CYP7A1) uses cholesterol as a substrate. It is exclusively expressed in liver and is down regulated by bile acids and up regulated by intracellular oxysterols, through nuclear receptors FXR and LXR respectively. The gene is 10kb long, containing six exons and localized to chromosome 8q21.13 [1096]. It encodes a 504 amino acids protein with a 57kD MW [1097]. CYP7A1 has been identified by sib-pair linkage analysis as a predisposing locus for high plasma LDL cholesterol in the general population [1098]. The second liver specific, 7α-hydroxylase (CYP39A1) or oxysterol 7α-hydroxylase uses 24-hydroxycholesterol as a substrate [1099]. It is poorly sensitive to bile acids or sterol feed back regulation, however it is more abundantly expressed in female mice. The CYP39A1 gene is 150kb long, located on human chromosome 6. The 470 amino acid protein is encoded by 12 exons. To date, no mutations were reported on either of these loci in humans. There is another oxysterol 7α-hydroxylase (CYP7B1), which uses 25- or 27-hydroxycholesterol as substrates. It has a more widespread expression, and is less sensitive to sterol or bile acid feedback regulations. The CYP7B1 gene is very similar (6 exons) and closely linked on chromosome 8q21.3 to that of CYP7A1 One homozygous mutation (Arg 388→ Stop) has been identified on CY7B1 in a newborn with a severe cholestatic liver disease resistant to cholic acid supplementation [1100].

27 hydroxylase (CYP27) is a mitochondrial enzyme of MW 57kD. Its primary sequence comprises 498 amino acids. It is part of cytochrome P-450, and ensures the hydroxylation of several cholesterol derivatives. It also converts vitamin D_3 into $25OH\text{-}D_3$ [1101]. Its expression, although predominant in liver, is rather diversified: duodenum, adrenals and lungs. It has been reported to promote an alternate pathway of cholesterol efflux in lung macrophages by producing 27-hydroxycholesterol able to be taken up by HDL [1102]. The CYP27 gene is 18.6kb long contains 9 exons and was

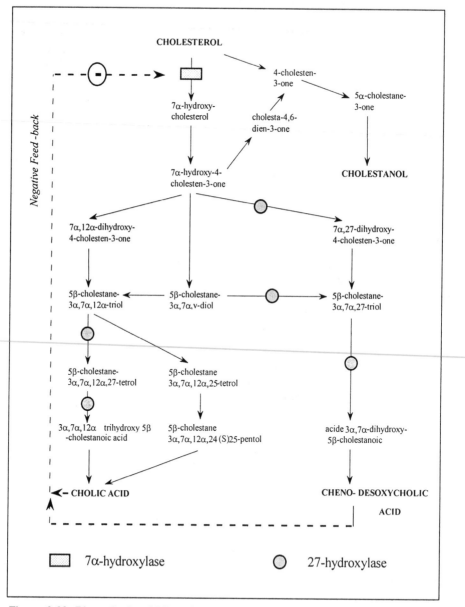

Figure 2-33. Biosynthesis of bile acids. Two enzymes are rate limiting: 7α-hydroxylase (CYP7) and 27-hydroxylase (CYP27). An enzymatic block on CYP27 results in the accumulation of toxic compounds like cholestanol.

localized to human chromosome 2, band q33-qter [1103,1104]. It encodes for a mRNA from 1.8 to 2.2kb. The first exon encodes for a signal peptide for integration into the mitochondrial membrane. The coding sequence comprises several highly conserved motifs in the family of microsomal enzymes of cytochrome P450. The promoter contains a motif recognized by Sp1 and a sequence of recognition for liver specific expression. Contrary to CYP7A1, there is no SRE upstream of the gene, which is in agreement with the insensitivity of CYP27 to sterol feed back regulation.

A monogenic disorder is associated with mutations of 27 hydroxylase: **Cerebrotendinous Xanthomatosis** [1105]. This recessively inherited disease is characterized by a progressive tissue accumulation of toxic oxysterol derivatives, particularly cholestanol and bile alcohols in urine (*figure 2-33*). Although patients are generally normolipidemic, they present with large tendinous xanthomas, corneal deposits and premature cataract. The outcome of the disease depends on disseminated, progressive and severe attacks of neurological deficiencies along with intellectual impairment and psychiatric disorders. The anatomical basis to this clinical presentation is visible by NMR imaging in the form of "xanthomatous deposits" of the white matter in the central nervous system [1106,1107]. Atherosclerosis is also reported incidentally. Although inconstantly reported, vitamin D metabolism does not seem disturbed. The severe outcome of the disease may be retarded by a treatment combining bile acid and HMGCoA reductase therapy. Many patients were described with mutations on the gene for CYP27 worldwide [1108]. Missense mutations cluster in highly conserved domains in the cytochrome P450 family of proteins. Some were involved in interactions with adrenodoxin, an electron transporter associated with mitochondrial enzymes of cytochrome P450. Others disrupt highly conserved residues interacting with haem. Disease causing mutations have been described worldwide. In Japan or Sephardic Jews apparently unrelated families were found with the same mutations as the result of a founder effect [1109]. In heterozygous carriers of CYP27 mutations, atherosclerosis and cardiovascular disease was reported, although inconstantly. The identification of naturally occurring mutations of CYP27 gave a molecular basis to this severely invalidating disease.

These first examples of genetic defects of the bile acid synthesis pathway although remotely connected with disorders of circulating lipoproteins, underline the biological significance of this pathway in lipid homeostasis. They emphasize the cellular toxicity of hydroxylated derivatives of cholesterol and their powerful influence on cholesterol metabolism.

2.4.3-   Other Apolipoproteins

Together with lipid soluble vitamins and antioxidants, which are bound to core lipids, other proteins may be bound to lipoproteins. These lipophilic proteins, initially named apolipoproteins, proved thereafter to be less specific in regulating lipoprotein circulation and to have more specialized tissue functions [1110].

Apolipoprotein D or lipocalin belongs to the family retinol binding proteins (RBP) or α2μ-microglobulin. It is a protein of MW 29kD, representing 5% of the protein content of HDL. It may be bound to LCAT, to CETP or to apo A-II with which it may dimerize. The gene of apo D (APOD) is 20kb long, organized in 5 exons. It is localized to human chromosome 3q26-2qter [1111]. It encodes for 169aa, which do not share homology with other apolipoproteins. Upstream regulatory sequences may bind to steroid hormone receptors and to acute phase response mediators [1112]. The protein is ubiquitously expressed, mainly in adrenals, spleen, kidney, pancreas, placenta, liver, intestine and white matter of the brain [1113]. More unusual it is secreted by axillary gland and has been identified as one of the axillary odorant carriers, which play a role of pheromones in mammals [1114]. In addition to its affinity for retinol and cholesterol this protein strongly binds bilirubin in mortified tissues. This last property would confer to apo D a role in the protection of tissues against oxidation, and in the anti-atherogenic role of HDL.

Apolipoprotein H or β2-glycoprotein-I binds acid phospholipids. It is found in VLDL, chylomicrons and HDL. This protein of 50 kD belongs to the family of the serum complement system, by the presence of 5 repeats of 60 aa in its primary sequence. The gene (APOH) is unique, encodes for 345 residues and is localized to chromosome 17 [1115,1116]. The role of apo H in lipoprotein metabolism is unclear although it is likely to enhance apo C-II activator activity for LPL. Its role in immunity seems a little more precise. It would be repressor in the acute phase response during infection. Its affinity for phospholipids seems required for the binding of anti-phospholipid antibodies to anionic phospholipids and the generation of auto-antigens in certain systemic diseases like lupus or the primary antiphospholipid syndrome. Common mutations (Cys306→Gly and Trp316→Ser) in the 5th hydrophobic domain of apo H, preventing phospholipid binding could represent protective alleles against the generation of anti-phospholipid antibodies [1117].

Apolipoprotein J or clusterin is a dimeric glycoprotein associated with HDL in plasma [1110]. It is synthesized as a propeptide of MW 70kD, which is cleaved in two sub-units of 34kD and 37kD bound by a disulfide bridge.

The gene (CLU) is 16.6kb long, containing 9 exons, localized on chromosome 8p21-p12 [1118]. It may be secreted as a native lipoprotein in the local cell environment. Its primary structure comprises amphipathic α-helices and heparin binding domains, encoded by 427 amino acids. Apo J is a natural ligand for megalin, which mediates its cellular uptake. It is preferably associated with damaged membranes, from which it could enhance cholesterol efflux. It has anti-cytolytic properties by inhibiting several complement factors. It is ubiquitously expressed, however predominantly in ovary, testis, brain, liver and in all cells facing with biological fluids or extracellular spaces. Apo J is thus allotted a role of extracellular chaperone, being cytoprotective by stabilizing proteins and various compounds at the interface between cell membranes and the external medium [1119]. This role was elegantly shown in cells undergoing apoptosis, among which only cells overexpressing apo J would survive, after induction of programmed cell-death [1120]. Apo J has been involved in neurodegenerative disease, in immune modulation and in the reproductive system. It has been suggested to be a part of anti-atherogenic properties of HDL.

Apolipoprotein L was recently identified as a novel apolipoprotein in small apo A-I-containing HDL [1121]. The protein is encoded by 383aa, including a 12aa signal peptide and a 371aa mature protein. The single chain protein has an apparent MW of 42kD, as the result of glycosylation. Apo L is secreted by the pancreas and is tightly bound to HDL. Although it contains four amphipathic α-helices, it does not share homologies with any known protein. Its physiological role is not known yet.

Serum Amyloid A proteins (SAA), are acute phase response proteins. After various tissue injuries such as this caused by microbial infections or burns, serum amyloid protein levels may increase by 100 to 1000 fold in plasma. SAA concentration may be permanently high in chronic inflammatory states and for unexplained reasons this protein accumulates in tissues forming the major component of the amyloid matter. Serum amyloid proteins are a family of proteins: SAA-1, SAA-2, SAA-3 and SAA-4. They are on average 104aa long, with a molecular weight comprised between 12 and 19kD. A single gene encodes each SAA protein, with an apolipoprotein-like organization in 4 exons. They are gathered in a cluster localized on the short arm chromosome 11, to 11p15.1 [1122]. Their structure is species specific. Indeed in the same species, they share more than 95% homology, however one particular isoform may be quite dissimilar from one species to another [1123]. This suggests a specific role closely related with the host immune system, rendering observations difficult to anticipate from one species to the other. The liver massively secretes these proteins in native HDL during inflammatory states. They may be involved in tissue-repair mechanisms by attracting phospholipids to the site of injury. Their great excess in HDL is likely to displace apo A-I and to inhibit LCAT activity and cellular cholesterol efflux. Moreover, their secretion is accompanied by a fall in plasma HDL concentrations. Thus, they could be involved in extracellular processes of atherogenesis. They are expressed in all the activated cellular components of the atherosclerotic plaque, whereas they are absent form normal arterial wall. Abnormal apo A-I or apo A-II isoforms were associated with amyloid deposits, which confirms SAA as a member of the family of small apolipoproteins [1124].

2.4.4- Antioxidative and Antitoxic Protection.

Since the 19th century the antitoxic role of lipoproteins had been suspected even before they were identified. This old concept has reunified with the "response to injury" hypothesis for atherosclerosis in which modified, and particularly oxidized lipoproteins retained in the arterial wall, could initiate foam cell formation, establishing a link between lipoprotein metabolism and atherosclerosis [1125]. Anti-oxidative properties of natural components of the lipid core (beta carotene, vitamin E etc.) and the specific anti-oxidative role of small apolipoproteins contained in HDL (apo A-I, Apo A-IV, apo E) were proposed as protective mechanisms against atherosclerosis. In addition, genes were identified in the anti-oxidative protection of lipoproteins.

Paraoxonase (PON/arylesterase), is an enzyme involved in detoxification of certain organophosphorous insecticides like parathion. Parathion is transformed into paraoxon by enzymes of the cytochrome P450 and is inactivated in plasma by paraoxonase or arylesterase. This protein of MW 43kD is associated with HDL, but its physiological role and its natural ligands are not known. It would have the capacity to protect LDL against oxidation [1126]. A functional polymorphism was known to determine two paraoxonase activities: one, common in populations determining weak PON activity (allele A) and the other less frequent, determining strong PON activity (allele B). A cDNA for paraoxonase (PON1) was cloned, and localized to chromosome 7q21-q22, near the locus for cystic fibrosis [1127]. The functional polymorphism was found to result from a genetic polymorphism. The "A" isoform results from the presence of a Glutamine at position 192 and the "B" results from an Arginine at the same position. Moreover, these functional mutations were found to be associated with plasma lipoprotein modifications and predisposition to atherosclerosis however with some inconsistency depending on study populations

[1128,1129]. The biological mechanisms underlying these modifications are unclear [1130,1131]. However, this first example shows that specific genes may contribute to protection against oxidative lipoprotein damage and tissue injury. They may help to identify new factors connecting lipoproteins metabolism and atherosclerosis. Other anti oxidative proteins have been identified, some having suggested to modulate atherosclerosis susceptibility [1132]

The identification of the first candidate genes of lipoprotein metabolism gave a molecular basis to inherited dyslipoproteinemia. They have shown great functional and structural diversity. The knowledge of the structure of different predisposing loci offered new ways to explore their expression and uncovered new therapeutic prospects. The metabolic consequences of natural mutants and of their expression in genetically engineered animals helped to elucidate novel functions and their coordinated regulations in vivo, which were hardly accessible at the protein level in humans. Genetic heterogeneity of lipid traits and molecular heterogeneity of disease causing mutations is a hallmark in familial dyslipidemia, like in other examples in molecular genetics. Candidate loci are polymorphic, offering an invaluable tool for analysis of their effects on normal and pathological lipid metabolism in individuals, in families or in populations.

III- THE GENETIC BASIS OF DYSLIPIDEMIAS

The idea that human disease, particularly cardiovascular disease and lipid disorders could be determined by some inherited component was already suspected in the past centuries. The recent emergence and expansion of molecular genetics has given plain and multiple evidence of its reality. Now genes may be used as single objects, dismantled by the direct analysis of their primary sequence on human DNA and grasped by the manipulation of their products (mRNA and proteins). The entire sequence of the human genome (and of other living organisms) readily accessible as universal patrimony, offers manifold opportunities to dissect the molecular basis of fundamental metabolic pathways and as a consequence of human disease.

In this chapter we will review through several examples, the increase in knowledge and the diversity of practical applications that came out from the genetic study of dyslipidemia. Genes described in the previous chapter directly control lipoprotein metabolism, defining them as candidate genes of lipoprotein metabolism. Several hundreds of others may be indirectly involved in dyslipidemia as part of their role in atherosclerosis or in energy balance. This increase in knowledge has led to the delineation of the genetic architecture of lipoprotein metabolism. Another consequence of the discovery of genes involved in lipoprotein metabolism was a unique opportunity to develop the candidate gene approach and other genetic strategies. Beyond a better understanding of the molecular basis of human dyslipidemia, novel medical responses may be defined to diagnose and treat these common disorders.

3.1- THE GENETIC ARCHITECTURE OF LIPOPROTEIN METABOLISM

More than 50 genes described in the previous chapter play a direct role in lipoprotein metabolism. However, their involvement may vary from one gene to the other depending of their function and on the way their product intervene among other proteins in the same metabolic pathway. Their role on a specific pathway may vary depending on their degree of functionality and on their functional diversity (multifunctionality). Their overall effect in vivo also depends on the presence of other proteins (i.e. genetic redundancy) which may compensate for possible loss of functions, like those resulting from spontaneous gene mutations. The molecular dissection of individual components of lipoprotein metabolism led to the understanding of a higher order of organization of its genetic architecture. As a consequence the limits of lipoprotein metabolism are envisioned outside of the bloodstream, now down to intracellular regulations (scale down), as well as up to tissue-specific and physiological regulatory mechanisms of lipid homeostasis (scale up).

3.1.1- The Functionality of Candidate Genes

3.1.11- Levels of Functional Significance of Candidate Genes

The concept of molecular disease (i.e. one gene/one disease) was already efficient in understanding the biological role of specific proteins long before DNA could be routinely analyzed. With recent progresses in molecular genetics, the identification of a naturally occurring gene mutation causing a particular phenotype in humans has broadened the spectrum of biological functions of proteins and their role in human disease. Candidate genes of lipoprotein metabolism have shown to display numerous loss-of-function mutations, or defective mutations.

According to the phenotypic consequences of a loss of function due to homozygous defective mutations in humans, candidate genes belong to three main classes (*Table 3-1*): The first class is the "**indispensable genes**" which play a fundamental role in biology. Their loss of expression is presumably embryonic lethal in humans. They are involved in very basic functions like lipid biosynthesis (e.g. HMGCoA reductase or Fatty Acid synthase), developmental processes at early stages of embryogenesis (LRP, Megalin) or act as important coordinators of lipid homeostasis (SREBPs, PPARs). A complete loss of function caused by a natural mutation on both alleles has hardly any chance to be observed as a living phenotype in humans. With the use of gene knock out (KO) technology it was possible to prove that this was true in mice. For example, embryonic lethality is constant in SREBP2 knock out mice and is observed in 85% of SREBP1 knock out mice. The role of SREBPs in coordinating several pathways of lipid homeostasis is mandatory for normal embryonic development. Only those SREBP1 KO mice, which overexpress SREBP2, and compensate for the loss of SREBP1 during embryogenesis will survive and look apparently healthy [1061]. Cellular lipid uptake is a very important process during early embryogenesis, particularly during brain development. Gene disruption of megalin, a member of the LDL receptor gene family causes severe forebrain malformation [769]. Among the loss of other functions, the defective uptake of protease/protease-inhibitor complexes causes early embryonic failure in LRP KO mice [441]. In contrast, heterozygous KO mice are generally healthy, due to a level of expression of the normal copy of the gene, sufficient to restore physiological functions. Only in the case of PPARγ, a dominant negative germ-line mutation was reported as a cause

Table 3-1. **Candidate genes of lipoprotein metabolism belong to three functionality classes.** Functional effects are those resulting from natural mutations in humans. * Effect may differ in other species.

Indispensable Genes		Rate-Limiting Genes		Modulator Genes	
Locus	Function	Locus	Function	Locus	Function
		LDLR	LDL	Lp(a)	Extracellular
HMGCoA-R	Lipid	APOB*	metabolism	MSR	Lipid
HMGCoA-S	Biosynthesis	MTP*		CD36	Retention &
Fatty Acid-S				SRB1	Scavenging
		LPL*	TG-rich		
LRP	Embryonic	APOC2	Lipoprotein	HL	
Megalin	Development	APOE	Metabolism	APOC1	
				APOC3	Lipoprotein
SREBP-1	Regulation	APOA1		APOA4	Remodeling &
SREBP-2	of Lipid and	LCAT	HDL	APOA2	Trafficking
PPAR-γ	of Energy	CETP	Metabolism	VLDLR*	
RXR	Homeostasis	PLTP		ASP	
		Cubilin			
				PPAR-α	Intracellular
		ABC1		FABP	Lipids
		NPC1	Cholesterol		
		LAL	Trafficking	APOD	
		HSL*		APOJ	Lipoprotein
				APOH	& Lipid
		SCD	Fatty Acids	SAA	Protection
		OCTN2	Trafficking	PON	
		Lamins			
		LXR			
		FXR	Bile Acids		
		CYP7B1	Metabolism		
		CYP27			

of insulin resistance and hypertension humans [1078]. Like seen for mutations in multimeric proteins, the dominant negative effect resulted from an inhibitory interaction of the mutated PPARγ protein with its natural partner RXR. This situation contrasts with the activation normally induced by the wild type allele of PPARγ - a situation found in heterozygous PPARγ KO mice, which express only one copy of the normal allele. This example underscores the fact that KO experimental models do not always reproduce a pathological situation found in humans. Moreover, phenotypes observed in knock out mice may be more severe than those observed in humans. In mice, apo B or MTP gene invalidation is embryonic lethal due to defective lipid uptake in the embryo, whereas it causes serious diseases in infant humans (familial hypobetalipoproteinemia and abetalipoproteinemia respectively). LPL KO mice, die soon after birth as the result of massive lung capillaries invasion by chylomicrons after their first sucking [508]. Human newborns with complete genetic LPL deficiency survive, although they have massive chylomicronemia and may suffer from acute pancreatitis during the first year of life.

Interestingly, if indispensable candidate genes have hardly a chance to cause a human disease, at least in homozygotes, this class of genes holds very interesting targets for plasma lipid-lowering or anti-atherogenic therapy. HMGCoA Reductase inhibitors (i.e. statins) represent the paradigm of such drug targets at the root of lipid metabolism and atherosclerosis, later exemplified by PPAR activators (fibrates, glitazones). Other drug targets might be found among factors regulating gene expression, particularly those regulating intracellular metabolism of lipids and their derivatives like bile acids, oxysterols, prostanoids etc. Moreover the basic functions of indispensable genes might have other therapeutic applications outside of the cardiovascular system. Statins through their action on the mevalonate pathway may modulate gene expression in tumor cell lines [32], may be immunomodulatory in heart-transplanted patients [1133] or increase bone formation and may be protective against osteoporosis [1134,1135].

The second class is the "**rate-limiting genes**". It is also the most represented by candidate genes for which the loss of function significantly disrupts a specific pathway of lipoprotein metabolism and causes a familial lipoprotein disorder. Defective mutations may manifest as autosomal dominant, like in cases of the genes for LDL receptors or ABC1, for apolipoproteins A-I, B, or E, for CETP or lamins. Familial lipoprotein disorders may be as well recessive like in the case of MTP, LPL, LCAT or CYP27 genes. In line with indispensable genes, several genes may be rate limiting in other animal species, whereas they have lost a crucial function in humans. For example, the chicken homologue of the VLDL receptor gene plays an important role in yolk formation and in lipoprotein metabolism in this bird species [782]. In contrast, in mammals and particularly in humans, the changes in tissue distribution and levels of expression have given this locus a minor role in lipoprotein metabolism. The identification of natural disease-causing mutations in humans has also helped to dissect the sequence of gene interventions in specific metabolic pathways. For example, lipolysis by LPL of triglycerides in apo B-containing lipoproteins is a prerequisite to LDL formation. A family originating from Quebec was identified with carriers of a homozygous LPL gene mutation together with a heterozygous LDL receptor gene mutation [1136]. As expected fasting chylomicronemia was observed like in subjects homozygous for LPL deficiency only. However, contrary to subjects with heterozygous LDL receptor deficiency, subjects with digenic deficiency (LPL and LDLR) had low levels of plasma LDL. This confirmed that lipolysis of triglyceride-rich particles is a major step for the generation of circulating LDL in humans. Thus, rate-limiting genes represent the best candidates for monogenic disease causing genes. They constitute the base on which to built up specific diagnostic and therapeutic strategies, most appropriate to prevent morbid complications of familial lipoprotein disorders.

The third class is the "**modulator genes**", for which loss of function mutations have a mild or no apparent effect at baseline in a specific individual, however their effects may be evident in larger number of individuals or in particular physiological, pathological or environmental conditions. A classical example is Lp(a) which is not expressed in many animal species, and may produce null alleles that are not expressed in humans without any major change in lipoprotein metabolism. However, this locus is a strong modulator of the atherogenic potential of LDL to which apo(a) is bound. Hepatic lipase has a modest lipolytic effect in plasma, however it is an important modulator of lipoprotein remodeling and of HDL metabolism [592]. The modulator effect may not result only from the primary structure of the protein. It may rather depend on the level of gene expression and on protein abundance in plasma or in extracellular spaces like in the case of small exchangeable apolipoproteins apo C-I, apo C-III, apo A-II or apo A-IV [646]. In fact, functional genetic variants were found in regulatory regions controlling gene expression at these loci. The modulator effect might also depend on specific physiological conditions like diet [818,819], age, hormonal regulations or another genetic condition. PPARα deficient mice develop progressive weight gain and hyperlipidemia when they grow in age. They also display a sexual dimorphism: females accumulate lipids in adipose tissue while males accumulate lipids in liver [1071]. Combined human CETP and HL deficiency result in high plasma HDL and increased atherosclerosis as the result of impaired reverse cholesterol transport [944]. In a broader definition modulator genes may also include genes that play a major role in other metabolic systems with repercussions on lipoprotein metabolism. Because the effects of modulator genes are moderate, their quantification may require the study of large human cohorts. However, they represent interesting targets for lifestyle and environmental modifications in common dyslipidemia.

3.1.12-   The Redundancy of Candidate Genes

Despite the fundamental role of several candidate genes like ACAT (intracellular cholesterol esterification) or DGAT (triglyceride biosynthesis), their disruption by naturally occurring or engineered mutations does not result in a major phenotype. This is due to genetic redundancy, a characteristic of vertebrate genes [1137]. Genetic redundancy could be a consequence of polyploidy (i.e. several copies of the individual's genomic information). Genomic duplications and fusions might have occurred several million years ago and would have generated several copies of an ancestral genome, later giving rise to multiple gene copies, creating multigene families. Candidate genes of lipoprotein metabolism all belong to multigene families (see *figure 2-20* page 100 and *figure 2-27* page 117) which have generally diversified across the evolution of living species through exon shuffling and mutation accumulation overtime. This supposes that genes share domains of structural homologies, generally corresponding to common functional protein motifs. The LDL receptor gene family is characterized by modules containing ligand-binding domains (or class A cysteine-rich motifs), EGF-like domains involved in receptor recycling, a single hydrophobic transmembrane domain and a cytoplasmic tail containing a signalling motif (NPXY) necessary for receptor endocytosis. The gene family of small exchangeable apolipoproteins share in common amphipathic α-helices, involved in lipid loading and unloading as well as in interactions with circulating or cell-surface ligands (e.g. enzymes, receptors). The apo B gene belongs to the vitellogenin family of large-capacity lipid transporters during egg formation. Proteins of the LPL gene family (including LPL, HL, PL, EL, LCAT and yolk proteins) belong to a family of lipid binding proteins containing a hydrophobic pocket trapping triglycerides and/or phospholipids. Transfer proteins (CETP, PLTP) belong to a family of secreted lipid transfer/lipopolysaccharide binding proteins, sharing a common boomerang shaped structure containing two barrel-like arms mediating lipid transfer. Scavenger receptors class B (CD36 and SR-B1) share a common U-shaped primary structure, with a highly glycosylated extracellular loop involved in multiligand binding and two membrane spanning domains and two cytoplasmic tails. Redundancy may be limited to specific functional domains. ABC1 belongs to a large family of membrane proteins mediating active translocation of a wide variety of molecules across membranes through a common pore-like general structure containing two blocks each containing 6 transmembrane domains and one ATP-binding cassette [1138]. However, ABC1 is specialized in cellular cholesterol efflux. Sterol sensing domains are also a common feature of a number of intracellular proteins mediating important cholesterol-dependent responses: SCAP (SREBP activation), NCP1 (intracellular cholesterol transport), HMGCoA reductase (cholesterol biosynthesis), *Patched* (membrane topology of *Sonic Hedgehog* during embryonic development).

Structural redundancy may go with functional redundancy. This is particularly manifest for genes involved in intracellular lipid metabolism like nuclear factors controlling gene expression (SREBPs, PPARs, etc.), or genes involved in intracellular lipid trafficking (ACAT), or biosynthesis like DGAT. For example, gene disruption of DGAT, which mediates the last step of triglyceride biosynthesis, is accompanied by normal triglyceride biosynthesis in knock out mice, suggesting that other biological pathways relay the DGAT defective pathway for triglyceride biosynthesis [1045]. A similar example led to the identification of a second ACAT gene, after the ACAT1 was silenced by gene knock out in mice [1033]. Another example is the redundant role of VLDLR and apo ER2 (both very similar in their general structure) in mediating *Reelin* signalling for neuronal positioning during brain development. Either gene disruption has no effect, whereas both gene defects result in cerebellar dysplasia [785]. Thus, gene redundancy could be a natural security mechanism (like an extra engine on airplanes), leading to healthy phenotypes despite the loss of expression of a protein involved in vital cellular functions. Redundancy may also account for partial functional compensation. This was shown in the case of small exchangeable apolipoproteins in their capacity to modulate lipoprotein metabolism. Apo A-I deficiency is partially replaced by apo E for reverse cholesterol transport and by apo A-IV for anti-oxidative properties of HDL. The role of LRP as a remnant receptor in the liver could be demonstrated after LDL receptor gene knock-out in mice transiently expressing RAP [759]. Only in the absence of remnant clearance by LDL receptors, LRP inhibition by RAP resulted in plasma accumulation of remnants. Redundancy may be functional only. Heparan sulfate proteoglycans (HSPG) represent high capacity binding sites for lipoproteins at the cell surface, favoring their remodeling and mediating their endocytosis. That structural and functional redundancy characterizes genes of lipoprotein metabolism is not surprising, when one considers the importance of lipid transport and its regulation in living species. Moreover, because tissue specific expression and protein structure and functions somehow differ, gene redundancy allows a broader diversity of biological responses on different pathways of lipoprotein metabolism in response to variable environmental conditions.

3.1.13-   The Multifunctionality of Candidate Genes

Another important finding emerged from the exploration of candidate genes of lipoprotein metabolism: multifunctionality. A common fact in molecular genetics is that a single gene product may have different functions (multidomain protein). Multidomain proteins are composed of sets of modules structurally and functionally independent. They were generated throughout evolution by duplication or transposition of discrete genomic regions comprising single exons or groups of exons (i.e. exon shuffling) [1139]. Multifunctionality is obvious in cases of large multifunctional proteins like macrophage class A scavenger receptor, cubilin, LRP, megalin and other members of the LDL receptor gene family. Moreover specialized proteins may combine independent functions within the same molecule. For example SREBPs combine the characteristics of a resident ER-membrane protein with those of a mobile transcription factor after sterol-sensitive proteolysis [1056]. The VLDL receptor combines functions of a plasma membrane receptor mediating ligand endocytosis with those of a signalling molecule in the brain [785]. CD36 combines the function of signalling-receptor binding thrombospondin, with those of a lipid transporter across the plasma membrane [475]. The diversity of functions is also found in small exchangeable apolipoproteins or lipases. Depending on the location of apo A-I mutations, amyloidosis or deficient LCAT activation may be observed (see *Table 2-11* page 131). This was shown to be critical in cases of missense mutations, which may disrupt one function while others are preserved on the expressed protein. LPL is an important factor of lipoprotein anchoring to HSPG and to cellular receptors, independently from its lipolytic activity. Defective mutations impairing lipolytic activity while preserving protein secretion were associated with increased susceptibility to atherosclerosis, presumably as the result of increased lipoprotein retention in the arterial wall [557]. Opposite phenotypes might be observed depending on the location of the mutation on the primary sequence of apo B (hypo- or hyperapobetalipoproteinemia) indicating that the same protein combines domains essential for LDL structure and domains essential for LDL catabolism. They may even lead to different tissue-specific diseases. In the case of lamins, cardiomyopathy or insulin resistance and hyperlipidemia may be observed depending on the location of mutations on the primary sequence. This suggests that depending on tissue-specific factors interacting with specific domains of the protein, different organ-specific diseases may be observed.

3.1.2-   The Driving Forces of Lipoprotein Metabolism

Lipoproteins are transient transporters of large amounts of neutral lipids. Lipids are available for immediate uptake by any cell in the organism. Apo A-I containing lipoproteins may as well upload cholesterol or oxysterols for their return to the liver. Their size, smaller than that of circulating cells, allows lipoproteins to rapidly circulate in extracellular spaces, which include the blood stream or the cerebrospinal fluid in the case of HDL. Their high plasticity allows a permanent remodeling favorable to rapid lipid exchanges between themselves or with cells. They are also used as shuttles to transport antioxidative products and lipophilic compounds like vitamins (retinol, vitamin E, beta-carotene etc.). Large triglyceride-rich particles may buffer toxic substances (e.g. bacterial endotoxins), driving them back to the liver where they are eventually degraded and eliminated [1140]. Specific proteins like human apo A-I may even have direct antitoxic or antiviral activities [799]. Contrary to circulating cells, their fragile structure and lipid-rich composition is also their Achilles heel making them highly susceptible to degradation and oxidation in extracellular spaces, with no possibilities of self-repair or regeneration. Their high capacity to interact with components of the cell surface makes them also vulnerable to their entrapment in extracellular spaces. This phenomenon called lipoprotein retention in the arterial wall, transforms degraded lipoprotein into alien bodies triggering inflammatory and other pathogenic responses typical of atherosclerosis [509]. Several ways to interrupt the chain of events have been proposed: 1°) decrease the amount of secreted atherogenic lipoproteins; 2°) decrease the residence time of lipoproteins by increasing their clearance; 3°) decrease their vulnerability to degradation and their retention in extracellular spaces; 4°) increase their capacity to eliminate potentially toxic lipids in the liver (reverse transport). This supposes that forces that drive these mechanisms are understood and controlled.

Several driving forces determine the metabolic fate of lipoproteins. First, abundant sources of lipids are needed for **cellular membranes**: synthesis, maintenance, repair or any modification in their lipid composition. Cells finely control the levels and the nature of lipids contained in their membranes or in lipid stores in their cytoplasm. Membrane lipid composition is crucial for protein activity (integral proteins or proteins neighboring membranes), for organelle and cellular structural and functional integrity. For example, this is critical in neurons for which cellular integrity is highly dependent on that of their membranes. Therefore, it was not that much surprising that apo E dependent triglyceride-rich lipoprotein endocytosis could influence neuronal outgrowth [685]. Moreover, humans and several species of primates are the rare species in which large apo B-rich lipoproteins (i.e. LDL) circulate in fasting plasma, allowing a permanent cholesterol uptake in peripheral cells through the LDL receptor pathway. This is particularly critical for neurons, which require large amounts of cholesterol and long chain fatty acids for their membrane maintenance. In addition, any situation of cell injury (infection, inflammation, ischaemia, necrosis etc.) or renewal will induce lipoprotein recruitment at least for membrane repair and synthesis. Therefore, that cytokines or cell-cycle regulatory proteins profoundly influence lipoprotein metabolism could be expected. Transient modifications in the lipid composition (cholesterol, fatty acids in phospholipids) of membranes modulate extracellular or intracellular signalling (e.g. cell-cell interactions, microdomain formation: rafts, caveolae, etc.), the trafficking of active substances along metabolic pathways within cells or across the plasma membrane, or the shape of the cell itself (cytoskeleton proteins anchoring). Moreover, cells establish permanent exchanges with the extracellular milieu though lipid uptake from the core of lipoproteins or through lipid efflux towards lipoproteins or the external milieu. These exchanges may modulate the lipid composition of cellular membranes. For example in Niemann Pick disease, the lipid composition of membranes is significantly altered in this disorder of intracellular cholesterol trafficking [1008]. On the other hand, lipids from the environment may modulate cellular membrane composition. Diets enriched in polyunsaturated fatty acids induce an increase in the amount of unsaturated fatty acids found in lipids contained in circulating lipoproteins and in cellular membranes [1141].

The second driving force is the control of cellular and whole body **energy balance**. Lipids and particularly fatty acids are highly efficient sources of energy. Large apo B-containing lipoproteins are high-capacity transporters, which deliver neutral lipids to cells sparing the energy they would spend for their de novo synthesis. Large lipoproteins are designed during their assembly in hepatocytes or enterocytes, for the transport of significant amounts of triglycerides together with cholesteryl esters. A potent inducer of apo B-rich lipoprotein secretion is cellular lipid accessibility during their assembly. Defective carnitine-mediated lipid transport to the mitochondrion for degradation results in increased liver VLDL secretion in OCTN2 deficiency [1054]. Interestingly, it was shown that cardiomyocytes (high energetic consumers) are also able to produce apo B-containing lipoproteins to export excessive amounts of lipids, which could otherwise accumulate in their cytosol [211]. Lipolysis of triglycerides circulating in lipoproteins is a fast way to deliver energy to peripheral cells. Contrary to other energetic compounds (carbohydrates, amino acids, medium and short chain fatty acids), which go to the liver through the portal vein after their intestinal absorption, long chain fatty acids and cholesterol are directly transported by chylomicrons into the general circulation allowing immediate uptake by peripheral tissues. Because in the Western lifestyle, the post-prandial period encompasses most of a 24h period, direct energy provisioning to peripheral tissues is an important determinant of lipoprotein trafficking. Interestingly, nuclear factors controlling the expression of candidate genes of lipoprotein metabolism (SREBPs, PPARs, HNFs, AP-1 etc.) have also other gene targets involved in carbohydrate or energetic metabolism. The control of lipoprotein metabolism could be part of a broader system of energetic control having strong impact on aging and life span. For example, it was shown that calorie restriction could prolong life span in mice, worms (*Caenorhabditis elegans*) and in yeast (*Saccharomices cerevisiae*) through mechanisms influencing the genomic control of chromatin silencing, which in turn influences gene expression and genetic recombination [1142]. Interestingly, this basic biological fact meets the classical observation in clinical practice that energetic restriction (e.g. low fat diet) could significantly modify plasma lipoprotein profile in humans.

The third driving force is the generation and the degradation of **active molecules** derived from neutral lipids. Cholesterol and fatty acids contained in triglycerides are important sources of signalling molecules. The classic example is steroid hormones derived from cholesterol. However, many other lipid derivatives like oxysterols or eicosanoids act as strong signalling molecules regulating vital cellular processes. The concept that lipids were signalling molecules and important actors of cellular activity has recently gained more credit with the identification of PPARs, a family of nuclear hormone receptors binding lipid derivatives. Moreover, oxysterols bind liver nuclear hormone receptors (e.g. LXR), which drive the level of bile acids secretion - an important regulatory mechanism of lipoprotein metabolism [1143]. Oxygenated lipid species may also result from active intracellular energetic metabolism (e.g. in mitochondria). A corollary of active lipid species generation is their degradation or elimination. Unesterified cholesterol may become by itself an active molecule, for at least the reason that its molecular backbone resists degradation within cells. Transfer proteins (i.e. CETP or PLTP) may reorient the fate of cholesterol transported by lipoproteins, by accelerating its transport back to the liver where it may be secreted into bile. Oxidized lipids may be formed at the surface of lipoproteins especially in extracellular spaces in the presence of activated macrophages, which secrete reactive oxygen species and other lipid modifying molecules (phospholipases, sphingomyelinase, metalloproteases etc). Sterols may become oxysterols and long chain fatty acids may generate oxygenated species and/or shorter chain fatty acids. For example, the short chain fatty acid butyrate (C4:0) has a strong impact on genomic activity. Butyrate is an inhibitor of histone deacetylases. Through the modulation of chromatin activity, butyrate could modulate genetic expression and cell survival with important implications in cancer and aging [1144,1145]. Oxidized lipoproteins may activate nuclear hormone receptors (e.g. PPARγ) in immuno-competent cells through their binding and endocytosis by specific receptors (scavenger receptors), which recognize negatively charged lipids [472]. On the other hand, lipoprotein susceptibility to oxidation may be buffered by the level of circulating small apolipoproteins. For example the dramatic progression of atherosclerosis in apo E knock out mice was suggested to result from hypercholesterolemia as well as from the loss of antioxidative protection [681]. Thus mechanisms aimed at the control of lipid-derived active molecules may not only influence the production and residence time of lipoproteins, but also their potential to become atherogenic.

In multicellular organisms where a lipoprotein metabolism exists, the coordination of these mechanisms reside within the cellular control of internal metabolic needs, in harmony with neighboring cells and other organs involved in broader physiological systems, themselves in **permanent response to environmental conditions**. These mechanisms may use the same molecular relays with an impact in a single cell together with an impact in the whole organism. For example, ABC1 expression in macrophages and in enterocytes induced by LXR enhances cellular cholesterol efflux towards HDL and cholesterol excretion into the intestinal lumen [976]. In this case reverse cholesterol transport is coordinated at the level of a single cell and at the level of whole body cholesterol homeostasis. Interestingly, preferential connections were established between lipoprotein metabolism and the central nervous system and the immune system, which mediate and coordinate interactions of the whole body with the environment. The understanding of molecular, cellular and physiological mechanisms controlling lipid homeostasis is at its very beginning. Their discovery might give new clues to the molecular basis of multifactorial and complex disease like dyslipidemia and atherosclerosis.

3.2-   THE CANDIDATE GENE APPROACH

With the advent of molecular biology technology, it became evident that genes recently identified could represent novel tools to study and to treat human disease. The sequencing of the human genome is close to completion, however little is known about its significance. Despite the fact that a few thousand human genes are now identified including some recognized as disease-causing genes [1146], the total number of functional genes is still debated, varying between 30 000 and 35 000 [1147-1149]. Thus, at present ways to study the genetic basis of disease could be partitioned into two approaches depending on whether a gene is known or not. If a gene is known to play a role on a specific metabolic pathway, it may be tested for its role as a disease-causing gene, namely defining the candidate gene approach. If a disease-causing gene is unknown, other strategies may be used depending on the assumption made on the genetic mechanisms of the disease, that is if the disease has Mendelian inheritance or not.

3.2.1-   The Candidate Gene Approach in Human Disorders of Lipoprotein Metabolism

Because gene identification was easier in the early eighties for those genes encoding known proteins and preferably proteins which could be purified in significant amounts, the first candidate genes identified were candidate genes for plasma proteins (apolipoproteins, apo(a), LPL) and for the LDL receptor [9,210,357,511,795]. Therefore one of the first fields of human disease to be explored by the candidate gene approach was lipoprotein metabolism. This approach has proven very efficient in the identification of the molecular basis of many inherited metabolic diseases and of novel susceptibility alleles to common polygenic disorders such as diabetes, high blood pressure or certain forms of cancers. The functions and the regulations of candidate genes could be dissected by recombinant DNA technologies in experimental models of human disease. It has also opened the way to novel diagnostic or therapeutic strategies in medical practice.

The candidate gene approach consists in testing whether a known gene encoding for a protein playing a specific biological role, is involved in the genetic predisposition to a human disease. However, the study design and the kind of results expected would vary depending if the disease (i.e. qualitative trait) has Mendelian inheritance or not. In the first case, one may hypothesize that a single gene determines most of the phenotype and that the disease is caused by a mutation on that gene. If the disease or the trait, usually quantitative (e.g. plasma lipid or lipoprotein profile) is non-mendelian (or complex), one can suppose that the disease may be caused by several factors including genetic factors, requiring strategies of genetic epidemiology to identify disease risk alleles and quantify their effects in populations. Therefore, a careful phenotyping and a clear definition of the trait to be studied are a prerequisite to the use of this approach.

3.2.11-   Identifying the Disease Causing Mutation in Inherited Dyslipidemia

Several forms of human dyslipidemia may have the characteristics of metabolic diseases with a Mendelian mode of inheritance. They are usually defined by a set of clinical and biological symptoms present in about half of the relatives when dominantly inherited. They are usually rare, with early onset and often associated with inbreeding in the case of recessive diseases. The identification of the disease causing mutation may require different strategies either direct or indirect. The **direct approach** may be used when one is oriented form the beginning by specific disease characteristics. These may be typical clinical signs of rare or recessive diseases [117,1146]. For example planar (or cutaneous xanthomas) in young children are highly suggestive of homozygous familial hypercholesterolemia due to LDL receptor gene mutations. Large yellow-orange tonsils are evocative of Tangier disease caused by ABC1 mutations. Fish eye-like corneal opacities are evocative of LCAT or apo A-I deficiency. Biological signs may also orient towards a specific disease-causing locus. Enzymatic or protein mass and activity (LCAT, LPL, CETP etc.) may be directly quantified by biochemical techniques in human plasma. For example, a plasma-LPL activity below 20% of normal is highly suggestive of a primary defect caused by mutations on both alleles of the LPL gene [506]. Cholestanol is an intermediate by-product of bile acids biosynthesis. Its plasma accumulation is highly suggestive of a defect of CYP27 causing cerebrotendinous xanthomatosis [1105]. On the other hand, the absence of a small apolipoprotein (A-I, C-II, etc.) or the presence of a truncated apolipoprotein B on isoelectric focusing is highly suggestive of a primary defect on the encoding gene. The gene itself may be directly analyzed in cases of mutations with a high prevalence in certain populations. This is the case for the Arg3500Gln apo B mutation in populations of European ancestry (prevalence 1/250 to 1/1200) or for founder mutations in specific populations (e.g. LDL receptor gene mutations in Afrikaners, in Finns, in French Canadians or in Lebanese, or CETP gene mutations in Japanese). Another advantage offered by direct analysis of genomic DNA is the detection of heterozygous mutations, which might have been harder to identify at the protein level.

However there are many cases, where a monogenic disease may be caused by different genes, justifying an **indirect approach**. A classical example is the case of heterozygous familial hypercholesterolemia and defective apolipoprotein B-100. Both diseases express with the same phenotype of primary type IIa hypercholesterolemia resulting from defective LDL catabolism (*figure 3-1*). Both genes are of average size and display high allelic heterogeneity (i.e. a high number of genetic variations in the coding sequence). Sequencing of the several thousand nucleotides encoding each gene and looking for large gene rearrangements would be tedious and expensive. Moreover nothing ensures that a third morbid locus is not involved in a particular patient. In such cases, genetic linkage analysis may spare a long and uncertain search for a causative mutation.

The concept of **genetic linkage** is based on the positive relationship existing between genetic distance and the frequency of meiotic recombination events [1150]. The closer two loci (e.g. a locus predisposing to a disease and a locus identified by a genetic marker) are physically located on a chromosomal region, the more likely marker alleles at these loci will be transmitted to the offspring both on the same chromosomal fragment (segregation). Marker alleles were historically protein serotypes detected in human plasma. At present, they have been replaced by DNA polymorphisms due to their extreme abundance and their precise ordering on the human genome [1151]. DNA polymorphisms are defined by between-individual variations of a portion of DNA sequence, with a frequency exceeding 1% in populations. These variations may be qualitative (single nucleotide substitution) then called Single Nucleotide Polymorphisms (SNPs). They may be sometimes detected by the modification of a cutting site for a restriction enzyme then called Restriction Fragment Length Polymorphisms (RFLPs). These variations may be quantitative, due to the variable repetition of simple motifs of two to several nucleotides, then called microsatellites or Short Tandem Repeats (STRs). Genetic polymorphisms are precious tools in

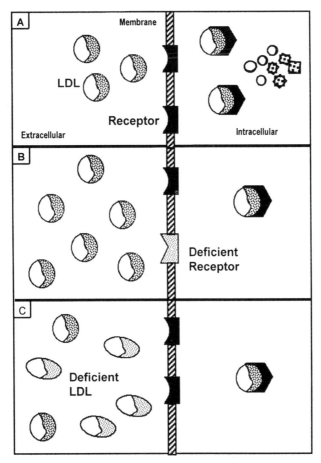

Figure 3-1. **Two loci for a single phenotype.** Circulating LDL are taken-up by cells through their interaction with LDL receptors, followed by their degradation within cells (A). Primary type IIa hypercholesterolemia may be caused by impaired LDL catabolism resulting from defective receptors (B) or from defective ligand (C). In the first case the disease locus (familial hypercholesterolemia) is on chromosome 19 whereas in the second case the disease locus (familial defective apo B-100) is on chromosome 2.

the search for disease causing mutations, because their location is precisely identified on the human genome, some of them being located within genes or in their flanking regions. These markers are used in linkage analysis to follow the segregation of a particular polymorphic allele with the symptoms of a monogenic disease. The strength of the segregation or genetic linkage may be evaluated by the Lod Score method [1152,1153]. This statistical method estimates the probability of genetic linkage between a monogenic disease and a locus identified by genetic markers. The calculation of this probability depends on the degree of knowledge on phenotypic classes (homozygote/heterozygote/unaffected), of their frequency in the population, on trait penetrance and on the relative genetic distances between the markers. Its power increases with the size of the family and on informativeness of markers (i.e. the frequency of the rare allele in the referral population). When the ratio of the probability of linkage over the probability of non-linkage exceeds 1000:1 (Lod Score >+3) the locus may be considered causative of the disease. Thus genetic linkage may indicate precisely, the gene on which the causal mutation should be sought. Genetic linkage analysis proved very powerful to identify or exclude predisposing loci for familial dyslipoproteinemia. Historically, J. Ott and colleagues had shown in 1974 that the locus predisposing to familial hypercholesterolemia was genetically close to loci for complement component C3 and for erythrocyte Luther group [1154]. At that time, the LDL receptor gene was unknown a fortiori it's chromosomal location. Ten years later, these data were confirmed by gene cloning, followed by the assignment on chromosome 19 of all the three genes [1155]. Genetic linkage analysis has shown to be robust in pointing candidate genes of lipoprotein metabolism as disease causing loci. In the case of the LDL receptor gene linkage analysis may lead to a positive diagnosis in 65 to 95% of cases [146]. The drawback of this approach is that it's power depends on the capacity to collect clinical and genetic data in extended families with a significant number of affected and unaffected individuals.

In some instances, Lod Score estimation may conclude to a probability of non-linkage greater than the probability of linkage of a specific locus with a specific disease. **Exclusion of a locus** defined by a negative Lod Score below -2 (100 more chances that the locus is not disease predisposing than that it is) may reorient towards the search of novel loci also predisposing to the same phenotype. For example, abetalipoproteinemia has clinical features very similar to homozygous hypobetalipoproteinemia, however linkage analysis could exclude apolipoprotein B as the disease-causing locus [325]. This finding helped to reorient towards a novel locus for abetalipoproteinemia later identified as encoding for a protein involved in lipoprotein assembly and secretion: MTP. More recently, linkage analysis has identified novel loci for dominantly inherited hypercholesterolemia [1156], hypobetalipoproteinemia [1157] or hypoalphalipoproteinemia [1158].

Once the disease locus is pointed either by a direct or an indirect approach, the genetic mutation may be validated as causative [1159]. Several criteria may provide evidence that a DNA sequence variation is a disease causing mutation: it was found in the affected proband and family members and absent in a significant number of unaffected individuals. In addition, it has biological effects at the protein or gene expression level and modifies an evolutionarily conserved nucleotide or amino-acid residue (i.e. an amino acid residue found identical or structurally similar in other animal species).

3.2.12-   A Redifinition of Inherited Dyslipidemia

The identification of candidate genes of lipoprotein metabolism offered an opportunity to better understand the mechanisms of **natural mutagenesis** resulting in monogenic dyslipidemia. Once a new gene was cloned in a particular species the characterization of its homologue in other species was generally straightforward due to high sequence homology particularly in coding sequences. By computerized sequence comparison, it was possible to identify the most highly conserved domains in candidate genes across the evolution of living species, thereby indicating domains of high functional significance. Because natural mutagenesis is supposed to occur randomly, a mutation profoundly impairing a gene function would have hardly a chance to perpetrate along the time scale of evolution. A corollary is that areas of lower degree of evolutionary conservation may be prone to common genetic variation with moderate functional significance at the protein level. The analysis of natural mutagenesis causing human dyslipidemia has revealed unusual mutagenic mechanisms like uniparental isodisomy for chromosomes 4 and 8 (see page 68). Moreover, the list of naturally occurring mutations may reveal specific regions for a higher risk of mutagenesis. In theory, genetic mutations may be point mutations: missense, nonsense or frameshift mutations in the coding sequence and in intronic, 5'- or 3' non-coding sequences. They may be large gene rearrangements, (e.g. insertions or deletions). However, depending on the gene structure and functions a specific kind of mutation or a specific clustering of mutations may be found. For example, apo B mutations causing hypobetalipoproteinemia described so far are confined to nonsense or frameshift mutations resulting in truncated apo B isoforms (see *table 2-4*, page 62). Missense mutations of the LPL gene are preferably clustered in the hydrophobic pocket surrounding the catalytic site (see *table 2-6* page 90). CpG dinucleotides are hot spots for C→T nucleotidic transition [165]. Exon 4 of the apo E gene contains a CpG island encoding for the receptor binding domain, where the majority of Apo E mutations causing dyslipidemia are described (see *table 2-10*, page 113). Large gene rearrangements have shown to result from unequal crossovers between homologous *Alu* sequences in the LDLR, the LPL gene or in the A1-C3-A4 gene cluster. Therefore, one could predict from the primary gene structure and its functional domains that there are sites for a higher risk of deleterious mutations.

The study of defective mutations in series of affected individuals allowed a **functional characterization** of these genes. In the case of the LDL receptor gene, a map of functional gene domains (ligand binding, endocytosis etc) was drawn from the identification of natural mutants and of their biological expression in cell cultures of skin-fibroblasts from subjects with homozygous familial hypercholesterolemia [108]. Following this pioneer example, functional domains were identified by the natural mutagenesis of genes encoding for enzymes (LPL [506] or LCAT [916]), for apolipoproteins (apo A-I [797] and apo B [273]), or is underway for the recently identified ABC1 gene [70,1138]. The identification of natural mutants has disentangled the structure-function relationships that remained difficult to study at the protein level. For example apo B is a high molecular weight and very hydrophobic protein whose direct study had remained delicate. The existence of truncated mutants circulating in plasma of patients with familial hypobetalipoproteinemia has shown that the post-transcriptional machinery accommodates with the assembly and the secretion of lipoproteins containing abnormal apo B species, truncated in their C-terminal end. In some instances, studies of genotype-phenotype relationships have shown unexpected pathogenic consequences of naturally occurring mutations. In CETP deficiency, despite spontaneously high plasma HDL concentrations (a supposedly protective factor), the risk of cardiovascular disease may increase as a consequence of defective reverse cholesterol transport. On the other hand, Apo A-I Milano, a natural mutant of Apo A-I was found associated with a low HDL/high triglyceride trait and decreased susceptibility to atherosclerosis as the result of increased capacity of HDL to mediate cellular cholesterol efflux [1160]. Novel functions for candidate genes outside of lipoprotein metabolism like in the case of apo E (e.g. neurodegenerative disease), or novel functions in lipoprotein metabolism for candidate genes in other systems (e.g. cubilin a cobalamin transporter or CD36 a thrombospondin receptor) were uncovered. It may be anticipated that other examples of unexpected structure-function relationships will be identified in the future.

In addition to these biological findings a conceptual turmoil was brought about by the identification of natural mutations in several forms of monogenic dyslipidemia. The delineation of **novel disease entities** defined by their molecular basis has dismantled previous clinical classifications. A single disease may be genetically heterogeneous. The most illustrative evidence came again from primary hypercholesterolemia. Familial hypercholesterolemia (FH) and familial defective apolipoprotein B-100 (FDB) appear to have a similar clinical expression of dominantly inherited hypercholesterolemia. However, some differences are discernible. Homozygous forms are very different: severe hypercholesterolemia in LDL receptor deficiency, whereas moderate in FDB. Moreover in heterozygotes plasma LDL cholesterol concentration is lower and more variable in FDB than in FH. Cardiovascular complications would be retarded with a lower prevalence [312]. These differences in clinical expression could also appear in a differential response to lipid-lowering therapy [188]. These differences might reflect the heterogeneity of LDL receptor mutations, some having a more pronounced effect on receptor activity than others [177]. Though, it could also reflect a different pathogenesis. In familial defective apo B-100, the clearance of LDL precursors is preserved through apo E binding to normal LDL receptors (see *figure 1-11* page 38). The production rate of LDL may be more variable and even remain normal, whereas this mechanism is also impaired in LDL receptor deficiency. This metabolic difference was shown to contribute to the contrasted levels of plasma cholesterol observed in homozygotes for FH or FDB [315]. With the increasing number of disease-causing genes identified, diseases themselves will be progressively redefined by their molecular basis. These novel clinical entities may be the basis of novel diagnostic and therapeutic approaches for monogenic dyslipidemia.

Another important finding was that **clinical expression** of a monogenic dyslipidemia, could be determined by the type and the location of genetic mutations. For example in FH, class 1 LDLR mutations (i.e. null alleles) are associated with a more precocious and more severe form of the disease particularly in homozygotes, than mutations preserving protein expression [108]. Depending on their type and location, LCAT or LAL gene mutations may be associated with a milder or more severe form of disease (complete LCAT deficiency versus Fish Eye disease, Wolman disease versus Cholesteryl Ester Storage disease). Depending on the location of ABC1 mutations along the gene sequence, the disease might have a more pronounced expression in the reticulo-endothelial system or in the cardiovascular system [1138]. Moreover, hypoalphalipoproteinemia may be dominantly or recessively inherited [70]. Therefore, monogenic dyslipidemia do not escape from a general rule in molecular genetics, in that the type of defective allele may contribute to the variation in disease expression.

3.2.13- The Evidence of Genetic Interactions

The identification of functional mutations or polymorphisms on candidate genes brought new insights into genetic interactions modulating disease expression. For example, hypobetalipoproteinemia was found to attenuate hypercholesterolemia resulting from a LDLR mutation in a large pedigree [195]. Moreover, through the same kind of compensatory effect, a novel cholesterol-lowering locus was identified in FH families on chromosome 13q [192,194]. Conversely, combined heterozygous familial hypercholesterolemia and FDB displays a more pronounced hypercholesterolemia than either disorder alone however milder than homozygous familial hypercholesterolemia [196,197]. The particular clinical expression of this form of digenic hypercholesterolemia reflects distinct actions of the receptor and its natural ligand Apo B-100 on LDL metabolism. In addition, epigenetic interactions have been reported between mutations causing monogenic dyslipidemia and functional polymorphisms. The E2E2 genotype could favor type III dysbetalipoproteinemia in heterozygotes for LDL receptor deficiency [727]. However, in heterozygous carriers of the E2 allele, lipid-lowering effects were found sex specific in a series of FH subjects with a single LDL receptor gene deletion [198]. In a large cohort of FH families, common LPL mutations were shown to aggravate the lipoprotein profile by lowering HDL cholesterol [199]. The «magnification» effect of defective mutations with strong functional impact reveals gene-gene interactions at the individual level with minor mutations, which would remain otherwise undetectable in normolipidemic individuals.

These interactions provide a first measurable appreciation of expressivity of candidate genes for dyslipoproteinemia. For example, heterozygous LPL gene mutations may remain unsuspected in women, until their first pregnancy, during which massive chylomicronemia and acute pancreatitis may suddenly occur [568]. The knowledge of the genotype gives a measurable appreciation of environmental interactions. Despite the fact that no major changes are observed at baseline, apo A-IV polymorphism is associated with differential responses to cholesterol-lowering diet [875]. A few studies were undertaken on the effects of lipid lowering drugs in the face of particular alleles of candidate genes of lipoprotein metabolism. For example, Apo E4 carriers whom are at higher risk of cardiovascular disease benefit from lipid lowering therapy by recovering a risk level similar to that of non-carriers [1161].

3.2.14-   Identifying Risk Alleles for Complex Diseases in Populations

Although the majority of dyslipidemia appear with an inherited component, only a few are authentically monogenic. Morphometric studies have suggested that heritability of quantitative traits like plasma lipid or lipoprotein parameters generally comply with a model consisting in major genes modulated by other genetic or environmental factors [1162]. On the basis of these observations it is possible to test candidate loci for their effects on the variance of quantitative and/or complex (or multifactorial, or polygenic) lipid traits in populations [111]. Like for other complex traits (hypertension, diabetes, etc.), because the Mendelian model is inapplicable, it is then a question of epidemiological estimation of a disease risk with respect to a genetic locus defined by genetic markers, with no assumption made on the underlying mechanisms [1163]. In human, these strategies are based on the study of DNA polymorphisms, located within candidate loci or in their vicinity. There are two categories of polymorphic markers:

functional and neutral polymorphisms. Functional polymorphisms have a proven biological or pathogenic effect, even if it is mild, because they induce measurable changes in protein structure (e.g. non synonymous SNP), function or level of expression. In contrast, neutral polymorphisms, are defined by the absence of measurable effect on protein structure or expression (synonymous mutations on the third base of a codon, nucleotidic variations in non-coding regions with low evolutionary sequence conservation) or assumed by default. Two main strategies were proposed for the study of complex dyslipoproteinemia: association studies in cohorts of unrelated subjects in search for linkage disequilibrium and studies of related individuals (sib-pairs) in search for a disease-susceptibility locus.

Linkage disequilibrium is a statistical concept defined by a preferential allelic association between two loci, found more frequently in the population than if it would be by random [1150]. It is more precisely defined by the "difference (D) between the frequency of an allelic association between two non-allelic loci (P_{AB}) and the product of allelic frequencies at each locus (P_A and P_B)":

$$D = P_{AB} - P_A \times P_B$$

Therefore this parameter depends on the studied loci and population, which determine the allelic "frequency", and does not assume anything on the biological mechanisms underlying the association [1164]. The study design is generally a case-control study in cohorts of unrelated individuals. When a preferential association is observed between certain polymorphic alleles and a disease in various populations, it suggests by its universality a strong biological effect. These novel disease-risk markers may contribute to define sub-groups of subjects among which, some are carrying authentic Mendelian diseases, and some are carriers of a specific genetic risk. This strategy proved successful in two classical examples: cystic fibrosis and multifactorial diseases linked to the HLA locus. Two anonymous (i.e. neutral) polymorphic markers (KM19 and XV2c) were in strong linkage disequilibrium with cystic fibrosis, an autosomal recessive disease. The strength of linkage disequilibrium had allowed predictive diagnosis even before the disease causing gene was identified [1165]. These markers flanking the locus proved physically distant (> 200 KB) from the CFTR (Cystic Fibrosis Transmembrane Conductance Regulator) gene, a member of the superfamily of ABC transporters. Strong linkage disequilibrium and the high prevalence of the disease in Caucasian populations suggest that an intense selection pressure stabilized this chromosomal area. This phenomenon of genomic stability authenticated by linkage disequilibrium was found in gene clusters of small exchangeable apolipoproteins (A1-C3-A4 and E-C1-C4-C2), suggesting that selective pressure would have influenced the pattern of concerted gene expression, more than the gene structure itself [815,816]. These indirect observations give a glimpse on the functional significance of poorly known genomic regions, which are not directly involved in protein synthesis or tissue-specific expression.

The second classical example was that of HLA associated diseases. Only association studies showing a preferential link between HLA antigens and diseases with an autoimmune component (e.g. type I diabetes, rheumatoid arthritis, multiple sclerosis, psoriasis, hæmochromatosis etc.) proved consistent across populations. Thus linkage disequilibrium with a multifactorial disease may be easier to authenticate when guided by a candidate gene approach. The case of HLA polymorphism is exemplary, because it relates with essential actors of the immune system subjected by definition to strong environmental pressure [1166]. Several of these polymorphisms were shown to be functional, suggesting that in multifactorial disease, functional variants may have greater chances to be directly validated as risk alleles than neutral polymorphisms. This was verified in the case of lipoprotein metabolism.

Functional polymorphisms in candidate genes of lipoprotein metabolism were authenticated as independent risk factors for complex dyslipidemia or other multifactorial diseases. Association studies (case-control design) were efficient to demonstrate a biological effect at the population level, which would remain undetectable in single individuals. Apo E isoforms were shown to contribute to 8-10% of the variance of plasma cholesterol levels in populations. Likewise, apo(a) length polymorphism could determine a significant part of plasma Lp(a) variance, however with some variability between European, Asian and African populations. Risk estimates were performed for late onset or complex diseases as a function apo E polymorphism. Isoform E4 is associated with a relative risk of about 2 for cardiovascular disease. This level of risk is comparable with that of common biological risk factors, like high plasma cholesterol or blood pressure in the general population [729]. Moreover, it was shown that apo E polymorphism has very high positive predictive value (>95%) for the diagnosis of Alzheimer's disease [737]. Other common functional polymorphisms were found through a systematic examination of heterozygous relatives of individuals with an autosomal recessive disease. An interesting example was heterozygous LPL deficiency, which is consistently associated with lower plasma HDL cholesterol and higher triglycerides than non-carrier relatives. Moreover, carriers of common functional LPL gene variants (frequency 2-6%) were associated with a higher risk of atherosclerosis in Caucasians [581]. Novel susceptibility alleles for common plasma lipid traits were found in heterozygous carriers of diseases distinct from dyslipidemia. For example, Gaucher's disease is a lysosomal recessive disease caused by mutations in the β-glucocerebrosidase gene. Heterozygous carriers had dominant hypoalphalipoproteinemia, the locus accounting for about 19% of plasma HDL variance [1167]. Because mutated alleles may be highly prevalent in Askenazi Jews (1 in 13.5) and in mixed populations (1 in 83), mutations at this locus could predispose to hypoalphalipoproteinemia in up to 2% of the general population, which is remarkably high for a single genetic locus. Many other functional polymorphisms were reported in exons (e.g. CD36, Apo A-IV, PON, PPARα) and regulatory regions (e.g. HL, MTP, apo A-I and C-III, CETP), of candidate genes for lipoprotein metabolism in association with various lipid traits or atherosclerosis. They await confirmation by the analysis of large human cohorts worldwide to quantify their actual effect on the variance of plasma lipoproteins and on their associated risk for common dyslipidemia. For example, it took the analysis of several thousands of subjects to authenticate the Pro12Ala variant of PPARγ as a risk factor for type II diabetes [1079].

Studies aimed at finding preferential associations between common multifactorial diseases and **neutral polymorphisms** may be more delicate to interpret. The high abundance of neutral SNPs in the genome (about 1 every 500 bp) is also their weak point; low level of evolutionary sequence conservation and high allele frequency being a sign of low level of functionality [541]. The power of these studies depends on the good adequacy between characteristics of genetic markers (genomic location, rare allele frequency) and the selection of subjects in the study sample (size, phenotypic selection criteria, etc.). For example, a neutral TaqI RFLPs in intron 1 of the CETP gene was associated with lower plasma HDL and higher progression of atherosclerosis on angiography over a two-year period [937]. In this case, the choice of a dynamic phenotype precisely measurable could show the effect of the CETP locus. However, these results do not allow drawing any pathophysiological conclusion, the genetic marker reflecting only linkage disequilibrium with some functional sequence somewhere in the genomic region containing CETP. A well-recognized weak point of association studies is the unsuspected stratification in the study population leading to artefactual associations or on the contrary masking real associations [1163]. Geographical, cultural, or genetic factors (migrations, genetic drift, preferential mating, consanguinity, etc.) within the population may have favored the prevalence of certain alleles in sub-groups in a way completely independent from that of the disease. Moreover, the required size of study sample to detect linkage disequilibrium may vary depending on the frequency of the rare allele, and on the intensity of the expected effect of the marker locus. Altogether, this may explain discrepancies often reported from one association study to another between neutral markers at candidate loci of lipid metabolism and common forms of dyslipidemia. The study of a short stretch (9.7kb) of the LPL genomic sequence, showed that neutral SNPs were abundant and that there were recombination hot-spots. On the basis of these findings it was suggested that a careful analysis of the structure of the candidate genomic region in terms of linkage disequilibrium, would be very useful before starting association studies at a candidate gene locus with combinations of genetic polymorphisms (i.e. haplotypes, see page 207) in a particular population [542].

A way to increase study robustness is to analyze related individuals. In between genetic linkage analysis in large families, which is appropriate for Mendelian disorders and association studies in unrelated individuals, is **sib-pair linkage analysis**, which combines the study of related individuals (parent-offspring or siblings) in large series of unrelated nuclear families. The advantage of this kind of study design is that it decreases the effects of population stratification between affected and non-affected individuals, because of a greater number of shared genetic and environmental determinants. The frequency of alleles at a candidate locus shared by phenotypically concordant or discordant sib-pairs is compared with the expected frequency of these alleles in the general population. If there is an excess of shared alleles in concordant pairs compared with random allele distribution, then this locus may contribute to some extent to the variance of this trait. In this case no particular assumption is made on any mechanism underlying the trait, but it may point to a genetic locus, which could contribute to disease susceptibility. This strategy was successful in showing that the apo(a) locus contributes to 90% of the variance of the circulating Lp(a) [389]. Plasma levels of HDL in normolipidemic subjects, were found to be linked with variations on the AI-CIII-AIV gene cluster and on the hepatic lipase gene [610], whereas in relatives of coronary heart diseased patients, apo A-II and CETP loci were found to determine plasma levels of HDL cholesterol [901]. Sib pair linkage analysis could identify the CYP7A1 gene involved in bile acid synthesis, as a modifier locus for fasting-plasma LDL cholesterol in the general population [1098].

3.2.15-   Genetic   Determinants   of   Dyslipidemia   in   Populations

Although cardiovascular diseases are observed in every part of the world, their incidence and their prevalence may differ from one country to another. For example, a North-South decreasing gradient of cardiovascular disease prevalence was reported in Europe [99]. In comparative population studies, the risk was correlated with plasma level of cholesterol and the level of saturated fat in the diet. However differences in the distribution of various alleles of candidate genes for atherosclerosis could also contribute to these differences [1168].

Different **patterns of monogenic disease distribution** were reported in certain populations. Certain human populations, for cultural or geographical reasons, underwent few admixture during their evolution, favoring the concentration of specific mutations and a higher prevalence of certain monogenic diseases. Here, the same mutation may be identified in numerous apparently unrelated subjects. Segregation analysis of polymorphic markers flanking that mutation allows the construction of haplotypes (i.e. a combination of alleles of polymorphic markers located in Cis on the same chromosome). Thus haplotyping provides with a kind of identification "bar-code" of the chromosomal region or gene of interest. If the haplotype in Cis of the disease causing mutation is identical in two apparently unrelated individuals, one may infer that both subjects have a common ancestor (founder effect). If it is different, one may deduce that two independent mutational events occurred at different time points (recurrent mutation) in different populations. Familial hypercholesterolemia has a high prevalence in the Canadian province of Quebec: 1/200 to 1/100. A founder effect was reported for 5 mutations in the LDL receptor gene, among which one is a large gene deletion frequently found in subjects living in the region of Saguenay-Lac-Saint-Jean. This deletion was found in a French subject from the Nantes region in France, an area left by the first settlers in Quebec in the 17th and 18th centuries [1169]. One can thus date this mutation back to at least three centuries ago. By tracing FH families in Italy, it was also possible to date a large LDLR gene rearrangement back to the 17th century [176]. A founder effect for several LDL receptor mutations was reported in Finland, Druses and Ashkenazi Jews in Israel, in Afrikaners from South Africa, and in Lebanon [154]. A common ancestry was also reported for three LPL gene mutations in Quebec; for the Arg3500Gln mutation of apo B-100 in Western Europe; for several CETP gene mutations in Japan and for 27-hydroxylase mutations in Sephardic Jews. Conversely several mutations are very rare or absent form certain populations: the Arg3500Gln mutation of apo B-100 was neither detected in Finland, nor in Japan, nor in Israel, whereas it accounts for 2 to 3% of type IIa hypercholesterolemia in other countries.

Monogenic diseases resulting form mutations of candidate genes of lipoprotein metabolism are remarkably frequent, in comparison with other monogenic diseases. About 1/500 individuals may carry a defective LDL receptor or LPL gene mutation in mixed populations of European ancestry. In Japan CETP deficiency is found in 1/200 subjects. By comparison the frequency of cystic fibrosis is 1/2000 (recessive disease), that of type 1 Neurofibromatosis is 1/3000 (dominant disease), that of Duchenne myopathy 1/3500 of male births (X related disease) [1170]. Thus several millions of subjects worldwide could carry a defective allele for one of the major candidate genes of lipoprotein metabolism. Most of the deleterious effects of mutations appear in adults and do not alter reproduction fitness. Up to recent time they would not limit survival before the average life expectancy, which was about three decades lower than at present in developed countries. In addition with molecular heterogeneity of mutations (e.g. several hundreds of defective mutations at the LDLR locus), cases of recurrent mutations in the LDL receptor or the LPL genes suggest that mutagenesis may naturally drift at these loci [145,1171]. Neo-mutations were anecdotally documented [565,832,1172]. Therefore a mild selection pressure might influence mutagenesis at candidate loci for human dyslipidemia.

Genetic polymorphisms also have a variable allelic distribution depending on the referral population. The apo E4 allele is grossly under-represented in Far-Eastern populations and in Southern Europe, in comparison with Scandinavian populations (particularly Finland) in which it is over-represented compared with the world average [700]. Differences in the distribution of apo(a) or LPL alleles were also reported between European, Asian or African populations [385,582]. Isoform E2 is associated with hypocholesterolemia and longevity in normolipidemic subjects, whereas it predisposes to type III hyperlipidemia in hyperlipidemic subjects. Isoform E4 was found a worsening factor for atherosclerosis and Alzheimer's disease however protective against age related macular degeneration. This could represent an example of balanced selection in lipoprotein metabolism, favoring either phenotype, according to the predisposing allele. This type of selection was reported in Drosophila [1173] and in the HLA gene cluster. Type I diabetes is associated with DR3 and DR4 alleles, whereas multiple sclerosis or narcolepsy are associated with the DR2 allele [1166].

The frequency of functional variants of lipoprotein metabolism raises an interesting issue. Hypocholesterolemia contemporary of inflammatory and infectious conditions and the anti-toxic role of lipoproteins were noticed since the last century. In infectious conditions, it is not unexpected that lipoproteins transport "fuels" required for cell renewal and activity. They also behave as anti-toxic agents, collect bacterial endotoxins and direct them towards the liver, shunting macrophages, which are prone to secrete cytokines activating cellular death. We saw that small apolipoproteins themselves had anti-oxydative and anti-inflammatory functions. Components of the extracellular matrix or multifunctional receptors intervening against infectious agents are also involved in lipoprotein metabolism. A comparative study between populations showed that an apo B haplotype was more frequent in populations from Northern Europe than in Africa, suggesting that it would have expanded by selective advantage in temperate climate [56]. Thus it may be hypothesized that mutations modifying the average lipoprotein profile towards moderate hypercholesterolemia would have represented a selective advantage, which would have contributed to resist infections and food shortages during past centuries. In sedentary populations in which infectious diseases are fought effectively from childhood on, these alleles in the context of average high saturated fat diet would become unfavorable and predispose to cardiovascular disease. This assumption was put forth in the case of the HLA multigenic cluster, for which multifactorial pathologies such as diabetes could have emerged, by selective advantage against infectious diseases [1166].

3.2.2- The Candidate Gene Approach In Experimental Models of Dyslipidemia

3.2.21   In Vitro Studies and Cell Cultures

Cell cultures are a powerful tool for physiological and biochemical analysis of cells isolated in a controlled environment. Primary cell-cultures from diseased patients or spontaneously mutant cell lines were invaluable tools to identify defective proteins or decipher metabolic pathways. Successful examples were discoveries of the LDL receptor, MTP, NPC1 or ABC1 genes by this approach. Primary hepatocytes in cell culture were used in the first ex-vivo gene therapy experiments, for the treatment of homozygous familial hypercholesterolemia (see below). In addition, the expression of mutations produced by site directed mutagenesis in permanent cell-lines might provide evidence of a functional defect resulting from a newly described mutation. This approach was used to map functional domains of candidate genes and to identify specific regulatory motifs in their promoters. However, lack of detectable protein defect *in vitro* does not necessarily indicate lack of functionality *in vivo*. For example, the common Ser447Stop LPL mutation was long considered neutral, and reported with higher levels of expression and normal LPL activity in vitro [581]. However, it is consistently found with higher levels of plasma HDL and lower TG in different populations with no clear functional explanation at present [582,590]. Moreover, cell cultures are invaluable tools to explore numerous intracellular mechanisms controlling lipoprotein metabolism: interactions at the cell surface (cell-cell, cell-lipoproteins, lipoproteins-extracellular matrix etc.), cellular signalling, lipid transport and trafficking, nuclear regulations etc. The development of permanent cell-lines may also prove useful. For example, a cell line expressing uncoupling proteins in brown fat tissue was obtained by causing tumors (hibernoma) in transgenic mice overexpressing the simian virus SV40 downstream of a promoter with adipocyte-specific expression [1174]. Cell-lines may orient towards novel therapeutic targets, by the identification of differential patterns of expression under particular physiological or pharmacological conditions. Gene expression arrays are now available in the form of several thousands of small DNA fragments bound to a small solid support (glass, silicon, nylon etc) or "DNA-chips" to which mRNAs or cDNAs extracted from a cell may be hybridized [1175]. Thus a change in throughput is operating from single to several gene transcripts analyzed by Northern blotting up to several thousands of them at a time. With the progress of mass spectrometry this may even be translated into the analysis of a set of thousands of proteins (proteomics). A corollary will be the necessity to develop appropriate systems to analyze such overwhelming and complex data sets.

3.2.22-   In Vivo Studies and Animal Models

Animal strains with natural mutations are interesting models of human disease: apart from the classical Watanabe rabbit with a 12 nucleotide deletion in the LDL receptor gene [58], cat or mink strains were identified with point mutations in the LPL gene [61,62]. There are natural mouse strains for NPC1 [1012], ACAT1 [1025], or SCD1 [1048]. These models have proven useful to study physiological regulations at the whole body level and may be invaluable for future trials in gene therapy.

However, as soon as recombinant DNA technologies became common, **genetically modified animals** (e.g. transgenic and knock-out mice) were generated. By caricaturing the expression of a transgene or on the contrary by completely abolishing its expression by homologous recombination, it is possible to individualize the effects of a gene, in vivo. At present, as soon as a novel gene is identified as a possible candidate in lipoprotein metabolism, genetically modified animals are generated, so that there is now at least one model of over-expression and one model of invalidation for every candidate gene. These models have been invaluable to unravel subtle gene effects in a particular metabolic context. This was the case for small exchangeable apolipoproteins like apo A-II or the family of apo Cs [646]. Physiological conditions under which genes contribute to resistance or on the contrary to susceptibility to atherosclerosis may be monitored in vivo. For instance apo E knock out mice, worsen their hypercholesterolemia and develop a severe atherosclerosis under high fat-high cholesterol diet [676]. Conversely, transgenic mice overexpressing apo E spontaneously resist to dietary induced hypercholesterolemia, as the result of increased catabolism of large lipoproteins [775]. Crossing experiments allows the study of genetic interactions in vivo. ACAT1 knock-out mice appear apparently healthy, unless they are crossed with hypercholesterolemic mice (LDLR or apo E deficient), then developing severe tissue cholesterol deposition [1025]. Moreover, the genetic background of the invalidated gene may not be favorable to provide evidence for an effect on atherosclerosis. For example, C57BL/6 mice are susceptible to atherosclerosis, whereas strains currently used for inducing homologous recombination (SJ129) in gene knock out experiments are not. Therefore additional back-cross experiments with C57BL/6 mice may be needed to yield strains with the modified gene in a predisposing genetic background.

In spite of their success, these models may give a distorted vision of natural physiological mechanism. Observations made on genetically modified animals are not always transposable to human for several reasons. The in vivo effects of a transgene depend on the way it was constructed. For example, human apo A-I is not expressed in the intestine of transgenic mice if the transgene does not comprise sequences located downstream of the apo C-III gene [813]. The same type of observation was made for the hepatic expression of apo E and apo C-I [662]. The interpretation of the metabolic effects caused by a human transgene may be confused by persisting endogenous expression of the murine gene, like was initially observed in human apo A-II transgenic mice [884].

The speciation of gene products may be a handicap. For example human apo (a) expressed in transgenic mice, is unable to bind covalently murine apo B-100 to form Lp(a), whereas the infusion of human LDL or the crossing with human apo B transgenics, results in the formation of Lp(a) [122,948]. In transgenic mice human apo B is poorly edited to form human apo B-48, whereas murine apo B-48 is normally produced [923,948]. Because the site and number of copies integrated into the mouse genome is unpredictable, the level of transgene expression may be variable together with the expression of other genes in the vicinity of the integration site. Moreover, the degree of methylation and thus the level of expression of certain transgenes may vary according to the genomic region in which it is integrated. In knock out mice, these difficulties are circumvented by the locus-specificity of homologous recombination. When viable mice are impossible to obtain because gene invalidation is embryonic lethal, conditional knock out restricted to gene invalidation after birth or in a tissue-specific manner may circumvent this difficulty.

These examples emphasize the limits of murine models in lipoprotein metabolism. In this species cholesterol transported mainly in HDL with a monodisperse distribution, low endogenous CETP activity and apo B editing in the liver; all features which profoundly distinguishes murine lipoprotein metabolism from that of human. Genetically modified rabbits were generated in order to study the expression of candidate genes in animal models, which transport cholesterol in LDL and are more susceptible to atherosclerosis [793]. Despite these limitations genetically modified animals are promising models for future physiological studies and therapeutic developments.

3.2.3- Clinical Issues of the Candidate Gene Approach

3.2.31- Applications in Current Patient Care

Possibilities of **molecular diagnosis** of human dyslipoproteinemia have immediate practical consequences. In the case of monogenic diseases, molecular diagnosis provides certainty on the cause of the disease. Not only, will it give a precise mechanism to explain the disease but also it dates and therefore quantifies precisely the intensity of the morbid risk in a particular individual. For example, that a heterozygous LDL receptor gene mutation is identified in an individual aged 40 allows quantifying its risk exposure to 1.5 to 2 folds the normal level of LDL cholesterol during 40 years, if remained untreated. Moreover, knowing the pathogenesis prompts to target lipid-lowering therapy towards LDL metabolism. There may be some added benefit to know the molecular defect, and more particularly if there is a residual expression of the mutated gene (defective allele) or not (null allele). For example in FH, defective alleles may have a milder expression than null alleles. In familial chylomicronemia caused by LPL gene mutations, a residual expression and preserved enzymatic mass may be more atherogenic if present than if not. In the case of a dominant disease, half of the family members may carry the same disease risk. Therefore, screening and therapy should be extended to high-risk relatives. Specific preventive measures may be taken as a function of age and other risk factors. Indeed, like in common forms of dyslipidemia cardiovascular risk factors (smoking, high blood pressure, diabetes etc.) are multiplicative in FH, which has particular significance in the face of this already high-risk disorder. Moreover, because monogenic lipid disorders are prevalent in all populations specific screening strategies need to be defined. Specific and adapted therapeutic strategies may then be evaluated by appropriately designed trials to these well-defined and recognized high-risk individuals.

A second issue provided by molecular genotyping, is the individualization of novel **genetic risk factors** (e.g. apo E or LPL gene polymorphisms), which do not cause the disease by themselves but contribute to its expression and severity. For instance, the screening of CETP deficient individuals could be used as a tool for assessing individuals with a higher cardiovascular risk and a paradoxically average to high plasma HDL cholesterol [926]. In these cases, like for other risk factors, genetic risk factors should be validated in large human cohorts due to their generally modest contribution to individual plasma lipoprotein variance (a few percent). Again, because they point to a particular pathogenesis, tailored preventive measures could be designed. For example, because it was shown that partial LPL deficiency delays the post-prandial response to oral fat load [581], appropriate dietary intervention aimed at lowering dietary fat might prove more useful in these individuals.

In cases of **severe and/or incurable disorders** such as homozygous LDL receptor deficiency, the need for genotyping may be a prerequisite to gene therapy. First experiments of LDL receptor gene transfer in hepatocytes were attempted in familial hypercholesterolemia. Moreover, prenatal diagnosis and the identification of heterozygous carriers have been proposed in severely malformative diseases (e.g. Smith Lemli Opitz).

3.2.32-   Corrective Gene Therapy

From the start, the potential for the use of recombinant DNA technologies was considered in the correction of inherited diseases remaining beyond any curative treatment. The development of such therapies shortly raised novel ethical and technological issues. In regard to numerous unknown factors on the function of the human genome and on risks inherent to the handling of hereditary material, gene therapy was confined to genetic modification of somatic cells in humans (as opposed to germ-line cells).

Severe and rare forms of dyslipoproteinemia, whose deficient genes are primarily expressed in the liver, could benefit from this type of therapy. Among them, homozygous familial hypercholesterolemia resulting from complete LDL receptor deficiency was found an interesting candidate. Although LDL receptors are ubiquitous, hepatic receptors ensure most of LDL daily clearance. Pioneer experiments of gene therapy were initially applied to Watanabe rabbits [1176], then to baboons [1177] and finally to a young adult carrying one French-Canadian missense mutation [1178]. Hepatocytes (1/10 of the liver mass) removed surgically were put in culture and transfected by a recombinant retrovirus containing the cDNA for the human LDL receptor. After transformation, cells were injected through the portal vein to disseminate into hepatic lobules. The genetic correction restored plasma LDL levels down to those of heterozygotes, remaining sensitive to lipid lowering drugs for six months. Preliminary experiments using alternate genes (VLDLR) or chimeric proteins (LDLR ligand binding domain for LDL binding + transferrin receptor binding domain for liver endocytosis) were attempted in non-human models [779,1179].

However promising were these first experiments, several problems remain unsolved and represent major obstacles to the diffusion of these approaches in practice. There are still incompatibilities between efficiency, durability and safety of gene therapy with existing techniques. Currently, techniques based on the manipulation of recombinant viruses offer the best yield in gene correction. The ex-vivo handling of human cells has drawbacks. Beside the invasive character and moderately dilapidating surgical sampling, cell-culture may induce phenotypic changes, potentially deleterious in the long run, after cellular reimplantation in the target organ. Strategies of in vivo gene therapy using pseudovirus may offer alternative to in vitro transfections on cells in primary culture. Because of their liver tropism, modified adenoviruses containing the target sequence downstream of the natural promoter were used to target the liver. Here, the effect did not last more than several weeks, not to mention the major drawback of strong immune reaction associated with adenoviral gene therapy [1180]. Retroviral vectors were shown most effective in their cellular transfection and genomic integration capacities. However, as opposed to the adeno-associated virus, which is integrated in a preferential region of the short arm of chromosome 19, retrovirus genomic integration may be at random. Moreover they best transfect replicating cells, in which cell-cycle mechanisms are activated. There is a risk of activation of oncogenic sequences especially when the inactivated virus has the opportunity to recombine with a wild-type viral strain. Besides, a risk of inactivation of physiologically significant sequences persists, as the result of random integration into the human genome. This integration may not be stable on the long run, erasing the effects of corrective gene therapy. Finally retroviruses only accept short DNA fragments, which may limit their use to cDNAs.

An alternative would be homologous recombination or single nucleotide repair correction, which target the deficient gene or mutation itself [1181,1182]. These approaches would have the advantage of leaving the corrected gene under the control of natural regulatory sequences, and to correct the right part of the gene, thus preventing from adverse effects of viral vectors. Unfortunately, these approaches have presently a very low yield. Moreover, they do not prevent from any undesirable modification of homologous sequences when the target sequence belongs to a gene family.

Many problems currently remain unsolved before considering a diffusion of gene therapy for the treatment of severe inherited dyslipoproteinemia. Safety and efficiency of methods, may improve in pace with increased knowledge on human genome and on biology of the cell. The development of less invasive methods, using non-viral vectors allowing a single, safe and lasting correction of the gene in "orthotopic" position, will provide full rise to these treatments of a new kind: biological therapies, early initiated by vaccination, blood transfusion or organ grafting.

3.2.4-   Present   Limitations   to   the   Candidate Gene Approach

The identification of candidate genes was a major breakthrough providing unprecedented molecular explanations to those clinically silent and heterogeneous metabolic disturbances that characterize dyslipidemia. Physiological concepts underlying the organization of lipoprotein metabolism and their connections with atherosclerosis were refined and sometimes questioned. They may definitely modify the way these disorders are managed in common practice. Beyond the performances of the candidate gene approach, difficulties have to be faced. Genetic heterogeneity of dyslipidemia would require family-based explorations, a strategy difficult to transfer in patient-based approach characterizing present medical practice in Western societies. Moreover, most disease-causing mutations have high allelic heterogeneity. Several hundreds of mutations may alter LDLR gene function [145,155]. Methodologies for a more rapid and efficient genotyping of extended gene sequences are rapidly evolving although remaining costly. Additional experiments in animals or humans will be needed to refine the relationships between genotypic changes and observed disease expression or drug responses. Moreover, numerous intracellular mechanisms regulating lipoprotein metabolism remain poorly understood, suggesting that many more candidate genes need to be identified. Finally there are common lipoprotein disorders like familial combined hyperlipidemia for which the genetic and molecular bases remain unknown.

3.3- NOVEL GENES, NOVEL APPROACHES

So far, the candidate gene approach to human dyslipidemia has been efficient in elucidating the molecular basis of common forms of primary hypercholesterolemia, of certain unusual forms of hypertriglyceridemia and is underway for disorders of reverse cholesterol transport. However, a vast area of lipoprotein metabolism remaining to be understood is that of combined hyperlipidemia. Combined hyperlipidemia represents an heterogeneous group of disorders, among which some are monogenic, however with a variable level of expression, highly sensitive to environmental factors including diet, hormones or age, and to their association with other polygenic conditions such as diabetes, hypertension or obesity. They challenge efforts for novel gene identification and for the definition of novel strategies to identify their molecular basis.

3.3.1- Identifying Novel Genes of Lipoprotein Metabolism

3.3.11- Positional Cloning of Novel Genes

In some instances, when a human disease is clearly monogenic, positional cloning of unknown genes has been most powerful to identify their molecular basis. Positional cloning is based on the "one gene-one disease" hypothesis. With the densification of the human genome map, large sets of highly polymorphic markers spanning the whole genome (genome wide scan) were available for linkage analysis (see page 196). In families with monogenic dyslipidemia, strong linkage (Lod Score >+3) may pinpoint a candidate genomic region in which the novel gene may be found. The subsequent analysis of the corresponding set of cloned chromosomal fragments (YAC clones: Yeast Artificial Chromosomes, or BAC clones: Bacterial Artificial Chromosomes) and sets of ESTs (Expressed Sequence Tags) may identify putative regions encoding for a protein. A genetic mutation causing the disease on one of these gene/protein(s) will eventually validate this novel locus as disease causing. Positional cloning was used to identify the previously unknown disease causing-loci for NPC1 [1010] or Tangier Disease [966] disclosing novel protein or metabolic regulations. Sitosterolemia, a rare atherogenic lipoprotein disorder, has been found to result from mutations of ABCG5, a member of ABC transporters involved in intestinal cholesterol absorption [1183]. Autosomal recessive hypercholesterolemia (ARH), a rare form of type IIa hypercholesterolemia named also "pseudo-homozygous familial hypercholesterolemia", has been found to result from mutations of an adaptor protein possibly involved in LDL receptor endocytosis in the liver [1184]. Other disease-causing genes may be soon identified for rare recessively inherited diseases. Positional cloning was also used to indicate putative genomic regions involved in more common lipid disorders like dominant hypoalphalipoproteinemia [1158], dominantly inherited hypercholesterolemia [1156] or familial combined hyperlipidemia. In the latter case, genome wide scan has identified several loci, which could contribute to various components of the disease: high plasma TG (10p11.2) or apo B (21q21). Moreover, depending on the sample population, linkage was either found on chromosome 11p in Dutch families [1185] or on chromosome 1q21 in Finnish families [1186]. Interestingly, the locus at chromosome 1q21-q23 found to predispose to the disease in the Finnish isolate, is syntenic with a locus *Hyplip1* predisposing to combined hyperlipidemia on chromosome 3 in mice [1187]. This indicates that a novel gene with a function conserved in mice and human controls VLDL metabolism, which disturbances are a hallmark of familial combined hyperlipidemia.

This example underscores the usefulness of positional cloning in animal models of human dyslipidemia to dissect the molecular components of complex or genetically heterogeneous disease. Murine species and specially mice is an excellent model for genetic studies. These animals are small, reproduce easily, have many offsprings and their generations expand shortly [119]. Many polymorphic markers are localized on mouse genome, which is the best-mapped mammalian genome after that of human [1188]. Moreover regions of synteny with the human genome are known, so that it is possible, when a gene is localized in mouse, to look for its counterpart in human [1189]. The "*jvs*", "*asebia*" and "*ald*" strains of mice were classical models of lipid disorders later found to be carriers of OCTN2, SCD1 and ACAT1 mutations [1053,1048,1030]. Recently, the "*fld*" mouse, which develops lipodystrophy, fatty liver and high plasma VLDL a condition reminiscent of human hypertriglyceridemia, was found to bear mutations of a novel nuclear protein "lipin" highly expressed in adipose tissue [1190]. There are several mouse models of human dyslipidemia for which the disease-causing gene remains to be identified [1191]. However, a drawback of "pure" strains is that linkage analysis may be difficult to perform due to high within-strain genetic homogeneity. Recombinant inbred strains (RIB) of mice were used to overcome this problem. Genetic admixture offers the possibility to trace the phenotype together with polymorphic genetic markers in offspring and to localize a disease susceptibility locus. Moreover, this method offers the possibility to map complex traits or modifier genes in multifactorial diseases, and to identify environmental factors (diet, age, exercise etc), which contribute to gene expression.

A limitation to mouse models is that lipoprotein metabolism in mouse rather differs from that in human and that natural strains of mice are spontaneously resistant to dietary-induced atherosclerosis. Other animal models with spontaneous forms of metabolic disorders have been explored. By studying the spontaneously hypertensive and insulino-resistant SHR rat, a predisposing locus neighboring CD36 could be identified. A line of rabbits the "St Thomas rabbit" is a model of familial combined hyperlipidemia in which turnover studies have confirmed a primary defect in VLDL production [119]. Primates have a lipoprotein metabolism closest to that of humans. In these species, particularly Old World monkeys, the central nervous system and the immune system have great impact on basic metabolism in general and more particularly on lipoprotein metabolism [119]. These species have proven useful for physiological or gene-diet interactions studies. Along with mapping and sequencing of the genome in other living species, it will be possible to identify novel loci for phenotypes more closely resembling that found in humans.

3.3.12-   Expression Cloning of Novel Genes

Strategies based on differential expression in cell lines or in genetically modified animals were used to identify novel genes of lipoprotein metabolism. For example SR-B1 was identified using cDNAs obtained from a mutant CHO (Chinese Ovary) cell line deficient for native LDL endocytosis however with preserved capacities to take up modified LDL [983]. When transfected into fibroblast cell lines naturally deficient for modified LDL endocytosis, several cDNAs including that encoding for SR-B1 conferred these cell lines the ability to take up modified LDL. Another approach uses differential patterns of expression on DNA micro-arrays between cells from diseased or non-diseased individuals. For example, fibroblasts from Tangier diseased patients exhibited different patterns of expression as compared with normal fibroblasts [965]. This helped to point ABC1 as the Tangier disease-causing gene. In the same line, adipose cells from genetically hypertensive and insulino-resistant SHR rats displayed more than 90% reduction in the expression of a clone previously identified, to be encoded by a genomic region on rat chromosome 4 [491]. Because this region was previously shown a candidate for the disease locus in the SHR rat, CD36 was investigated as a gene involved in insulin resistance in the SHR rat. Gene targeting may be used as an alternate strategy to identify novel genes regulating lipoprotein metabolism. Physically, or chemically induced mutagenesis may generate novel lipoprotein disorders in mice which may be further mapped and characterized by positional cloning. For example, genetic targeting of an extended genomic region in mouse has uncovered an abnormal lipid phenotype, pointing OCTN2 as regulator of VLDL synthesis and secretion in the liver [1054].

3.3.13-   In Silico Cloning of Novel Genes

Most of the sequence of the human genome is now accessible through public databases [1192]. This is also the case for complete genomic sequences for other living species like the baker's yeast (*Saccharomyces Cerevisiae*) [1193], or for the microscopic worm (*Caenorhabditis Elegans*) [1194], or close to complete sequencing for the mouse. At present a large amount of these sequences comprise regions with putatively expressed sequences, also called ESTs (Expressed Sequence Tags). Novel genes may be found among ESTs, because they could be genes encoding for expressed and functional proteins. By computing homologies between human ESTs and genes from other living species it may be possible to identify novel genes. For example, the human family of ACAT genes was identified through their sequence homology with the "ARE" family of genes, which were found important for sterol esterification in yeast [1025].

Despite being apparently distinct, a combination of positional, expression or homology cloning strategies was used in many instances in the quest for novel gene identification. With the expansion of genomic resources and increased throughput of molecular and computing technologies, it may be anticipated that many more examples of newly identified genes of lipoprotein metabolism are expected in the near future using these combined strategies.

3.3.2-   Novel Approaches to Identify Novel Mechanisms of Human Dyslipidemia

Despite the significant accomplishments of the past decades in uncovering the molecular basis of dyslipidemia and their connections with atherosclerosis and other human diseases much remains to be explored. If diseases with relatively defined and constant phenotypes under "baseline" conditions have now a molecular basis, those variable phenotypes like combined hyperlipidemia, strongly dependent on environmental conditions (e.g. diet, physical exercise, microbial pressure, climate etc.) or those evolving over time like atherosclerosis remain to be investigated. This may require novel phenotypic definitions of the disease, either by developing novel tools to approach molecular processes closer to gene expression, by exploring dynamic phenotypes (evolving over time) or inducible phenotypes (by drug, diet etc.). Because human populations have evolved in divergent environments, several alleles might have been selected overtime, which may open to novel gene identification. The exploration of various forms of dyslipidemia and their genetic components worldwide is therefore needed to refine the understanding of the genetics of human dyslipidemia. Moreover, novel models to define a phenotype departing from the classical additive set of features determined by a single mechanism, could open the poorly understood field of polygenic and multifactorial diseases.

The understanding of the cellular architecture and the compartmentation of metabolic pathways, which strongly impact on lipoprotein trafficking, are at their very beginning. For example, ABC1 is not only an important factor of reverse cholesterol transport but plays a crucial role in the structure of caveolae thereby modulating other signalling or metabolic pathways within cells. Microsomal and cytosolic modifications of proteins need also a better understanding. Mammals, which "edit" apo B-48 in the liver redistribute cholesterol transport towards HDL particles, whereas when apo B is exclusively edited in the intestine, this transport is ensured by LDL [233]. Apo B degradation in the proteasome is an important regulatory mechanism of lipoprotein secretion, so that any disturbances in proteasomal activity may profoundly impact on lipoprotein secretion.

In addition, most remains to be understood of the human genome. More than 90% of genomic sequences do not have the features of a sequence encoding for a protein (i.e. a gene) [1192]. Little is known about their function. Recently, it was shown by comparative sequence analysis that evolutionary conserved anonymous regions of the human genome, located remotely from gene sequences could have a regulatory influence and could coordinate the expression of a set of complementary genes. Such conserved noncoding sequences (CNS) were identified in the intergenic region between interleukin 4 and 13, in the interleukin gene cluster on human chromosome 5 [1195]. More than acting as classical transcriptional regulatory elements, these elements would act as modulators of chromatin structure towards the coordinated expression of particular genes in a genomic region extending over 120kb. These data are more than reminiscent of remote tissue-specific control regions in the A1-C3-A4 gene cluster on chromosome 11 [813] or in the E-C1-C4-C2 gene cluster on chromosome 19 [662]. If such observations were made in genes controlling lipoprotein metabolism this would be of great importance for the search for therapeutic targets, which would trigger these elements to correct complex dyslipidemia. Moreover, such mechanisms in combination with the higher polymorphism of noncoding sequences could contribute to the molecular basis of polygenic dyslipidemia or atherosclerosis.

They give a glimpse on poorly understood mechanisms regulating genomic topology and function in the nucleus. For example genomic imprinting [338] or nucleolar dominance [1196] in which preferential allelic expression is driven by changes in chromatin structure, or age-related changes in genomic structure (e.g. telomere shortening, premature aging in cloned ovine) suggest that much remains to be understood about the human genome. Thus, at the turn of the millennium, if the first and second dimensions of genomic and cellular functions have uncovered several of their secrets, the third dimension (spatial organization) if not the fourth (time related events) remain to be explored. There, might be important clues to the understanding of multifactorial, environmental and age-dependent disorders, like lipoprotein disorders and atherosclerosis.

CONCLUSION

Dyslipidemia, like other risk factors of atherosclerosis are multifactorial diseases for which, genetic and phenotypic heterogeneity only reflects the complexity of lipoprotein metabolism. Lipoproteins are essential components of a vital function: lipid homeostasis in living species. It is basically determined by physiological demands in permanent response to environmental conditions. The diversity of genes involved, their polymorphism and the requirement of environmental input for their proper coordination determine the intrinsic adaptability of this system. Alike for candidate genes in hematology, the identification of the first candidate genes of lipoprotein metabolism has benefited from a phenotypic advantage due to the presence of many regulatory proteins in the bloodstream, as opposed to those for diabetes or hypertension for example. Soon after these genes could be manipulated like any other experimental object by rapidly evolving molecular biology technologies. Major advances in the understanding of previously obscure physiological mechanisms determining quantitative and qualitative levels of circulating lipids were obtained. They led to the development of novel drugs, designed to specifically target a molecular mechanism (e.g. HMGCoA reductase inhibitors), which later proved extremely powerful to prevent or retard the course of atherosclerosis in populations at large. It may be anticipated that many other drugs or therapeutic strategies will come up soon. The identification of candidate genes of lipoprotein metabolism has profoundly modified the definition of dyslipidemia shifting from a phenotypic towards a molecular definition of disease. The emergence of novel patho-physiological entities and risk (or protective) factors in turn may soon modify the medical approach to lipid disorders, atherosclerosis or other age-related diseases and any disorder connected with lipid homeostasis. They have given clues to the multifaceted evolution of this system in human populations. It may be anticipated that with the recent release of the human genome sequence, the pace of discoveries will accelerate. This increase in knowledge will eventually delineate novel methodological and theoretical barriers needed to pass over for the new millennium.

1. FREDRICKSON DS : Phenotyping. On reaching base camp (1950-1975). *Circulation* 1993; 87 sup III: III1-III15

2. FELTGEN K : *Le cholestérol : 1758-1913*. Essai historique sur l'intérêt qu'il a suscité en médecine depuis sa découverte au milieu du XVIIIe siècle jusqu'à l'aube du XXe siècle. MD Thesis 1993. Faculté Mixte de Médecine et de Pharmacie de Rouen.

3. FOURCROY AF de : Calculs biliaires. *Encyclopédie Méthodique-Médecine*. Paris. 1792; tome IV 280-291

4. FOURCROY AF de : *Chimie. Encyclopédie Méthodique-Chimie*. Paris, An IV, tome III, 303-781

5. CHEVREUL ME : Des corps que l'on a appelés adipocires, c'est à dire, de la substance cristallisée des calculs biliaires humains, du spermaceti et de la substance grasse des cadavres. *Annales de Chimie* 1815; 2: 339-372

6. MAC INTYRE N and HARRY DS : *Lipids and lipoproteins in clinical practice*. Wolfe Publishing London 1992

7. OBERMAN A, KREISBERG R, HENKIN Y : *Principles and management of lipid disorders. A primary care approach*. Baltimore, Maryland. Williams and Wilkins 1992

8. ANITSCHKOW N : Über die veranderungen der kaninchenaorta bei experimenteller cholesterinsteatose. *Beitrage zur pathologishen Anatomie und zur allgemeinen Pathologie* 1913; 80: 379-404

9. BROWN MS and GOLDSTEIN JL : A receptor-mediated pathway for cholesterol homeostasis. *Science* 1986; 232: 34-47

10. VIRCHOW R : *Cellular pathology based on the physiological and patholgical study of tissues*, 2nd edition. Paris. Baillière ed. 1861

11. QUINEY JR and WATTS GF : *Classic papers in hyperlipidæmia*. London, Science Press Ltd 1989

12. RAYER PFO : *Traité théorique et pratique des maladies de la peau*. Paris, Baillière 1835

13. BRUCHET P : Xanthélasma In : *Nouveau Dictionnaire de Médecine et de Chirurgie Pratiques*. Paris, Baillière 1886 Tome 39: 639-664

14. CHAUFFARD A : Les dépôts locaux de cholestérine et leurs rapports avec la cholestérinémie. *Revue de Médecine* 1911; 176-194

15. THANNHAUSER SJ and MAGENDANTZ H : The different clinical groups of xanthomatous diseases : a clinical physiological study of 22 cases. *Ann Int Med* 1938; 11: 1662

16. MÜLLER C : Xanthomata, hypercholesterolemia, angina pectoris. *Acta Med Scand* 1938; 89 sup: 75-84

17. BOUDET F : Nouvelles recherches sur la composition du sérum et du sang humain. *Annales de Chimie et de Physique* 1833; 52: 337-348

18. MACHEBOEUF MA : Recherches sur les phosphoaminolipides et les stérides du sérum et du plasma sanguins : II Etude physicochimique de la fraction protéique la plus riche en phospholipides et en stérides. *Bull Soc Chim Biol* 1929; 11: 485- 503

19. GOFMAN JM, LINDGREN FT, ELLIOTT H : Ultracentrifugal studies of lipoproteins of human serum. *J Biol Chem* 1949; 179: 973-979

20. FREDRICKSON DS, LEVY RI, LEES RS : Fat transport in lipoproteins - an integrated approach to mechanisms and disorders. *N Engl J Med* 1967; 276 34-44, 94-103, 148-156, 215-225, 273-281

21. DE GENNES JL: Les hyperlipidémies idiopathiques. Proposition d'une classification simplifiée. *La Presse Médicale* 1971; 79: 791-795

22. UTERMANN G : Apo E polymorphism in health and disease. *Am Heart J* 1987; 113: 433-440

23. GARROD AE : Inborn errors of metabolism (Croonian lectures). *Lancet* 1908; 2: 1, 73, 142, 214

24. BEADLE GW, TATUM EL : Genetic control of biochemical reactions in Neurospora. *Proc Natl Acad Sci USA* 1941; 27: 499

25. PAULING L, ITANO HA, SINGER SJ, WELLS IC : Sickle cell anemia : a molecular disease. *Science* 1949; 110: 543

26. KHACHADURIAN AK : The inheritance of essential familial hypercholesterolemia. *Am J Med* 1964; 37: 402-407

27. KEYS A, ANDERSON JT, GRANDE F : Prediction of serum-cholesterol responses of man to changes in fats in the diet. *Lancet* 1957; ii: 959-966

28. KEYS A, KIMURA N, KUSUKAWA A, BRONTE-STEWART B, LARSEN N, HANEY KEYS M : Lessons from serum cholesterol studies in Japan, Hawaii and Los Angeles. *Ann Int Med* 1958; 48: 83-94

29. GOLDSTEIN JL, HAZZARD WR, SCHROTT HG, MOTULSKY AG, BIERMAN EL : Hyperlipidemia in coronary heart disease. 1. Lipid levels in 500 survivors of myocardial infarction. *J Clin Invest* 1973; 52: 1533-1543

30. GOLDSTEIN JL, HAZZARD WR, SCHROTT HG, MOTULSKY AG, BIERMAN EL : Hyperlipidemia in coronary heart disease. 2. Genetic analysis of lipid levels in 176 families and delineation of a new inherited disorder, combined hyperlipidemia. *J Clin Invest* 1973; 52: 1544-1568

31. ALBERTS B, BRAY D, LEWIS J, RAFF M, ROBERTS K, WATSON JD : *Molecular biology of the cell.* 3rd ed. New York, NY. Garland Publishing, 1994; 1 vol

32. GOLDSTEIN JL, BROWN MS : Regulation of the mevalonate pathway. *Nature* 1990; 343: 425-430

33. GLOMSET J, GELB M, FARNSWORTH C : The prenylation of proteins. *Curr Opin Lipidol* 1991; 2: 118-124

34. SINENSKY M and LUTZ RJ : The prenylation of proteins. *Bioessays* 1992; 14: 25-31

35. SEABRA MC, BROWN MS, GOLDSTEIN JL : Retinal degeneration in choroideremia: deficiency of Rab geranylgeranyl transferase. *Science* 1993; 259: 377-381

36. ROBINSON GW, TSAY YH, KIENZLE BK, SMITH-MONROY CA, BISHOP RW : Conservation between human and fungal squalene synthetases : similarities in structure, function, and regulation. *Mol Cell Biol* 1993; 13: 2706- 2717

37. OSBORNE TF, ROSENFELD JM : Related membrane domains in proteins of sterol sensing and cell signaling provide a glimpse of treasures still buried within the dynamic realm of intracellular metabolic regulation. *Curr Opin Lipidol* 1998; 9: 137-140

38. COOPER MK, PORTER JA, YOUNG KE, BEACHY PA : Teratogen-mediated inhibition of target tissue response to SHH-signaling. *Science* 1998; 1603-1607

39. TINT GS, IRONS M, ROY ELLIAS E, BATTA AK, FRIEDEN R, CHEN TS, SALEN G : Defective cholesterol biosynthesis associated with the Smith-Lemli-Opitz syndrome. *N Engl J Med* 1994; 330: 107-113

40. FITZKY BU, GLOSSMANN H, UTERMANN G, MOEBIUS FF : Molecular genetics of the Smith-Lemli-Opitz syndrome and postsqualene sterol metabolism. *Curr Opin Lipidol* 1999; 10: 123-131

41. SEGREST JP, JONES MK, DE LOOF H, BROUILLETTE CG, VENKATACHALAPATHI YV, ANANTHARAMAIAH GM : The amphipathic helix in the exchangeable apolipoproteins: a review of secondary structure and function. *J Lipid Res* 1992; 33: 141-166

42. SOMERVILLE C, BROWSE J : Plant lipids : metabolism, mutants, and membranes. *Science* 1991; 252: 80-87

43. BOWNES M : Why is there sequence similarity between insect yolk proteins and vertebrate lipases. *J Lipid Res* 1992; 33: 777-790

44. BAKER ME : Is vitellogenin an ancestor of apolipoprotein B-100 of human low-density lipoprotein and human lipoprotein lipase ? *Biochem J* 1988; 255: 1057-1060

45. CHAPMAN MJ : Animal lipoproteins : chemistry, structure and comparative aspects. *J Lipid Res* 1980; 21: 789-853

46. RYAN RO : Dynamics of insect lipophorin metabolism. *J Lipid Res* 1990; 31: 1725-1739

47. COLE KDG, FERNANDO-WARNAKULASURIYA GJP, BOGUSKI MS, FREEMAN M, GORDON JI, CLARK WA, LAW JH, WELLS MA : Primary structure and comparative sequence analysis of an insect apolipoprotein : apolipophorin III from manduca sexta. *J Biol Chem* 1987; 262: 11794-11800

48. KANOST MR, BOGUSKI MS, FREEMAN M, GORDON JI, WYATT G, WELLS MA : Primary structure of apolipophorin-III from the migratory locust, Locusta migratoria. Potential amphipathic structures and molecular evolution of an insect apolipoprotein. *J Biol Chem* 1988; 263: 10568-10573

49. DANTUMA NP, POTTERS M, DE WINTHER MP, TENSEN CP, KOOIMAN FP, BOGERD J, VAN DER HORST DJ : An insect homolog of the vertebrate very low density lipoprotein receptor mediates endocytosis of lipophorins. *J Lipid Res* 1999; 40: 973-978

50. BABIN PJ, VERNIER JM : Plasma lipoproteins in fish. *J Lipid Res* 1989; 30: 467-489

51. LAMON-FAVA S, SASTRY R, FERRARI S, RAJAVASHISTH TB, LUSIS AJ, KARATHANASIS SK : Evolutionary distinct mechanisms regulate apolipoprotein A-I gene expression: differences between avian and mammalian apo A-I gene transcription control regions. *J Lipid Res* 1992; 33: 831-842

52. LUSIS AJ : Genetic factors affecting blood lipoproteins : the candidate gene approach. *J Lipid Res* 1988; 29: 397-429

53. EGGEN DA, ABEE R, MALCOLM GT, STRONG JP : Survey of serum cholesterol and triglyceride concentration and lipoprotein electrophoretic pattern in rhesus monkeys (Macaca mulatta). *J Med Primatol* 1982; 11: 1-9

54. PAIGEN B, ISHIDA BY, VERSTUYFT J, WINTERS RB, ALBEE D : Atherosclerosis susceptibility differences among progenitors of recombinant inbred strains of mice. *Arteriosclerosis* 1990; 10: 316-323

55. RAPACZ J, HASLER-RAPACZ J, TAYLOR C, CHECOVICH WJ, ATTIE AD : Lipoprotein mutations in pigs are associated with elevated plasma cholesterol and atherosclerosis. *Science* 1986; 234: 1573-1577

56. RAPACZ J, CHEN L, BÜTLER-BRUNNER E, WU MJ, HASLER-RAPACZ JO, BÜTLER R, SCHUMAKER VN : Identification of the ancestral haplotype for apolipoprotein B suggests an African origin of Homo sapiens sapiens and traces their subsequent migration to Europe and the Pacific. *Proc Natl Acad Sci USA* 1991; 88: 1403-1406

57. SPILMAN CH, HART KL, DINH DM, VIDMAR TJ : Spontaneous hypercholesterolemia in cynomolgus monkeys : evidence for defective low-density lipoprotein catabolism. *Biochim Biophys Acta* 1992; 1128: 26-34.

58. YAMAMOTO T, BISHOP RW, BROWN MS, GOLDSTEIN JL, RUSSELL DW : Deletion in cystein-rich region of LDL receptor impedes transport to cell surface in WHHL rabbit. *Science* 1986; 232: 1230-1237

59. HUMMEL M, LI Z, PFAFFINGER D, NEVEN L, SCANU AM : Familial hypercholesterolemia in a rhesus monkey pedigree: molecular basis of low density lipoprotein receptor deficiency. *Proc Natl Acad Sci USA* 1990; 87: 3122-3126

60. GRUNWALD KA, SCHUELER K, UELMEN PJ, LIPTON BA, KAISER M, BUHMAN K, ATTIE AD : Identification of a novel Arg-->Cys mutation in the LDL receptor that contributes to spontaneous hypercholesterolemia in pigs. *J Lipid Res* 1999; 40:475-485

61. GINZINGER DG, LEWIS ME, MA Y, JONES BR, LIU G, JONES SD : A mutation in the lipoprotein lipase is the molecular basis of chylomicronemia in a colony of domestic cats. *J Clin Invest* 1996; 97: 1257-1266

62. SAVONEN R, NORDSTOGA K, CHRISTOPHERSEN B, LINDBERG A, SHEN Y, HULTIN M , OLIVECRONA T, OLIVECRONA G : Chylomicron metabolism in an animal model for hyperlipoproteinemia type I. *J Lipid Res* 1999; 40: 1336-1346

63. DOMINICZAK MH : Apolipoproteins and lipoproteins in human plasma. In : Rifai N, Warnick RG, Dominiczak MH. *Hanbook of lipoprotein testing*. AACC Press, Washington. 1997; 1-24

64. HAVEL R, KANE JP : Structure and metabolism of plasma lipoproteins. In : Scriver CR, Beaudet AL, Sly WS, Valle D (eds). *The Metabolic Basis of Inherited Disease*, 6th edn. Mac Graw Hill, New York. 1989; 1129-1138

65. GLOMSET JA : The plasma lecithins:cholesterol acyltransferase reaction. *J Lipid Res* 1968; 9: 155-167

66. ROTHBLAT GH, MAHLBERG FH, JOHNSON WJ, PHILLIPS MC : Apolipoproteins, membrane cholesterol domains, and the regulation of cholesterol efflux. *J Lipid Res* 1992; 33: 1091-1097

67. IKONEN E : Molecular mechanisms of intracellular cholesterol transport. *Curr Opin Lipidol* 1997; 8: 60-64

68. LANGER C, HUANG Y, CULLEN P, WIESENHUTTER B, MAHLEY RW, ASSMANN G, VON ECKARDSTEIN A : Endogenous apolipoprotein E modulates cholesterol efflux and cholesteryl ester hydrolysis mediated by high-density lipoprotein-3 and lipid free apolipoproteins in mouse peritoneal macrophages. *J Mol Med* 2000; 78: 217-227

69. ANDERSON RG. The caveolae membrane system. *Ann Rev Biochem* 1998; 67:199-225

70. HAYDEN MR, CLEE SM, BROOKS-WILSON A, GENEST JJr, ATTIE A, KASTELEIN JJP : Cholesterol efflux regulatory protein, Tangier disease and familial high-density lipoprotein deficiency. *Curr Opin Lipidol* 2000; 11: 117-122

71. ORSÓ E, BROCCARDO C, KAMINSKI WE, BÖTTCHER A, LIEBISCH G, DROBNIK W, GÖTZ A, CHAMBENOIT O, DIEDERICH W, LANGMANN T, SPRUSS T, LUCIANI M-F, ROTHE G, LACKNER KJ, CHIMINI G, SCHMITZ G : Transport of lipids from Golgi to plasma membrane is defective in Tangier disease patients and Abc-1 deficient mice. *Nat Genet* 2000; 24: 192-196

72. MATVEEV S, VAN DER WESTHUYZEN DR, SMART EJ : Co-expression of scavenger receptor-BI and caveolin-1 is associated with enhanced selective cholesteryl ester uptake in THP-1 macrophages. *J Lipid Res* 1999; 40: 1647-1654

73. TRIGATTI B, RIGOTTI A, KRIEGER M : The role of the high-density lipoprotein receptor SR-BI in cholesterol metabolism. *Curr Opin Lipidol* 2000; 11: 123-131

74. MOESTRUP SK, KOZYRAKI R. Cubilin, a high-density lipoprotein receptor. *Curr Opin Lipidol* 2000; 11: 133-140

75. TALL AR : Metabolic and genetic control of HDL cholesterol levels. *J Int Med* 1992; 231: 661-668

76. SNIDERMAN AD, ZHANG X-J, CIANFLONE K : Governance of the concentration of plasma LDL : a reevaluation of the LDL receptor paradigm. *Atherosclerosis* 2000; 148: 215-229

77. SACKS FM, WALSH BW : Sex hormones and lipoprotein metabolism. *Curr Opin Lipidol* 1994; 5: 236-240

78. RUDLING M, ANGELIN B : Loss of resistance to dietary cholesterol in the rat after hypophysectomy: importance of the presence of growth hormone for hepatic low density lipoprotein-receptor expression. *Proc Natl Acad Sci USA* 1993; 90: 8851-8855

79. SPARKS JD, SPARKS CE : Hormonal regulation of lipoprotein assembly and secretion. *Curr Opin Lipidol* 1993; 4: 177-186

80. FEINGOLD KR, HARDARDOTTIR I, GRUNFELD C. Beneficial effects of cytokine induced hyperlipidemia. *Z Ernahrungswiss* 1998; 37 supp 1: 66-74

81. ZECHNER R, NEWMAN TC, SHERRY B, CERAMI A, BRESLOW JL : Recombinant human cachectin/Tumor necrosis factor but not Interleukin-1α downregulates lipoprotein lipase gene expression at the transcriptional level in mouse 3T3-L1 adipocytes. *Mol Cell Biol* 1988; 8: 2394-2401

82. SNIDERMAN AD, MASLOWSKA M, CIANFLONE K : Of mice and men (and women) and the acylation-stimulating protein pathway. *Curr Opin Lipidol* 2000; 11: 291-296

83. GREWAL T, BOUDREAU M, ROY M, CHAMBERLAND A, LEFEBVRE C, LAVIGNE J, DAVIGNON J, MINNICH A. Expression of γ-IFN responsive genes in scavenger receptor over-expressing monocytes is associated with xanthomatosis. *Atherosclerosis* 1998; 138: 335-345

84. BROWN MS, GOLDSTEIN JL : A proteolytic pathway that controls the cholesterol content of membranes, cells, and blood. *Proc Natl Acad Sci USA* 1999; 96: 11041-11048

85. SCHOONJANS K, MARTIN G, STAELS B, AUWERX J : Peroxisome proliferator-activated receptors, orphans with ligands and fuctions. *Curr Opin Lipidol* 1997; 8: 159-166

86. BRUN R, KIM JB, HU E, SPIEGELMAN BM : Peroxisome proliferator-activated receptor gamma and the control of adipogenesis. *Curr Opin Lipidol* 1997; 8: 212-218

87. ROSS R : The pathogenesis of atherosclerosis : a perspective for the 1990s. *Nature* 1993 362: 801- 809

88. SAXENA U, GOLDBERG I : Endothelial cells and atherosclerosis: lipoprotein metabolism, matrix interactions, and monocyte recruitment. *Curr Opin Lipidol* 1994; 5: 316-322

89. LIBBY P : Current concepts in cardiovascular pathology: the role of LDL cholesterol in plaque rupture and stabilization. *Am J Med* 1998: 104: 14S-18S

90. BENNETT MR, EVAN GI, SCHWARTZ SM : Apoptosis of human vascular smooth muscle cells derived from normal vessels and coronary atheroscleorotic plaques. *J Clin Invest* 1995; 95: 2266-2274

91. SOLBERG LA, STRONG JP : Risk factors and atherosclerotic lesions : a review of autopsy studies. *Arteriosclerosis* 1983; 3: 187-198

92. NAPOLI C, D'ARMIENTO FP, MANCINI FP, POSTIGLIONE A, WITZTUM JL, PALUMBO G, PALINSKI W : Fatty streak formation occurs in human fetal aortas and is greatly enhanced by maternal hypercholesterolemia. *J Clin Invest* 1997; 100: 2680-2690

93. VAUGHAN CJ, GOTTO AM Jr, BASSON CT : The evolving role of statins in the management of atherosclerosis. *J Am Coll Cardiol* 2000; 35: 1-10

94. CASTELLI WP, GARRISON RJ, WILSON PWF, ABBOTT RD, KALOUSDIAN S, KANNEL WB : Incidence of coronary heart disease and lipoprotein cholesterol levels. The Framingham study. *JAMA* 1986; 256: 2835-2838

95. PEKKANEN J, LINN S, HEISS G, SUCHINDRAN CM, LEON A, RIFKIND B, TYROLER H : Ten-year mortality from cardiovascular disease in relation to cholesterol level among men with and without preexisting cardiovascular disease. *N Engl J Med* 1990; 322: 1700-1707

96. Mac MURRY, CERQUEIRA MT, CONNOR SL, CONNOR WE : Changes in lipid and lipoprotein levels and body weight in Tarahumara Indians after consumption of an affluent diet. *N Engl J Med* 1991; 325: 1704-1708

97. SØRENSEN TIA, NIELSEN GG, ANDERSEN PK, TEASDALE TW : Genetic and environmental influences on premature death in adult adoptees. *N Engl J Med* 1988 318: 727-732

98. SIMONS LA : Interrelations of lipids and lipoproteins with coronary artery disease mortality in 19 countries. *Am J Cardiol* 1986 57: 5G-10G

99. KUULASMAA K, TUNSTALL-PEDOE H, DOBSON A, FORTMANN S, SANS S, TOLONEN H, EVANS A, FERRARIO M, TUOMILEHTO J : Estimation of contribution of changes in classic risk factors to trends in coronary-event rates across the WHO MONICA project populations. *Lancet* 2000; 355: 675-687

100. SYTKOWSKI PA, KANNEL WB, D'AGOSTINO : Changes in risk factors and the decline in mortality from cardiovascular disease. *N Engl J Med* 1990; 322: 1635-1641

101. ROSSOUW JE, RIFKIND BM : Does lowering serum cholesterol level lower coronary heart disease risk ? *Endoc Metab Clin North Amer* 1990; 19: 279- 297

102. BROWN G, ZHAO XQ, SACCO DE, ALBERS JJ : Regression of arteriosclerosis, plaque disruption and cardiovascular events : a justification to lipid lowering therapies in coronary artery disease. *Annu Rev Med* 1993; 44: 365- 376

103. SCANDINAVIAN SIMVASTATIN SURVIVAL STUDY GROUP : Randomized trial of cholesterol lowering in 4444 patients with coronary heart disease : the scandinavian simvastatin survival study (4S). *Lancet* 1994; 344: 1383-1389

104. SHEPHERD J, COBBE SM, FORD I, ISLES CG, LORIMER AR, Mac FARLANE PW, Mc KILLOP JH, PACKARD CJ: Prevention of coronary heart disease with parvastatin in men with hypercholesterolemia. West of Scotland coroanry prevention study group. *N Engl J Med* 1995; 333: 1301-1307

105. GOTTO AM : Assessing the benefits of lipid-lowering therapy. *Am J Cardiol* 1998; 82: 2M-4M

106. KWITEROVICH PO, CORESH J, BACHORIK PS : Prevalence of hyperapobetalipoproteinemia and other lipoprotein phenotypes in men (aged ≤50 years) and women (≤60 years) with coronary artery disease. *Am J Cardiol* 1993; 71: 631-639

107. DAVIGNON J, GENEST JJ : Genetics of lipoprotein disorders. *Endocrinol Metab Clin North Amer* 1998; 27: 521-550

108. GOLDSTEIN JL, HOBBS HH, BROWN MS : Familial Hypercholesterolemia In : Scriver CR, Beaudet AL, Sly WS, Valle D. *The metabolic and molecular bases of inherited disease* 7th ed. Mc Graw Hill, New York. 1995. 1981-2030

109. LINTON MRF, FARESE RV, YOUNG SG : Familial Hypobetalipoproteinemia. *J Lipid Res* 1993; 34: 521-541

110. WITZTUM JL, STEINBERG D : Role of oxidized low density lipoprotein in atherogenesis. *J Clin Invest* 1991; 88: 1785-1792

111. SING CF, MOLL PP : Genetics of atherosclerosis. *Annu Rev Genet* 1990; 24: 172-187

112. RIFAI N, WARNICK GR, DOMINICZAK MH : *Handbook of lipoprotein testing.* AACC press. Washington, DC. 1997.

113. GRUNDY SM, VEGA GL : Causes of high blood cholesterol. *Circulation* 1990; 81: 412-427

114. BROOKS-WILSON A, MARCIL M, CLEESM, ZHANG L-H, ROOMP K, VAN DAM M, YU L, BREWER C, COLLINS JA, MOLHUIZEN HOF, LOUBSER O, OUELETTE FBF, FICHTER K, ASHBOURNE-EXCOFFON KJD, SENSEN CW, SCHERER S, MOTT S, DENIS M, MARINDALE D, FROHLICH J, MORGAN K, KOOP B,

PIMSTONE S, KASTELEIN JJP, GENEST JJr, HAYDEN MR : Mutations in ABC1 in Tangier disease and familial high density lipoprotein deficiency. *Nat Genet* 1999; 22: 336-345

115. BODZIOCH M, ORSÓ E, KLUCKEN J, LANGMANN T, BÖTTCHER A, DIEDERICH W, DROBNICK W, BARLAGE S, BÜCHLER C, PORSCH-ÖZCÜRÜMEZ M, KAMINSKI WE, HAHMANN HW, OETTE K, ROTHE G, ASLANIDIS LACKNER KJ, SCHMITZ G : The gene encoding ATP-binding cassette transporter 1 is mutated in Tangier disease. *Nat Genet* 1999; 22: 347-351

116. RUST S, ROSIER M, FUNKE H, REAL J, AMOURA Z, PIETTE J-C, DELEUZE J-F, BREWER HB, DUVERGER N, DENEFLE P, ASSMANN G : Tangier disease is caused by mutations in the gene encoding ATP-binding cassette transporter 1. *Nat Genet* 1999; 352-355

117. SCRIVER CR, BEAUDET AL, SLY WS, VALLE D. *The metabolic and molecular bases of inherited disease.* 7th ed. Mc Graw Hill. New York. 1995

118. GENEST JJ, MARTIN-MUNLEY SS, Mac NAMARA JR, ORDOVAS JM, JENNER J, MYERS RH, SILBERMAN SR, WILSON PWF, SALEM DN, SCHAEFER EJ: Familial lipoprotein disorders in patients with premature coronary artery disease. *Circulation* 1992; 85: 2025-2033

119. LUSIS AJ, SPARKES RS : *Genetic factors in atherosclerosis: approaches and model systems. Monographs in human genetics.* KARGER Ed, BASEL 1989; 12: 227p

120. UTERMANN G : Apo E polymorphism in health and disease. *Am Heart J*; 1987 113: 433-440

121. BLANGERO J, MAC CLUER JW, KAMMERER CM, MOTT GE, DYER TD, MC GILL HC : Genetic analysis of apolipoprotein A-I in two dietary environments. *Am J Hum Genet* 1990; 47: 414-428

122. BRESLOW JL : Apolipoprotein genetic variation and human disease. *Physiol Rev* 1988; 68: 85-132

123. BROWN MS, GOLDSTEIN JL : Familial hypercholesterolemia : defective binding of lipoproteins to cultured fibroblasts associated with impaired regulation of 3-hydroxy-3-methylglutaryl coenzyme-A reductase activity. *Proc Natl Acad Sci USA* 1974; 71: 788-792

124. SCIENTIFIC STEERING COMMITTEE ON BEHALF OF THE SIMON BROOME REGISTER GROUP : Mortality in treated heterozygous familial hypercholesterolaemia: implications for clinical management. *Atherosclerosis* 1999; 142: 105-112.

125. RUDLING M, REIHNER E, EINARSSON K, EWERT S, ANGELIN B : Low density lipoprotein receptor-binding activity in human tissues : quantitative importance of hepatic receptors and evidence for regulation of their expression in vivo. *Proc Natl Acad Sci USA* 1990; 87: 3469-3473

126. LINDGREN V, LUSKEY KL, RUSSELL DW, FRANCKE U : Human genes involved in cholesterol metabolism: chromosomal mapping of the loci for the low density lipoprotein receptor and 3-hydroxy-3-methylglutaryl-coenzyme A reductase with cDNA probes. *Proc Natl Acad Sci USA* 1985; 82: 8567-8571

127. YAMAMOTO T, DAVIS CG, BROWN MS, SCHNEIDER WJ, CASEY ML, GOLDSTEIN JL, RUSSELL DW : The human LDL receptor : a cysteine-rich protein with multiple *Alu* sequences in its mRNA. *Cell* 1984; 39: 27-38

128. SÜDHOF TC, GOLDSTEIN JL, BROWN MS, RUSSELL DW : The LDL receptor gene: a mosaic of exons shared with different proteins. *Science* 1985; 228: 815-822

129. FASS D, BLACKLOW S, KIM PS, BERGER JM. Molecular basis of familial hypercholesterolaemia from structure of LDL receptor module. *Nature* 1997; 388: 691-693

130. DAVIS CG, GOLDSTEIN JL, SÜDHOF TC, ANDERSON RGW, RUSSELL DW, BROWN MS: Acid-dependent ligand dissociation and recycling of LDL receptor mediated by growth factor homology region. *Nature* 1987; 326: 760-765

131. PATHAK RK, YOKODE M, HAMMER RE, HOFMAN SL, BROWN MS, GOLDSTEIN JL : Tissue-specific sorting of the human LDL receptor in polarized epithelia of transgenic mice. *J Cell Biol* 1990 111: 347-359

132. MATTER K, WHITNEY JA, YAMAMOTO EM, MELLMEN I : Common signals control low density lipoprotein receptor sorting in endosomes and the golgi complex of MDCK cells. *Cell* 1993; 74: 1053-1064

133. YOKODE M, HAMMER RE, ISHIBASHI S, BROWN MS, GOLDSTEIN JL : Diet-induced hypercholesterolemia in mice : prevention by overexpression of LDL receptors. *Science* 1990; 250: 1273-1275

134. ISHIBASHI S, GOLDSTEIN JL, BROWN MS, HERZ J, BURNS DK : Massive xanthomatosis and atherosclerosis in cholesterol-fed low density lipoprotein receptor-negative mice. *J Clin Invest* 1994; 93: 1885-1893

135. MEHTA KD, CHEN W-J, GOLDSTEIN JL, BROWN MS : The low density lipoprotein receptor in xenopus laevis. Five domains that resemble the human receptor. *J Biol Chem 1991*; 266: 10406-10414

136. HOBBS HH, RUSSELL DW, BROWN MS, GOLDSTEIN JL : The LDL receptor locus in familial hypercholesterolemia : mutational analysis of a membrane protein. *Annu Rev Genet* 1990; 24: 133-170

137. FURUHASHI M, SEO H, MIZUTANI S, NARITA O, TOMODA Y, MATSUI N: Expression of low density lipoprotein receptor gene in human placenta during pregnancy. *Mol Endoc* 1989; 3: 1252-1256

138. DAWSON PA, HOFMAN SL, VAN DER WESTHUYZEN DR, SÜDHOF TC, BROWN MS, GOLDSTEIN JL : Sterol-dependent repression of low density lipoprotein receptor promoter mediated by 16-base pair sequence adjacent to binding site for transcription factor Sp1. *J Biol Chem* 1988 263: 3372-3379

139. WANG X, SATO R, BROWN MS, HUA X, GOLDSTEIN JL : SREBP-1, a membrane-bound transcription factor released by sterol-regulated proteolysis. *Cell* 1994; 77: 53-62

140. FOX JC, McGILL HC, DEE CAREY K, GETZ GS : In-vivo regulation of hepatic LDL receptor mRNA in the Baboon. Differential effects of saturated and unsaturated fat. *J Biol Chem* 1987; 262: 7014-7020

141. SORCI-THOMAS M, WILSON MD, JOHNSON FL, WILLIAMS DL, RUDEL LL : Studies on expression of genes encoding apolipoprotein B-100 and B-48 and the low density lipoprotein receptor in non human primates. *J Biol Chem* 1989; 264: 9039-9045

142. GOLDSTEIN JL, SOBHANI MK, FAUST JR, BROWN MS : Heterozygous familial hypercholesterolemia : failure of normal allele to compensate for mutant allele at a regulated genetic locus. *Cell* 1976; 9: 195-203

143. DAVIGNON J, LAAKSONEN R. Low-density lipoprotein-independent effects of statins. *Curr Opin Lipidol.* 1999; 10: 543-559

144. LEITERSDORF E, CHAKRAVARTI A, HOBBS HH : Polymorphic DNA haplotypes at the LDL receptor locus. *Am J Hum Genet* 1989; 44: 409-421

145. WILSON DJ, GAHAN M, HADDAD L, HEATH K, WHITTALL RA, WILLIAMS RR, HUMPHRIES SE, DAY IN. A World Wide Web site for low-density lipoprotein receptor gene mutations in familial hypercholesterolemia: sequence-based, tabular, and direct submission data handling. *Am J Cardiol* 1998; 81: 1509-1511.

146. MISEREZ AR, SCHUSTER H, CHIODETTI N, KELLER U : Polymorphic haplotypes and recombination rates at the LDL receptor locus in subjects with and without familial hypercholesterolemia who are from different populations. *Am J Hum Genet* 1993; 52: 808-826

147. KOIVISTO PVI, KOIVISTO UM, MIETTINEN TA, KONTULA K : Deletion of exon 15 of the LDL receptor gene is associated with a mild form of familial hypercholesterolemia FH-Espoo. *Arterioscler Thromb* 1993; 13: 1680-1688

148. SCHUSTER H, FISCHER HJ, KELLER C, WOLFRAM G, ZÖLLNER N : Identification of the 408 valine to methionine mutation in the low density lipoprotein receptor in a German family with familial hypercholesterolemia. *Hum Genet* 1993; 91: 287-289

149. BERKMAN N, WEIR BS, PRESSMAN-SCHWARTZ S, RESHEF A, LEITERSDORF E : Haplotype analysis at the low density lipoprotein receptor locus : application to the study of familial hypercholesterolemia in Israel. *Hum Genet* 1992; 88: 405-410

150. DAGA A, MATTIONI T, COVIELLO DA, CORTE G, BERTOLINI S : Use of three DNA polymorphisms of the LDL receptor gene in the diagnosis of familial hypercholesterolemia. *Hum Genet* 1990; 84: 412-416

151. YAMAKAWA T, YANAGI H, SAKU K, SASAKI J, OKAFUJI T, SHIMAKURA Y, KAWAI K, TSUCHIYA S, TAKADA K, NAITO S, ARAKAWA K, HAMAGUCHI H : Family studies of the LDL receptor gene of relatively severe hereditary hypercholesterolemia associated with Achilles tendon xanthomas. *Hum Genet* 1991; 86: 445-449

152. HUMPHRIES SE, KING-UNDERWOOD L, GUDNASON V, SEED M, DELATTRE S, CLAVEY V, FRUCHART JC : Six DNA polymorphisms in the low density lipoprotein receptor gene : Their genetic relationship and an example of their use for identifying affected relatives of patients with familial hypercholesterolemia. *J Med Genet* 1993; 30: 273-279

153. LOUX N, BENLIAN P, PASTIER D, BOILEAU C, CAMBOU JP, MONNIER L, PERCHERON C, JUNIEN C : Recurrent mutation at aa 792 in the LDL receptor gene in a French patient. *Hum Genet* 1991; 87: 373-375

154. HOBBS HH, BROWN MS, GOLDSTEIN JL : Molecular genetics of the LDL receptor gene in familial hypercholesterolemia. *Hum Mutat* 1992; 1: 445-466

155. VARRET M, RABES J-P, THIART R, KOTZE MJ, BARON H, CENARRO A, DESCAMPS O, EBHARDT M, HONDELIJN J-C, KOSTNER G, MIYAKE Y, POCOVI M, SCHMIDT H, SCHMIDT H, SCHUSTER H, STUHRMANN, YAMAKURA T, JUNIEN C, BÉROUD C, BOILEAU C : LDLR database (second edition): new additions to the database and the software, and results of the first molecular analysis. *Nucl Acids Res* 1998; 26: 248-252

156. PORTER CJ, TALBOT CC Jr, CUTICCHIA J : Central mutation databases: a review. *Hum Mutat* 2000; 15: 36-44

157. DAVIS CG, ELHAMMER A, RUSSELL DW, SCHNEIDER WJ, KORNFELD S, BROWN MS, GOLDSTEIN JL : Deletion of clustered O-linked carbohydrates does not impair function of low density lipoprotein receptor in transfected fibroblasts. *J Biol Chem* 1986; 261: 2828-2838

158. KAJINAMI K, MABUCHI H, ITOH H, MICHISHITA I, TAKEDA M, WAKASUGI T, KOIZUMI J, TAKEDA R : New variant of low density lipoprotein receptor gene FH-Tonami. *Arteriosclerosis* 1988; 8:187-192

159. KORENBERG JR and RYKOWSKI MC : Human genome organization : Alu, lines and the molecular structure of metaphase chromosome bands. *Cell* 1988; 53: 391-400

160. STOPPA-LYONNET D, CARTER PE, MEO T, TOSI M : Clusters of intragenic Alu repeats predispose the human C1 inhibitor locus to deleterious rearrangements. *Proc Natl Acad Sci USA* 1990; 87: 1551-1555

161. TOP B, KOELEMAN BPC, LEUVEN JAG, HAVEKES LM, FRANTS RR : Rearrangements in the LDL receptor gene in Dutch familial hypercholesterolemic patients. *Atheroscleosisr* 1990; 83: 127-136

162. SUN XM, WEBB JC, GUDNASON V, HUMPHRIES S, SEED M, THOMPSON GR, KNIGHT BL, SOUTAR AK : Characterization of deletions in the LDL receptor gene in patients with familial hypercholesterolemia in the United Kingdom. *Arterioscler Thromb* 1992; 12: 762-770

163. AALTO-SETÄLÄ K, HELVE E, KOVANEN PT, KONTULA K : Finnish type of low density lipoprotein receptor gene mutation (FH-Helsinki) deletes exons encoding the carboxy-terminal part of the receptor and creates an internalization-defective phenotype. *J Clin Invest* 1989; 84: 499-505

164. LEJEUNE P, DANCHIN A: Mutations in the BglY gene increase the frequency of spontaneous deletions in Escherichia coli K-12. *Proc Natl Acad Sci USA* 1990; 87: 360-363

165. COOPER DN, ANTONARAKIS SE, KRAWCZAK M : The nature and mechanisms of human gene mutation. In Scriver CR, Beaudet AL, Sly WS, Valle D, (eds). *The metabolic and molecular bases of inherited disease*, 7th ed. Mc Graw Hill, New York, 1995; 259-291.

166. KRAWCZAK M, BALL EV, COOPER DN : Neighboring-nucleotide effects on the rates of germ-line single-base-pair substitution in human genes. *Am J Hum Genet* 1998; 63: 474-488.

167. GUDNASON V, KING-UNDERWOOD L, SEED M, SUN X-M, SOUTAR AK, HUMPHRIES SE : Identification of recurrent and novel mutations in exon 4 of the LDL receptor gene in patients with familial hypercholesterolemia. *Arteriosclerosis Thrombosis* 1993; 13: 56-63

168. WEBB JC, PATEL DD, JONES MD, KNIGHT BL, SOUTAR AK : Characterization of two new point mutations in the low density lipoprotein receptor gene of an English patient with homozygous familial hypercholesterolemia. *J Lipid Res* 1992; 33: 689-698

169. KRAWCZAK M, COOPER DN : Gene deletions causing human genetic disease: mechanisms of mutagenesis and the role of the local DNA sequence environment. *Hum Genet* 1991; 86: 425-441

170. AALTO-SETÄLÄ K, KOIVISTO U-M, MIETTINEN TA, GYLLING H, KESÄNIEMI YA, SAVOLAINEN M, VIIKARI J, KONTULA K : Prevalence and geographical distribution of major LDL receptor gene rearrangements in Finland. *J Intern Med* 1992; 231: 227-234

171. GRAADT VAN ROGGEN JF, VAN DER WESTHUYZEN DR, MARAIS AD, GEVERS W, COETZEE GA : LDL receptor founder mutations in Afrikaner familial hypercholesterolemic patients: a comparison of two geographic areas. *Hum Genet* 1991; 88: 204-208

172. LEITERSDORF E, Van Der WESTHUYZEN DR, COETZEE GA, HOBBS HH : Two common low density lipoprotein receptor gene mutations cause familial hypercholesterolemia in Afrikaners. *J Clin Invest* 1989; 84: 954-961

173. LEITERSDORF E, TOBIN EJ, DAVIGNON J, HOBBS HH : Common low density lipoprotein receptor mutations in the French Canadian population. *J Clin Invest* 1990; 85: 1014-1023

174. MEINER V, LANDSBERGER D, BERKMAN N, RESHEF A, SEGAL P, SEFTEL HC, VAN DER WESTHUYZEN DR, JEENAH MS, COETZEE GA, LEITERSDORF E : A common Lithuanian mutation causing familial hypercholesterolemia in Ashkenazi Jews. *Am J Hum Genet* 1991; 49: 443-449

175. LEHRMAN MA, SCHNEIDER WJ, BROWN MS, DAVIS CG, ELHAMMER A, RUSSELL DW, GOLDSTEIN JL : The Lebanese allele at the LDL receptor locus : nonsense mutation produces truncated receptor that is retained in endoplasmic reticulum. *J Biol Chem* 1987 262: 401-410

176. BERTOLINI S, LELLI N, COVIELLO DA, GHISELLINI M, MASTURZO P, TIOZZO R, ELICIO N, GADDI A, CALANDRA S. A large deletion in the LDL receptor gene-- the cause of familial hypercholesterolemia in three Italian families: a study that dates back to the 17th century (FH-Pavia). *Am J Hum Genet* 1992; 51:123-134

177. SOUTAR AK : Update on low density lipoprotein receptor mutations. *Curr Opin Lipidol* 1998; 9:141-147

178. OSE L : An update on familial hypercholesterolaemia. *Ann Med* 1999; 31:13-18

179. JENSEN HK, JENSEN LG, MEINERTZ H, HANSEN PS, GREGERSEN N, FAERGEMAN O : Spectrum of LDL receptor gene mutations in Denmark: implications for molecular diagnostic strategy in heterozygous familial hypercholesterolemia. *Atherosclerosis* 1999;146: 337-344.

180. LOMBARDI MP, REDEKER EJ, DEFESCHE JC, KAMERLING SW, TRIP MD, MANNENS MM, HAVEKES LM, KASTELEIN JJ : Molecular genetic testing for familial hypercholesterolemia: spectrum of LDL receptor gene mutations in The Netherlands. *Clin Genet* 2000; 57:116-124

181. SCHMIDT H, KOSTNER GM : Familial hypercholesterolemia in Austria reflects the multi-ethnic origin of our country. *Atherosclerosis* 2000; 148: 431-432

182. KOTZE MJ, DE VILLIERS WJS, STEYN K, KRIEK JA, MARAIS AD, LANGENHOVEN E, HERBERT JS, GRAADT VAN ROGGEN JF, VAN DER WESTHUYZEN DR, COETZEE GA : Phenotypic variation among familial hypercholesterolemics heterozygous for either one of two Afrikaner founder LDL receptor mutations. *Arterioscler Thromb* 1993; 13: 1460-1468

183. MOORJANI S, ROY M, TORRES A, BETARD C, GAGNÉ C, LAMBERT M, BRUN D, DAVIGNON J, LUPIEN PJ : Mutations of low-density-lipoprotein-receptor gene, variation in plasma cholesterol, and expression of coronary heart disease in homozygous familial hypercholesterolemia. *Lancet* 1993; 341: 1303-1306

184. GAUDET D, VOHL MC, COUTURE P, MOORJANI S, TREMBLAY G, PERRON P, GAGNE C, DESPRES JP : Contribution of receptor negative versus receptor defective mutations in the LDL-receptor gene to angiographically assessed coronary artery disease among young (25-49 years) versus middle-aged (50-64 years) men. *Atherosclerosis* 1999; 143:153-161

185. UAUY R, VEGA GL, GRUNDY SM : Coinheritance of two mild defects in low density lipoprotein receptor function produces severe hypercholesterolemia. *J Clin Endocrinol Metab* 1991; 72: 179-187

186. JEENAH M, SEPTEMBER W, GRAADT VAN ROGGEN F, DE VILLIERS W, SEFTEL H, MARAIS D : Influence of specific mutations at the LDL-receptor gene locus on the response to simvastatin therapy in Afrikaner patients with heterozygous familial hypercholesterolæmia. *Atherosclerosis* 1993; 98: 51-58

187. LEITERSDORF E, EISENBERG S, ELIAV O, FRIEDLANDER Y, BERKMAN N, DANN EJ, LANDSBERGER D, SEHAYEK E, MEINER V, WURM M, BARD JM, FRUCHART JC, STEIN Y : Genetic determinants of responsiveness to the HMG-CoA reductase inhibitor fluvastatin in patients with molecularly defined heterozygous familial hypercholesterolemia. *Circulation* 1993; 87 sup III: III35-III44

188. SUN XM, PATEL DD, KNIGHT BL, SOUTAR AK : Influence of genotype at the low density lipoprotein (LDL) receptor gene locus on the clinical phenotype and response to lipid-lowering drug therapy in heterozygous familial hypercholesterolaemia. The Familial Hypercholesterolaemia Regression Study Group. *Atherosclerosis* 1998; 136:175-185

189. SUN XM, PATEL DD, WEBB JC, KNIGHT BL, FAN L-M, CAI H-J, SOUTAR AK : Familial hypercholesterolemia in China. Identification of mutations in the LDL-receptor gene that result in a receptor-negative phenotype. *Arterioscler Thromb* 1994 14: 85-94

190. PEREIRA E, FERREIRA R, HERMELIN B, THOMAS G, BERNARD C, BERTRAND V, MENDEZ Del CASTILLO D, BEREZIAT G, BENLIAN P : Recurrent and novel LDL receptor gene mutations causing heterozygous familial hypercholesterolemia in La Habana. *Hum Genet* 1995; 96: 319-322

191. SLIMANE MN, SUN X-M, LESTAVEL-DELATTRE S, MAATOUK F, SOUTAR A, BEN FAHRAT MH, CLAVEY V, BENLIAN P, HAMMAMI M : FH-Souassi: a founder frameshift mutation in exon 10 of the LDL receptor gene, associated with a mild phenotype in Tunisian families. *Atherosclerosis* 2001; 154: 557-565

192. HOBBS HH, LEITERSDORF E, LEFFERT CC, CRYER DR BROWN, MS GOLDSTEIN JL : Evidence for a dominant gene that suppresses hypercholesterolemia in a family with defective low density lipoprotein receptors. *J Clin Invest* 1989; 84: 656-664

193. VEGA GL, HOBBS HH, GRUNDY SM : Low density lipoprotein kinetics in a family having defective low density lipoprotein receptors in which hypercholesterolemia is suppressed. *Arterioscler Thromb* 1991; 11: 578-585

194. KNOBLAUCH H, MÜLLER-MYSHOK B, BUSJAHN A, BEN AVI L, BÄRHING S, BARON H, HEATH SC, UHLMANN R, FAULHABER H-D, SHPITZEN S, AYDIN A,

RESHEF A, ROSENTHAL M, ELIAV O, MÜHL A, LOWE A, SCHURR D, HARATS D, JESCHKE EN FRIEDLANDER Y, SCHUSTER H, LUFT FC, LEITERSDORF E : A cholesterol lowering gene maps to chromosome 13q. *Am J Hum Genet* 2000; 66: 157-166

195. EMI M, HEGELE R, HOPKINS PN, WU L, PLAETKE R, WILLIAMS RR, LALOUEL JM : Effects of three genetic loci in a pedigree with multiple lipoprotein phenotypes. *Arterioscler Thromb* 1991; 11: 1349-1355

196. RUBINSZTEIN DC, RAAL FJ, SEFTEL HC, PILCHER G, COETZEE GA, VAN DER WESTHUYZEN : Characterization of six patients who are double heterozygotes for familial hypercholesterolemia and familial defective apo B-100. *Arterioscler Thromb* 1993; 13: 1076-1081

197. BENLIAN P, DE GENNES JL, DAIROU F, HERMELIN B, GINON I, VILLAIN E, LAGARDE JP, FEDERSPIEL MC, BERTRAND V, BERNARD C, BEREZIAT G : Phenotypic expression in double heterozygotes for familial hypercholesterolemia and familial defective apo B-100. *Hum Mutat* 1996; 7: 340-345

198. FERRIÈRES J, SING CF, ROY M, DAVIGNON J, LUSSIER-CACAN S : Apolipoprotein E polymorphism and heterozygous familial hypercholesterolemia. Sex-specific effects. *Arterioscler Thromb* 1994; 14: 1553-1560

199. WITTEKOEK ME, MOLL E, PIMSTONE SN, TRIP MD, LANSBERG PJ, DEFESCHE JC, VAN DOORMAAL JJ, HAYDEN MR, KASTELEIN JJ : A frequent mutation in the lipoprotein lipase gene (D9N) deteriorates the biochemical and clinical phenotype of familial hypercholesterolemia. *Arterioscler Thromb Vasc Biol* 1999; 19: 2708-2713

200. GRUNDY SM : HMG-CoA reductase inhibitors for treatment of hypercholesterolemia. *N Engl J Med* 1988; 319: 24-33

201. KANE JP : Apolipoprotein B : structure and metabolic heterogeneity. *Annu Rev Physiol* 1983; 45: 637-650

202. CHEN SH, YANG CY, CHEN PF, SETZER D, TANIMURA M, LI WH, GOTTO AM, CHAN L : The complete cDNA and amino acid sequence of human apolipoprotein B-100. *J Biol Chem* 1986; 261: 12918-12921

203. KNOTT TJ, PEASE RJ, POWELL LM, WALLIS SC, RALL SC, INNERARITY TL, BLACKHART BD, TAYLOR WH, MARCEL YL, MILNE RW, JOHNSON D, FULLER M, LUSIS AJ, Mc CARTHY BJ, MAHLEY RW, LEVY-WILSON B, SCOTT J : Complete protein sequence and identification of structural domains of human apolipoprotein-B. *Nature* 1986; 323: 734-738

204. LAW SW, GRANT SM, HIGUCHI K, HOSPATTANKAR AV, LACKNER K, LEE N, BREWER HB : Human liver apolipoprotein B-100 cDNA: complete nucleic acid and derived amino acid sequence. *Proc Natl Acad Sci USA* 1986; 83: 8142-8146

205. YOUNG SG : Recent progress in understanding apolipoprotein B. *Circulation* 1990; 82: 1574-1594

206. BAKER ME : Is vitellogenin an ancestor of apolipoprotein B-100 of human low-density lipoprotein and human lipoprotein lipase ? *Biochem J* 1988; 255: 1057-1060

207. WEISGRABER KH and RALL SC : Human apolipoprotein B-100 heparin binding sites. *J Biol Chem* 1987; 262: 11097-11103

208. AVIRAM M, LUND-KATZ S, PHILLIPS MC, CHAIT A : The influence of the triglyceride content of low density lipoprotein on the interaction of apolipoprotein B-100 with cells. *J Biol Chem* 1988; 263: 16842-16848

209. PEASE RJ, MILNE RW, JESSUP WK, LAW A, PROVOST A, FRUCHART JC, DEAN RT, MARCEL YL, SCOTT J : Use of bacterial expression cloning to localize the epitope for a series of monoclonal antibodies against apolipoprotein B-100. *J Biol Chem* 1990; 265: 553-568

210. KNOTT TJ, RALL SC, INNERARITY TL, JACOBSON SF, URDEA MS, LEVY-WILSON B, POWELL LM, PEASE RJ, EDDY R, NAKAI,H, BYERS M, PRIESTLEY LM, ROBERTSON E, RALL LB, BETSHOLTZ C, SHOWS TB, MAHLEY RW, SCOTT J : Human apolipoprotein B: structure of carboxyl-terminal domains, sites of gene expression, and chromosomal localization. *Science* 1985; 230: 37-43

211. VENIANT MM, NIELSEN LB, BOREN J, YOUNG SG: Lipoproteins containing apolipoprotein B-100 are secreted by the heart. *Trends Cardiovasc Med* 1999; 9:103-107

212. KOENIG M, MONACO AP, KUNKEL LM : The complete sequence of dystrophin predicts a rod-shaped cytoskeletal protein. *Cell* 1988; 53: 219-228

213. BLACKHART BD, LUDWIG EM, PIEROTTI VR, CAIATI L, ONASCH MA, ALLIS SC, POWELL L, PEASE R, KNOTT TJ, CHU ML, MAHLEY RW, SCOTT J, Mc CARTHY BJ, LEVY-WILSON B : Structure of the human apolipoprotein B gene. *J Biol Chem* 1986; 261: 15364-15367

214. LEVY-WILSON B and FORTIER C : Tissue specific undermethylation of DNA sequences at the 5' end of the human apolipoprotein B gene. *J Biol Chem* 1989; 264: 9891-9896

215. YAO Z and MAC LEOD RS : Synthesis and secretion of hepatic apolipoprotein B-containing lipoproteins. *Biochim Biophys Acta* 1994; 1212: 152-166

216. KARDASSIS D, ZANNIS VI, CLADARAS C : Organization of the regulatory elements and nuclear activities participating in the transcription of human apolipoprotein B gene. *J Biol Chem* 1992; 267: 2622-2632

217. LADIAS JAA and KARATHANASIS SK : Regulation of the apolipoprotein AI gene by ARP-1, a novel member of the steroid receptor superfamily. *Science* 1991; 251: 561-565

218. PAULWEBER B and LEVY-WILSON B : The mechanisms by which a human apolipoprotein B gene enhancer and reducer interact with the promoter are different in cultured cells of hepatic and intestinal origin. *J Biol Chem* 1991; 266: 24161-24168

219. ZHUANG H, CHUANG SS, DAS HK : Transcriptional regulation of the apolipoprotein B100 gene : purification and characterization of trans-acting factor BRF-2. *Mol Cell Biol* 1992; 12: 3183-3191

220. BROOKS AR, BLACKHART BD, HAUBOLD K, LEVY-WILSON B : Characterization of tissue-specific enhancer in the second intron of human apolipoprotein B gene. *J Biol Chem* 1991; 266: 7848-7859

221. LEVY-WILSON B, PAULWEBER B, NAGY BP, LUDWIG EH, BROOKS AR : Nuclease-sensitive sites define a region with enhancer activity in the third intron of the human apolipoprotein B gene. *J Biol Chem* 1992; 267: 18735-18743

222. YOUNG SG, FARESE RV, PIEROTTI VR, TAYLOR S, GRASS DS, LINTON MR : Transgenic mice expressing human apoB100 and apo B48. *Curr Opin Lipidol* 1994; 5: 94-101

223. RAABE M, KIM E, VENIANT M, NIELSEN LB, YOUNG SG : Using genetically engineered mice to understand apolipoprotein-B deficiency syndromes in humans. *Proc Assoc Am Physicians* 1998; 110:521-530.

224. YOUNG SG, BERTICS SJ, SCOTT TM, DUBOIS BW, CURTISS LK, WITZTUM JL : Parallel expression of the MB19 genetic polymorphism in apoprotein B-100 and apoprotein B-48. *J Biol Chem* 1986; 261: 2995-2998

225. SCOTT J : A place in the world for RNA editing. *Cell* 1995; 81: 833-836.

226. DAVIES MS, WALLIS SC, DRISCOLL DM, WYNNE JK, WILLIAMS GW, POWELL LM, SCOTT J : Sequence requirements for apolipoprotein B RNA editing in transfected rat hepatoma cells. *J Biol Chem* 1989; 264: 13395-13398

227. CHAN L, CHANG BH, NAKAMUTA M, LI WH, SMITH LC : Apobec-1 and apolipoprotein B mRNA editing. *Biochim Biophys Acta* 1997; 1345: 11-26

228. SCOTT J, NAVARATNAM N, BHATTACHARYA S, MORRISON JR : The apolipoprotein B messenger RNA editing enzyme. *Curr Opin Lipidol* 1994; 5: 87-93

229. LELLECK H, KRISTEN R, DIEHL I, APOSTEL F, BUCK F : Purification and molecular cloning of a novel essential component of the apolipoprotein B mRNA editing enzyme-complex. *J Biol Chem* 2000; 275: 19848-19856

230. DAVIDSON NO, POWELL LM, WALLIS SC, SCOTT J : Thyroid hormone modulates the introduction of a stop codon in rat liver apolipoprotein B messenger RNA. *J Biol Chem* 1988; 263: 13482-13485

231. HIGUCHI K, KITAGAWA K, KOGISHI K, TAKEDA T : Developmental and age-related changes in apolipoprotein B mRNA editing in mice. *J Lipid Res* 1992; 33: 1753-1764

232. QIAN X, BALESTRA ME, YAMANAKA S, BORÉN J, LEE I, INNERARITY TL : Low expression of the apolipoprotein B mRNA-editing transgene in mice reduces LDL levels but does not cause liver dysplasia or tumors. *Arterioscler Thromb Vasc Biol* 1998; 18: 1013-1020

233. GREEVE J, ALTKEMPER I, JENS-HOLGER D, GRETEN H, WINDLER E : Apolipoprotein B mRNA editing in 12 different species: hepatic expression is reflected in low concentrations of apoB-containing plasma lipoproteins. *J Lipid Res* 1993; 34: 1367-1383

234. PULLINGER CR, NORTH JD, TENG BB, RIFICI VA, RONHILD DE BRITO AE, SCOTT J : The apolipoprotein B gene is constitutively expressed in HepG2 cells : regulation of secretion by oleic acid, albumin, and insulin, and measurements of the mRNA half life. *J Lipid Res* 1989; 30: 1065-1077

235. CHUCK SL, YAO Z, BLACKHART BD, Mc CARTHY BJ, LINGAPPA VR : New variation on the translocation of protein during early biogenesis of apolipoprotein B. *Nature* 1990; 346: 382-385

236. CHUCK SL and LINGAPPA VR : Pause transfert : a topogenic sequence in apolipoprotein B mediates stopping and retarting of translocation. *Cell* 1992; 68: 9-21

237. BORÉN J, WETTESTEN M, SJÖBERG A, THORLIN T, BONDJERS G, WIKLUND O, OLOFSSON SO : The assembly and secretion of apoB 100 containing lipoproteins in HepG2 cells. *J Biol Chem* 1990; 265: 10556-10564

238. GINSBERG HN : Role of lipid synthesis, chaperone proteins and proteasomes in the assembly and secretion of apoprotein B-containing lipoproteins from cultured liver cells. *Clin Exp Pharmacol Physiol* 1997; 24: A29-32

239. LIANG J-S, WU X, JIANG H, ZHOU M, YANG H, ANGKEOW P, HUANG L-S, STURLEY SL, GINSBERG H : Translocation efficiency, susceptibility to proteasomal degradation, and lipid responsiveness of apolipoprotein B are determined by the presence of β sheet domains. *J Biol Chem* 1998; 273: 35216-35221

240. DAVIS RA : Cell and molecular biology of the assembly and secretion of apolipoprotein B-containing lipoproteins by the liver. *Biochim Biophys Acta.* 1999; 1440:1-31

241. TRAN K, WANG Y, DE LONG CJ, CUI Z, YAO Z : The assembly of very low density lipoproteins in rat hepatoma McA-RH7777 cells is inhibited by phospholipase A2 antagonists. *J Biol Chem* 2000; 275: 25023-25030.

242. BROWN A-M, CASTLE J, HEBBACHI A-M, GIBBONS GF : Administration of n-3 fatty acids in the diets of rats or directly to hepatocyte cultures results in different effects on hepatocellular apo B metabolism and secretion. *Arterioscler Thromb Vasc Biol* 1999; 19: 106-114

243. HIRSCHFELD J and BLOMBÄCK M : A new anti-Ag serum (LL). *Nature* 1964; 202: 706-707

244. BÜTLER R, BRUNNER E, MORGANTI G : Contribution to the inheritance of the Ag groups. *Vox Sang* 1974; 26: 485-496

245. YOUNG SG and HUBL ST : An ApaLI restriction site polymorphism is associated with the MB19 polymorphism in apolipoprotein B. *J Lipid Res* 1989; 30: 443-449

246. WANG X, SCHLAPFER P, MA Y, BUTLER R, ELOVSON J, SHUMAKER VN : Apolipoprotein B : The Ag(a1/d) immunogenetic polymorphism coincides with a T-to-C substitution at nucleotide 1981, creating an AluI restriction site. *Arteriosclerosis* 1988; 8: 429-435

247. WU MJ, BÜTLER E, BÜTLER R, SCHUMAKER VN : Identification of the base substitution responsible for the Ag(x/y) polymorphism of apolipoprotein B-100. *Arterioscler Thromb* 1991; 11: 379-384

248. XU C, NANJEE N, TIKKANEN MJ, HUTTUNEN JK, PIETINEN P, BÜTLER R, ANGELICO F, BEN MD, MAZZARELLA B, ANTONIO R, MILLER NG, HUMPHRIES S, TALMUD P : Apolipoprotein B amino acid substitution from arginine to glutamine creates the Ag(h/i) epitope : the polymorphism is not associated with differences in serum cholesterol and apolipoprotein B levels. *Hum Genet* 1989; 82: 322-326

249. MA Y, SCHUMAKER VN, BUTLER R, SPARKES RS : Two DNA restriction fragment length polymorphisms associated with Ag(t/z) and Ag(c/g) antigenic sites of human apolipoprotein B. *Arteriosclerosis* 1987; 7: 301-305

250. BREGUET G, BÜTLER R, BÜTLER-BRUNNER E, SANCHEZ-MAZAS A : A worldwide population study of the Ag-System haplotypes, a genetic polymorphism of human low-density lipoprotein. *Am J Hum Genet* 1990; 46: 502-517

251. RAPACZ J, CHEN L, BÜTLER-BRUNNER E, WU MJ, HASLER-RAPACZ JO, BÜTLER R, SCHUMAKER VN : Identification of the ancestral haplotype for apolipoprotein B suggests an African origin of Homo sapiens sapiens and traces their

subsequent migration to Europe and the Pacific. *Proc Natl Acad Sci USA* 1991; 88: 1403-1406

252. SANCHEZ-MAZAS A and LANGANEY A : Common genetic pools between human populations. *Hum Genet* 1988 78: 161-166

253. YOUNG SG, BERTICS, CURTISS LK, CASAL DC, WITZTUM JL : Monoclonal antibody MB19 detects genetic polymorphism in apolipoprotein B. *Proc Natl Acad Sci USA* 1986; 83: 1101-1105

254. GAVISH D, BRINTON EA, BRESLOW JL : Heritable allele-specific differences in amounts of apoB and low-density lipoproteins in plasma. *Science* 1989; 244: 72-76

255. ARNOLD KS, BALESTRA ME, KRAUSS RM, CURTISS LK, YOUNG SG, INNERARITY TL : Isolation of allele-specific, receptor-binding-defective low density lipoproteins from familial defective apolipoprotein B-100 subjects. *J Lipid Res* 1994; 35: 1469-1476

256. FRIEDL W, LUDWIG EH, BALESTRA ME, ARNOLD KS, PAULWEBER B, SANDHOFER F, Mc CARTHY BJ, INNERARITY TL : Apolipoprotein B gene mutations in Austrian subjects with heart disease and their kindred. *Arterioscler Thromb* 1991; 11: 371-378

257. LINTON MRF, GISH R, HUBL ST, BÜTLER E, ESQUIVEL C, BRY WI, BOYLES JK, WARDELL MR, YOUNG SG : Phenotypes of apolipoprotein B and apolipoprotein E after liver transplantation. *J Clin Invest* 1991; 88: 270-281

258. YANG CY, GU ZW, WENG SA, KIM TW, CHEN SH, POWNALL HJ, SHARP PM, LIU SW, LI WH, GOTTO AM, CHAN L : Structure of apolipoprotein B-100 of human low density lipoproteins. *Arteriosclerosis* 1989; 9: 96-108

259. LUDWIG EH and Mc CARTHY BJ : Haplotype analysis of the human apolipoprotein B mutation associated with familial defective apolipoprotein B100. *Am J Hum Genet* 1990; 47: 712-720

260. DUNNING AM, RENGES HH, XU CF, PEACOCK R, BRASSEUR R, LAXER G, TIKKANEN MJ, BÜTLER R, SAHA N, HAMSTEN A, ROSSENEU M, TALMUD P, HUMPHRIES SE : Two amino acid substitutions in apolipoprotein B are in complete allelic association with the Ag group (x/y) polymorphism : evidence for little recombination in the 3' end of the human gene. *Am J Hum Genet* 1992; 50: 208-221

261. HASSTEDT SJ, WU L, WILLIAMS RR : Major locus inheritance of apolipoprotein B in Utah pedigrees. *Genet Epidemiol* 1987; 4: 67-76

262. AUSTIN MA, NEWMAN B, SELBY JV, EDWARDS K MAYER EJ, KRAUSS RM : Genetics of LDL subclass phenotypes in women twins. Concordance, heritability, and commingling analysis. *Arterioscler Thromb* 1993; 13: 687-695

263. HUMPHRIES SE, TALMUD PJ : Hyperlipidaemia associated with genetic variation in the apolipoprotein B gene. *Curr Opin Lipidol* 1995; 6:215-222

264. STURLEY SL, TALMUD PJ, BRASSEUR R, CULBERTSON MR, HUMPHRIES SE, ATTIE AD : Human apolipoprotein B signal sequence variants confer a secretion-defective phenotype when expressed in yeast. *J Biol Chem.* 1994; 269:21670-21675

265. BOERWINKLE E and CHAN L : A three codon insertion/deletion polymorphism in the signal peptide region of the human apolipoprotein B (APOB) gene directly typed by the polymerase chain reaction. *Nucleic Acids Res* 1989; 17: 4003

266. HUANG LS, RIPPS ME, BRESLOW JL : Molecular basis of five apolipoprotein B gene polymorphisms in noncoding regions. *J Lipid Res* 1990; 31: 71-77

267. HUANG LS, KAYDEN H, SOKOL RJ, BRESLOW JL : ApoB gene nonsense and splicing mutations in a compound heterozygote for familial hypobetalipoproteinemia. *J Lipid Res* 1991; 32: 1341-1348

268. JONES T, RAJPUT-WILLIAMS J, KNOTT TJ, SCOTT J : An MspI RFLP in the APO B promoter. *Nucleic Acids Res* 1989; 17: 472

269. NAVAJAS M, LAURENT AM, MOREEL JF, RAGAB A, CAMBOU JP, CUNY G, CAMBIEN F, ROIZES G : Detection by denaturing gradient gel electrophoresis of a new polymorphism in the apolipoprotein B gene. *Hum Genet* 1990; 86: 91-93

270. SHRIVER MD, SIEST G, BOERWINKLE E : Length and sequence variation in the apolipoprotein B intron 20 Alu repeat. *Genomics* 1992; 14: 449- 454

271. DEEB SS, FAILOR RA, BROWN BG, BRUNZELL JD, ALBERS JJ, MOTULSKY AG, WIJSMAN E : Association of apolipoprotein B gene variants with plasma apoB and low density lipoprotein (LDL) cholesterol levels. *Hum Genet* 1992; 88: 463-470

272. LASKAREWSKI PM, KHOURY P, MORRISON JA, KELLY K, MELLIES MJ, GLUECK CJ : Prevalence of familial hyper- and hypolipoproteinemias: the Princeton school district family study. *Metabolism* 1982; 31: 558-577

273. KANE JP, HAVEL RJ : Disorders of the biogenesis and secretion of lipoproteins containing the B apolipoprotein. In: Scriver CR, Beaudet AL, Sly WS, Valle D (eds). *The Metabolic and Molecular Bases of Inherited Disease*, 7th edn. Mac Graw Hill, New York, 1995 1853-1885

274. LEPPERT M, BRESLOW JL, WU L, HASSTEDT S, O'CONNELL P, LATHROP M, WILLIAMS RR, WHITE R, LALOUEL JM : Inference of a molecular defect of apolipoprotein B in hypobetalipoproteinemia by linkage analysis in a large kindred. *J Clin Invest* 1988; 82: 847-851

275. HUANG LS, RIPPS E, KORMAN SH, DECKELBAUM RJ, BRESLOW JL : Hypobetalipoproteinemia due to an apolipoprotein B gene exon 21 deletion derived by Alu-Alu recombination. *J Biol Chem* 1989; 264: 11394-11400

276. TALMUD PJ, KRUL E, PESSAH M, GAY G, SCHONFELD G, HUMPHRIES S E, INFANTE R : Donor splice mutation generates a lipid-associated apolipoprotein B-27.6 in a patient with homozygous hypobetalipoproteinemia. *J Lipid Res* 1994; 35: 468-477

277. COLLINS DR, KNOTT TJ, PEASE RJ, POWELL LM, WALLIS SC, ROBERTSON S, PULLINGER CR, MILNE RW, MARCEL YL, HUMPHRIES SE, TALMUD PJ, LLOYD JK, MILLER NE, MULLER D, SCOTT J : Truncated variants of apolipoprotein B cause hypobetalipoproteinemia. *Nucleic Acids Res* 1988; 16: 8361-8375

278. YOUNG SG, HUBL ST, SMITH RS, SNYDER SM, TERDIMAN JF : Familial hypobetaliproteinemia caused by a mutation in the apolipoprotein B gene that results in a truncated species of apolipoprotein B (B 31). *J Clin Invest* 1990; 85: 933-942

279. MAC CORMICK SPA, FELLOWES AP, WALMSLEY TA, GEORGE PM : Apolipoprotein B-32 : a new truncated mutant of human apolipoprotein B capable of forming particles in the low density lipoprotein range. *Biochim Biophys Acta* 1992; 1138: 290-296

280. YOUNG SG, PULLINGER CR, ZYSOW BR, HOFMANN-RADVANI H, LINTON MR, FARESE RV, TERDIMAN JF, SNYDER SM, GRUNDY SM, VEGA GL, MALLOY MJ, KANE JP : Four new mutations in the apolipoprotein B gene causing hypobetaliproteinemia, including two different frameshift mutations that yield truncated apoprotein B proteins of identical length. *J Lipid Res* 1993; 34: 501-507

281. YOUNG SG, NORTHEY ST, Mc CARTHY BJ : Low plasma cholesterol levels caused by a short deletion in the apolipoprotein B gene. *Science* 1988; 241: 591-593

282. GROENEWEGEN WA, AVERNA MR, PULAI J, KRUL ES, SCHONFELD G : Apolipoprotein B-38.9 mutation does not associate with apo(a) and forms two distinct HDL density particle populations that are larger than HDL. *J Lipid Res* 1994; 35: 1012-1025

283. OHASHI K, ISHIBASHI S, YAMAMOTO M, OSUGA J-I, YAZAKI Y, YUKAWA S, YAMADA N : A truncated species of apolipoprotein B (B38.7) in a patient with homozygous hypobetalipoproteinemia associated with diabetes mellitus. *Arterioscler Thromb Vasc Biol* 1998; 18: 1330-1334

284. TARUGI P, LONARDO A, BALLARINI G, GRISENDI A, PULVIRENTI M, BAGNI A, CALANDRA S : Fatty liver in heterozygous hypobetalipoproteinemia caused by a novel truncated form of apolipoprotein B. *Gastroenterology* 1996;111:1125-1133

285. KRUL ES, KINOSHITA M, TALMUD P, HUMPHRIES SE, TURNER S, GOLDBERG AC, COOK K, BOERWINKLE E, SCHONFELD G : Two distinct truncated apolipoprotein B species in a kindred with hypobetalipoproteinemia. *Arteriosclerosis* 1989; 9: 856-868

286. SRIVASTAVA N, NOTO D, AVERNA M, PULAI J, SRIVASTAVA RA, COLE TG, LATOUR MA, PATTERSON BW, SCHONFELD G : A new apolipoprotein B truncation (apo B-43.7) in familial hypobetalipoproteinemia: genetic and metabolic studies. *Metabolism* 1996; 45: 1296-1304

287. WELTY FK, ORDOVAS J, SCHAEFER EJ, WILSON PW, YOUNG SG : Identification and molecular analysis of two apoB gene mutations causing low plasma cholesterol levels. *Circulation* 1995; 92: 2036-2040

288. YOUNG SG, BIHAIN B, FLYNN LM, SANAN DA, AYRAULT-JARRIER M, JACOTOT B : Asymptomatic homozygous hypobetalipoproteinemia associated with apolipoprotein B45.2. *Hum Mol Genet* 1994; 3:741-744

289. YOUNG SG, HUBL ST, CHAPPELL DA, SMITH RS, CLAIBORNE F, SNYDER SM, TERDIMAN JF : Familial hypobetalipoproteinemia associated with a mutant species of apolipoprotein B (B-46). *N Engl J Med* 1989; 320: 1604-1610

290. RUOTOLO G, ZANELLI T, TETTAMANTI C, RAGOGNA F, PARLAVECCHIA M, VIGANO F, CATAPANO AL : Hypobetalipoproteinemia associated with apo B-48.4, a truncated protein only 14 amino acids longer than apo B-48. *Atherosclerosis* 1998; 137:125-131

291. HARDMAN DA, PULLINGER CR, KANE JP, MALLOY MJ : Molecular defect in normotriglyceridemic abetalipoproteinemia. *Circulation* 1989; 80 II: 466

292. GROENEWEGEN WA, KRUL ES, SCHONFELD G : Apolipoprotein B-52 mutation associated with hypobetalipoproteinemia is compatible with a misaligned pairing deletion mechanism. *J Lipid Res* 1993; 34: 971-981

293. WAGNER RD, KRUL ES, TANG J, PARHOFER KG, GARLOCK K, TALMUD P, SCHONFELD G : ApoB-548, a truncated apolipoprotein found primarily in VLDL, is associated with a nonsense mutation in the apo B gene and hypobetalipoproteinemia. *J Lipid Res* 1991; 32: 1001-1011

294. TALMUD PJ, CONVERSE C, KRUL E, HUQ L, MAC ILWAINE GG, SERIES JJ, BOYD P, SCHONFELD G, DUNNING A, HUMPHRIES S E : A novel truncated apolipoprotein B (apo B-55) in a patient with familial hypobetalipoproteinemia and atypical retinis pigmentosa. *Clin Genet* 1992; 42: 62-70

295. PULLINGER CR, HILLAS E, HARDMAN DA, CHEN GC, NAYA-VIGNE J, IWASA JA, HAMILTON RL, LALOUEL JM, WILLIAMS RR, KANE JP : Two apolipoprotein B gene defects in a kindred with hypobetalipoproteinemia, one of which results in a truncated variant, apo B-61, in VLDL and LDL. *J Lipid Res* 1992; 33: 699-710

296. WELTY FK, HUBL ST, PIEROTTI VR, YOUNG SG : A truncated species of apolipoprotein B (B67) in a kindred with familial hypobetalipoproteinemia. *J Clin Invest* 1991; 87: 1748-1754

297. GROENEWEGEN WA, KRUL ES, AVERNA MR, PULAI J, SCHONFELD G : Dysbetalipoproteinemia in a kindred with hypobetalipoproteinemia due to mutations in the genes for apo B (ApoB 70.5) and Apo E (Apo E2). *Arterioscler Thromb* 1994; 14: 1695-1704

298. KRUL ES, PARHOFER KG, BARRETT HR, WAGNER RD, SCHONFELD G : Apo B-75, a truncation of apolipoprotein B associated with familial hypobetalipoproteinemia : genetic and kinetic studies. *J Lipid Res* 1992; 33: 1037-1050

299. FARESE RV, GARG A, PIEROTTI VR, VEGA GL, YOUNG SG : A truncated species of apolipoprotein B, B-83 associated with hypobetalipoproteinemia. *J Lipid Res* 1992; 33: 569-577

300. LINTON MF, PIEROTTI VR, HUBL ST, YOUNG SG : An apo-B gene mutation causing familial hypobetalipoproteinemia analyzed by examining the apo-B cDNA amplified from the fibroblast RNA of an affected subject. *Clin Res* 1990; 38: 286A

301. TENNYSON GE, GABELLI C, BAGGIO G, BILATO C, BREWER HB : Molecular defect in the apolipoprotein B gene in a patient with hypobetalipoproteinemia and three distinct apoB species. *Clin Res* 1990; 38: 482A

302. SCHONFELD G : The hypobetalipoproteinemias. *Annu Rev Nutr* 1995; 15: 23-34

303. HERSCOVITZ H, HADZOPOULOU-CLADARAS M, WALSH M, CLADARAS C, ZANNIS VI, SMALL DM : Expression, secretion, and lipid-binding characterization of the N-terminal 17% of apolipoprotein B. *Proc Natl Acad Sci USA* 1991; 88: 7313-7317

304. MAC LEOD RS, ZHAO Y, SELBY SL, WESTERLUND J, YAO Z : Carboxy-terminal truncations impairs lipid recruitment by apolipoprotein B100 but does not affect secretion of the truncated apolipoprotein B-containing lipoproteins. *J Biol Chem* 1994; 269: 2852- 2862

305. PARHOFER KG, DAUFHERTY A, KINOSHITA M, SCHONFELD G : Enhanced clearance from plasma of low density lipoproteins containing a truncated apolipoprotein, B-89. *J Lipid Res* 1990; 31: 2001-2007

306. PARKER RC, ILLINGWORTH RD, BISSONNETTE J, CARR BR : Endocrine changes during pregnancy in a patient with homozygous familial hypobetalipoproteinemia. *N Engl J Med* 1986; 314: 557-560

307. KAHN JA and GLUECK CJ : Familial hypobetalipoproteinemia: absence of atherosclerosis in a postmorterm study. *JAMA* 1978; 240: 47-48

308. INNERARITY TL, MAHLEY RW, WEISGRABER KH, BERSOT TP, KRAUSS RM, VEGA GL, GRUNDY SM, FRIEDL W, DAVIGNON J, Mc CARTHY BJ : Familial defective apolipoprotein B-100 : a mutation of apolipoprotein B that causes hypercholesterolemia. *J Lipid Res* 1990; 31: 1337-1349

309. TYBJÆRG-HANSEN A, HUMPHRIES SE : Familial defective apolipoprotein B-100 : a single mutation that causes hypercholesterolemia and premature coronary artery disease. *Atherosclerosis* 1992; 96: 91-107

310. MYANT NB : Familial defective apolipoprotein B-100: a review, including some comparisons with familial hypercholesterolaemia. *Atherosclerosis* 1993; 1-18

311. HANSEN PS : Familial defective apolipoprotein B-100. *Dan Med Bull* 1998; 45: 370-382

312. TYBJÆRG-HANSEN A, STEFFENSEN R, MEINERTZ H, SCHNOHR P, NORDESTGAARD BG. Association of mutations in the apolipoprotein B gene with hypercholesterolemia and the risk of ischemic heart disease. *N Engl J Med* 1998; 338: 1577-1584

313. PIMSTONE SN, DEFESCHE JC, CLEE SM, BAKKER HD, HAYDEN MR, KASTELEIN JJP. Differences in the phenotype between children with familial defective apolipoprotein B-100 and familial hypercholesterolemia. *Arterioscler Thromb Vasc Biol* 1997; 17: 826-833

314. HANSEN PS, DEFESCHE JC, KASTELEIN JJP, GERDES LU, FRAZA L, GERDES C, TATO F, JENSEN HK, JENSEN LG, KLAUSEN IC, FAERGEMAN O, SCHUSTER H : Phenotypic variation in patients heterozygous for familial defective apolipoprotein B (FDB) in three european countries. *Arterioscler Thromb Vasc Biol* 1997; 17: 751-747

315. MÄRZ W, BAUMSTARK MW, SCHARNAGL H, RUZICKA V, BUXBAUM S, HERWIG J, POHL T, RUSS A, SCHAAF L, BERG A, BÖHLES HJ, USADEL KH, GROß W : Accumulation of "small dense" low density lipoproteins (LDL) in a homozygous patient with familial defective apolipoprotein B-100 results from heterogeneous interaction of LDL subfractions with the LDL receptor. *J Clin Invest* 1993; 92: 2922-2933

316. STALENHOEF AFH, DEFESCHE JC, KLEINVELD HA, DEMACKER PNM, KASTELEIN JJP : Decreased resistance against in vitro oxidation of LDL from patients with familial defective apolipoprotein B-100. *Arterioscler Thromb* 1994; 14: 489-493

317. BERSOT TP, RUSSELL SJ, THATCHER SR, POMERNACKI NK, MAHLEY RW, WEISGRABER KH, INNERARITY TL, FOX CS : A unique haplotype of the apolipoprotein B-100 allele associated with familial defective apolipoprotein B-100 in a Chinese man discovered during a study of the prevalence of this disorder. *J Lipid Res* 1993; 34: 1149-1154

318. RAUH G, SCHUSTER H, SCHEWE CK, STRATMANN G, KELLER C, WOLFRAM G, ZÖLLNER N : Independent mutation of arginine(3500)→glutamine associated with familial defective apolipoprotein B-100. *J Lipid Res* 1993; 34: 799-805

319. ABDEL-WARETH LO, PIMSTONE SN, LAGARDE JP, RAISONNIER A, BENLIAN P, PRITCHARD H, HAYDEN MR, FROHLIC JJ. Familial defective apolipoprotein B-100 in hypercholesterolemic Chinese Canadians: identification of a unique haplotype of the apolipoprotein B-100 allele. *Atherosclerosis* 1997; 135: 181-185

320. FISHER E, SCHARNAGL H, HOFFMANN MM, KUSTERER K, WITTMANN D, WIELAND H, GROSS W, MÄRZ W : Mutations in the apolipoprotein (apo) B-100 receptor-binding region: detection of apo B-100 (Arg3500→Trp) associated with two new haplotypes and evidence that apo B-100 (Glu3405→Gln) diminishes receptor-mediated uptake of LDL. *Clin Chem* 1999; 45: 1026-1038

321. BORÉN J, OLIN K, CHAIT A, WRIGHT TN, INNERARITY TL : Identification of the principal proteoglycan-binding site in LDL. A single-point mutation in apo-B100 severely affects proteoglycan interaction without affecting LDL receptor binding. *J Clin Invest* 1998; 101: 2658-2664

322. BASSEN FA and KRONZWEIG AL : Malformation of the erythrocytes in a case of atypical retinitis pigmentosa. *Blood* 1950; 5: 381-387

323. ROSS RS, GREGG RE, LAW SW, MONGE JC, GRANT SM, HIGUCHI K, TRICHE TJ, JEFFERSON J, BREWER HB : Homozygous hypobetalipoproteinemia: a disease distinct from abetalipoproteinemia at the molecular level. *J Clin Invest* 1988; 81: 590-595

324. LACKNER KJ, MONGE JC, GREGG RE, HOEG JM, TRICHE TJ, LAW SW, BREWER HB : Analysis of the apolipoprotein B gene and messenger ribonucleic acid in abetalipoproteinemia. *J Clin Invest* 1986; 78: 1707-1712

325. TALMUD PJ, LLOYD JK, MULLER DPR, COLLINS DR, SCOTT J, HUMPHRIES S E : Genetic evidence from two families that apolipoprotein B gene is not involved in abetalipoproteinemia. *J Clin Invest* 1988; 82: 1803-1806

326. GREGG RE, WETTEREAU JR : The molecular basis of abetalipoproteinemia. *Curr Opin Lipidol* 199; 5: 81-86

327. SHARP D, BLINDERMAN L, COMBS KA, KIENZLE B, RICCI B, WAGER-SMITH K, GIL CM, TURCK CW, BOUMA ME, RADER DJ, AGGERBECK LP, GREGG RE, GORDON DA, WETTEREAU JR : Cloning and gene defects in microsomal triglyceride transfer protein associated with abetalipoproteinemia. *Nature* 1993; 365: 65-69

328. SHOULDERS CC, BRETT DJ, BAYLISS JD, NARCISI TME, JARMUZ A, GRANTHAM TT, LEONI PRD, BHATTACHARYA S, PEASE RJ, CULLEN PM, LEVI S, BYFIELD PGH, PURKISS P, SCOTT J : Abetalipoproteinemia is caused by defects of the gene encoding the 97 kDa subunit of a microsomal triglyceride transfer protein. *Hum Mol Genet* 1993; 2: 2109- 2116

329. SHARP D, RICCI B, KIENZLE B, LIN MC, WETTERAU JR : Human microsomal triglyceride transfer protein large subunit gene structure. *Biochemistry* 1994; 33:9057-9061

330. HAGAN DL, KIENZLE B, JAMIL H, HARIHARAN N : Transcriptional regulation of human and hamster microsomal triglyceride transfer protein genes. *J Biol Chem* 1994; 269: 28737-28744

331. RICCI B, SHARP D, O'ROURKE E, KIENZLE B, BLINDERMAN L, GORDON D, SMITH-MONROY C, ROBINSON G, GREGG RE, RADER DJ, WETTERAU JR : A 30-amino acid truncation of the microsomal triglyceride transfer protein large subunit disrupts its interaction with protein disulfide-isomerase and causes abetalipoproteinemia. *J Biol Chem* 1995; 270:14281-14285

332. NARCISI TM, SHOULDERS CC, CHESTER SA, READ J, BRETT DJ, HARRISON GB, GRANTHAM TT, FOX MF, POVEY S, DE BRUIN TW, et al : Mutations of the

microsomal triglyceride-transfer-protein gene in abetalipoproteinemia. *Am J Hum Genet* 1995; 57: 1298-12310

333. REHBERG EF, SAMSON-BOUMA ME, KIENZLE B, BLINDERMAN L, JAMIL H, WETTERAU JR, AGGERBECK LP, GORDON DA : A novel abetalipoproteinemia genotype. Identification of a missense mutation in the 97-kDa subunit of the microsomal triglyceride transfer protein that prevents complex formation with protein disulfide isomerase. *J Biol Chem* 1996; 271: 29945-29952

334. YANG XP, INAZU A, YAGI K, KAJINAMI K, KOIZUMI J, MABUCHI H : Abetalipoproteinemia caused by maternal isodisomy of chromosome 4q containing an intron 9 splice acceptor mutation in the microsomal triglyceride transfer protein gene. *Arterioscler Thromb Vasc Biol* 1999; 19: 1950-1955

335. WANG J, HEGELE RA : Microsomal triglyceride transfer protein (MTP) gene mutations in Canadian subjects with abetalipoproteinemia. *Hum Mutat* 2000; 15: 294-295

336. OHASHI K, ISHIBASHI S, OSUGA JI, TOZAWA RI, HARADA K, YAHAGI N, SHIONOIRI F, IIZUKA Y, TAMURA Y, NAGAI R, ILLINGWORTH DR, GOTODA T, YAMADA N : Novel mutations in the microsomal triglyceride transfer protein gene causing abetalipoproteinemia. *J Lipid Res* 2000; 41: 1199-1204

337. BENLIAN P, FOUBERT L, GAGNÉ E, BERNARD L, DE GENNES JL, LANGLOIS S, ROBINSON W, HAYDEN MR : Complete paternal isodisomy for chromosome 8 unmasked by lipoprotein lipase deficiency. *Am J Hum Genet* 1996; 59: 431-436

338. HALL JG : Genomic imprinting : review and relevance to human diseases. *Am J Hum Genet* 1990 46: 857-873

339. LEDBETTER DH, ENGEL E. Uniparental disomy in humans: development of an imprinting map and its implications for prenatal diagnosis. *Hum Mol Genet* 1995; 4: 1757-1764

340. KARPE F, LUDHAHL B, EHRENBORG E, ERIKSSON P, HAMSTEN A : A common functional polymorphism in the promoter region of the microsomal triglyceride transfer protein gene influences plasma LDL levels. *Arterioscler Thromb Vasc Biol* 1998; 18: 756-761

341. HANK JUO S-H, HAN Z, SMITH JD, COLANGELO L, LIU K : Common polymorphism in the promoter of microsomal triglyceride transfer protein gene influence cholesterol, apoB, and triglyceride levels in young African American men. Results from the coronary artery risk development in young adults (CARDIA) study. *Arteriocler Thromb Vasc Biol* 2000; 20: 1316-1322

342. COUTURE P, OTVOS JD, CUPPLES A, WILSON PWF, SCHAEFER EJ, ORDOVAS JM : Absence of association between genetic variation in the promoter of microsomal triglyceride transfer protein gene and plasma lipoproteins in the Framingham offspring study. *Atherosclerosis* 2000; 148: 337-343

343. WHITE DA, BENNETT AJ, BILLETT MA, SALTER AM : The assembly of triacylglycerol-rich lipoproteins: an essential role for the microsomal triacylglycerol transfer protein. *Br J Nutr* 1998; 80: 219-229

344. GORDON DA, JAMIL H : Progress towards understanding the role of microsomal triglyceride transfer protein in apolipoprotein-B lipoprotein assembly. *Biochim Biophys Acta* 2000; 1486: 72-83

345. BENOIST F, GRAND-PERRET T : Co-translational degradation of apolipoprotein B-100 by the proteasome is prevented by microsomal triglyceride transfer protein. *J Biol Chem* 1997; 272: 20435-20442

346. LEUNG GK, VÉNIANT MM, KIM SK, ZLOT CH, RAABE M, BJÖRKEGREN J, NEESE RA, HELLERSTEIN MK, YOUNG SG. A deficiency of microsomal triglyceride transfer protein reduces apolipoprotein B secretion. *J Biol Chem* 2000; 275: 7515-7520

347. TIETGE UJF, BAKILLAH A, MAUGEAIS C, TSUKAMOTO K, HUSSAIN M, RADER DJ : Hepatic overexpression of microsomal triglyceride transfer protein (MTP) results in increased in vivo secretion of VLDL triglycerides and apolipoprotein B. *J Lipid Res* 1999; 40: 2134-2139

348. RAABE M, FLYNN LM, ZLOT CH, WONG JS, VÉNIANT MM, HAMILTON RL, YOUNG SG : Knock out of the abetalipoproteinemia gene in mice: reduced lipoprotein secretion in heterozygotes and embryonic lethality in homozygotes. *Proc Natl Acad Sci USA* 1998; 95: 8686-8691

349. CHANG BH-J, LIAO W, NAKAMUTA M, MACK D, CHAN L : Liver-specific inactivation of the abetalipoproteinemia gene completely abrogates very low density lipoprotein/low density lipoprotein production in a viable conditional knock out mouse. *J Biol Chem* 1999; 274: 6051-6055

350. WETTEREAU JR, GREGG RE, HARRITY TW, ARBEENY C, CAP M, CONOLLY F, CHU C-H, GEORGE RJ, GORDON DA, JAMIL H, JOLIBOIS KG, KUNSELMAN LK, LAN S-J, MACCAGAN TJ, RICCI B, YAN M, YOUNG D, CHEN Y, FRYSZMAN OM, LOGAN JVH, MUSIAL CL, POSS MA, ROBL JA, SIMPKINS LM, SLUSARCHYK WA, SULSKY R, TAUNK P, MAGNIN DR, TINO JA, LAWRENCE RM, DICKSON JK, BILLER SA : An MTP inhibitor that normalize atherogenic lipoprotein levels in WHHL rabbits. *Science* 1998; 282: 751-754

351. ALLISON AC, BLUMBERG BS: An isoprecipitation reaction distinguishing human serum protein types. *Lancet* 1961; i: 634-637

352. BERG K : A new serum type system in man : the Lp-system. *Acta Pathol Microbiol Scand* 1963; 59: 369-382

353. UTERMANN G : The mysteries of lipoprotein (a). *Science* 1989; 242: 904-910

354. HOBBS HH and WHITE AL : Lipoprotein(a):intrigues and insights.*Curr Opin Lipidol* 1999; 10: 225-236

355. SCANU AM and FLESS GM : Lipoprotein (a). Heterogeneity and biological relevance. *J Clin Invest* 1990; 85: 1709-1715

356. Mac LEAN JW, TOMLINSON JE, KUANG WJ, EATON DL, CHEN EY, FLESS GM, SCANU AM, LAWN RM : cDNA sequence of human apolipoprotein (a) is homologous to plasminogen. *Nature* 1987; 330: 132- 137

357. GAW A and HOBBS HH : Molecular genetics of lipoprotein (a) : new pieces to the puzzle. *Curr Opin Lipidol* 1994; 5: 149-155

358. KLEZOVITCH O, EDELSTEIN C, AND SCANU AM : Apolipoprotein(a) binds via its C-terminal domain to the protein core of the proteoglycan decorin. Implications for the retention of lipoprotein(a) in atherosclerotic lesions. *J Biol Chem* 1998; 273: 23856-65

359. TOMLINSON JE, Mc LEAN JW, LAWN RM : Rhesus monkey apolipoprotein (a). Sequence, evolution and sites of synthesis. *J Biol Chem* 1989; 264: 5957-5965

360. WHITE AL, LANFORD RE : Biosynthesis and metabolism of lipoprotein(a). *Curr Opin Lipidol* 1995; 6: 75-80

361. BRUNNER C, KRAFT HG, UTERMANN G, MÜLLER HJ : Cys4057 of apolipoprotein (a) is essential for lipoprotein assembly. *Proc Natl Acad Sci USA* 1993; 90: 11643-11647

362. KOCHINSKY ML, CÔTE GP, GABEL B, VAN DER HOEK YY : Identification of the cysteine residue in apolipoprotein (a) that mediates extracellular coupling with apolipoprotein B-100. *J Biol Chem* 1993; 268: 19819-19825

363. GAUBATZ JW, HEIDEMANN C, GOTTO AM, MORRISETT JD, DAHLEN G : Human plasma lipoprotein (a) : structural properties. *J Biol Chem* 1983; 258: 4582-4589

364. CHIESA G, HOBBS HH, KOSCHINSKI ML, LAWN RM, MAIKA SD, HAMMER RE : Reconstitution of lipoprotein (a) by infusion of human low density lipoprotein into transgenic mice expressing human apolipoprotein (a). *J Biol Chem* 1992; 267: 24369-24374

365. LINTON MRF, FARESE RV, CHIESA G, GRASS DS, CHIN P, HAMMER RE, HOBBS HH, YOUNG SG : Transgenic mice expressing high plasma concentrations of human apolipoprotein B-100 and lipoprotein (a). *J Clin Invest*; 1993; 92: 3029-3037

366. ALBERS JJ, MARCOVINA SM, LODGE MS : The unique lipoprotein(a): properties and immunochemical measurement. *Clin Chem* 1990; 36: 2019-2026

367. MILES LA, FLESS GM, LEVIN EG, SCANU AM, PLOW EF : A potential basis for the thrombotic risks associated with lipoprotein (a). *Nature* 1989; 339: 301- 303

368. HAJJAR KA, GAVISH D, BRESLOW JL, NACHMAN RL : Lipoprotein (a) modulation of endothelial cell surface fibrinolysis and its potential role in atherosclerosis. *Nature* 1989; 339: 303-305

369. HARPEL PC, GORDON BR, PARKER TS : Plasmin catalyzes binding of lipoprotein (a) to immobilized fibrinogen and fibrin. *Proc Natl Acad Sci USA* 1989; 86: 3847-3851

370. GRAINGER DJ, METCALFE JC : Transforming growth factor-beta: the key to understanding lipoprotein(a). *Curr Opin Lipidol* 1995; 6: 81-85

371. CUSHING GL, GAUBATZ JW, NAVA ML, BURDICK BJ, BOCAN TMA, GUYTON JR, WEILBAECHER D, De BAKEY ME, LAWRIE GM, MORRISETT JD : Quantitation and localization of apolipoprotein (a) and B in coronary bypass vein grafts resected at re-operation. *Arteriosclerosis* 1989; 9: 593-603

372. RATH M, NIENDORF A, REBLIN T, DIETEL M, KREBBER HJ, BEISIEGEL U : Detection and quantification of Lp(a) in arterial wall of 107 coronary bypass patients. *Arteriosclerosis* 1989; 9: 579-592

373. UTERMANN G : Lipoprotein(a). In: Scriver CR, Beaudet AL, Sly WS, Valle D (eds). *The Metabolic and Molecular Bases of Inherited Disease*, 7th edn. Mac Graw Hill, New York, 1995; 1887-1912

374. LAWN RM, WADE DP, HAMMER RE, CHIESA G, VERSTUYFT JG, RUBIN EM : Atherogenesis in transgenic mice expressing human apolipoprotein (a). *Nature* 1992; 360: 670-672

375. FRANK SL, KLISAK I, SPARKES RS, MOHANDAS T, TOMLINSON JE, Mc LEAN JW, LAWN RM, LUSIS AJ : The apolipoprotein (a) gene resides on human

chromosome 6q26-27, in close proximity to the homologous gene for plasminogen. *Hum Genet* 1988; 79: 352-356

376. DRAYNA DT, HEGELE RA, HASS PE, EMI M, WU LL, EATON DL, LAWN RM, WILLIAMS RR, WHITE RL, LALOUEL JM : Genetic linkage between lipoprotein (a) phenotype and a DNA polymorphism in the plasminogen gene. *Genomics* 1988; 3: 230-236

377. FRANK SL, KLISAK I, SPARKES RS, LUSIS AJ : A gene homologous to plasminogen located on human chromosome 2q11-p11. *Genomics* 1989; 4 449-451

378. BYRNE CD, SCHWARTZ K, CHENG JF, LAWN RM : The human apolipoprotein (a)/plasminogen gene cluster contains a novel homologue transcribed in liver. *Arterioscler Thromb* 1994; 14: 534-541

379. MAGNAGHI P, CITTERIO E, MALGARETTI N, ACQUATI F, OTTOLENGHI S, TARAMELLI R : Molecular characterization of the human apolipoprotein (a)-plasminogen gene family clustered on the telomeric region of chromosome 6 (6q26-27). *Hum Mol Genet* 1994; 3: 437-442

380. MALGARETTI N, ACQUATI F, MAGNAGHI P, BRUNO L, PONTOGLIO M, ROCCHI M, SACCONE S, DELLA VALLE G, D'URSO M, LEPASLIER D, OTTOLENGHI S, TARAMELLI R : Characterization by yeast artificial chromosome cloning of the linked apolipoprotein (a) and plasminogen genes and identification of the apolipoprotein (a) 5' flanking region. *Proc Natl Acad Sci USA* 1992; 89: 11584-11588

381. WADE DP, CLARKE JG, LINDAHL GE, LIU AC, ZYSOW BR, MEER K, SCHWARTZ K, LAWN RM : 5' control regions of the apolipoprotein (a) gene and members of the related plasminogen gene family. *Proc Natl Acad Sci USA* 1993; 90: 1369-1373

382. LACKNER C, BOERWINKLE E, LEFFERT CC, RAHMING T, HOBBS HH : Molecular basis of apolipoprotein (a) isoform size heterogeneity as revealed by pulse-field gel electrophoresis. *J Clin Invest* 1991; 87: 2153-2161

383. ERDEL M, HUBALEK M, LINGENHEL A, KOFLER K, DUBA H-C, UTERMANN G : Counting the repetitive kringle-IV repeats in the gene encoding human apolipoproein(a) by fibre-FISH. *Nature Genet* 1999; 21: 357-358

384. LACKNER C, COHEN JC, HOBBS HH : Molecular definition the extreme size polymorphism of apolipoprotein (a). *Hum Mol Genet* 1993; 7: 933-940

385. UTERMAN G : Genetic architecture and evolution of the lipoprotein(a) trait. *Curr Opin Lipidol* 1999; 10: 133-141

386. HOURCADE D, MIESNER DR, BEE C, ZELDES W, ATKINSON JP : Duplication and divergence of the amino-terminal coding region of the complement receptor 1 (CR1) gene. An example of concerted (horizontal) evolution within a gene. *J Biol Chem* 1990; 265: 974-980

387. KOCHINSKY M, BEISIEGEL U, HENNE-BRUNS D, EATON DL, LAWN RM : Apolipoprotein (a) size heterogeneity is related to variable number of repeat sequences in its mRNA. *Biochemistry* 1990; 29: 640-644

388. UTERMANN G, DUBA HC, MENZEL HJ : Inheritance of Lp(a) glycoprotein phenotypes. *Hum Genet* 1988; 78: 47-50

389. BOERWINKLE E, LEFFERT C, LACKNER C, CHIESA G, HOBBS HH: Apolipoprotein (a) gene accounts for greater than 90% of the variation in plasma lipoprotein (a) concentrations. *J Clin Invest* 1992; 90: 52-60

390. COHEN JC, CHIESA G, HOBBS HH : Sequence polymorphisms in the apolipoprotein (a) gene Evidence for dissociation between apolipoprotein (a) size and lipoprotein (a) levels. *J Clin Invest* 1993; 91: 1630-1636

391. AUSTIN MA, SANDHOLZER C, SELBY JV, NEWMAN B, KRAUSS RM, UTERMANN G : Lipoprotein (a) in women twins : heritability and relationships to apolipoprotein (a) phenotypes. *Am J Hum Genet* 1992; 51: 829-840

392. AZROLAN N, GAVISH D, BRESLOW JL : Plasma lipoprotein (a) concentration is controlled by apolipoprotein (a) (Apo(a)) protein size and the abundance of hepatic Apo(a) mRNA in cynomolgus monkey model. *J Biol Chem* 1991; 266: 13866-13872

393. OGORELKOVA M, GRUBER A, UTERMAN G : Molecular basis of Lp(a) deficiency: a frequent apo(a) "null" mutation in Caucasians. *Hum Mol Genet* 1999; 8: 2087-2096

394. THILLET J, DOUCET C, CHAPMAN J, HERBETH B COHEN D, FAURE-DELANEF L : Elevated lipoprotein(a) levels and small apo(a) isoforms are compatible with longevity: evidence from a large population of French centenarians. *Atherosclerosis* 1998; 136: 389-394

395. MAEDA S, ABE A, SEISHIMA M, MAKINO K, NOMA A, KAWADE M : Transient changes of serum lipoprotein (a) as an acute phase protein. *Atherosclerosis* 1989; 78: 145-150

396. PARK YB, LEE SK, LEE WK, SUH CW, LEE CH, SONG CH, LEE J : Lipid profiles in untreated patients with rheumatoid arthritis. *J Rheumatol* 1999; 26: 1701-1704

397. CONSTANS J, PELLEGRIN JL, PEUCHANT E, DUMON MF, SIMONOFF M, CLERC M, LENG B, CONRI C : High plasma lipoprotein (a) in HIV-positive patients. *Lancet* 1993; 341: 1099-1100

398. KAGAWA A, AZUMA H, AKAIKE M, KANAGAWA Y, MATSUMOTO T : Aspirin reduces apolipoprotein(a) production in human hepatocytes by suppression of apo(a) gene transcription . *J Biol Chem* 1999; 274: 34111-34115

399. MOOSER V, BERGER MM, CAYEUX C, MARCOVINA SM, DARIOLI R, NICOD P, CHIOLERO R : Major reduction in plasma Lp(a) levels during sepsis and burns. *Aterioscler Thromb Vasc Biol* 2000; 20: 1137-1142

400. GEISS HC, RITTER MM, RICHTER WO, SCWANDT P, ZACHOVAL R : Low lipoprotein(a) levels during acute viral hepatitis. *Hepatology* 1996; 24: 1334-1337

401. FRAZER KA, NARLA G, ZHANG JL, RUBIN EM : The apolipoprotein(a) gene is regulated by sex hormones and acute-phase inducers in YAC transgenic mice. *Nature Genet* 1995; 9: 424-431

402. EDEN S, WIKLUND O, OSCARSSON J, ROSEN T, BENGTSSON BA : Growth hormone treatment of growth-hormone deficient adults results in a marked increase in Lp(a) and HDL concentrations. *Arterioscler Thromb* 1993; 13: 296-301

403. TAO RX, ACQUATI F, MARCOVINA SM, HOBBS HH: Human growth hormone increases apo(a) expression in transgenic mice. *Arterioscler Thromb Vasc Biol* 1999; 19: 2439-2447

404. HENRIKSSON P, ANGELIN B, BERGLUND L : Hormonal regulation of serum Lp(a) levels. Opposite effects after estrogen treatment and orchidectomy in males with prostatic carcinoma. *J Clin Invest* 1992 89: 1166-1171

405. SOMA M, FUMAGALLI R, PAOLETTI R, MESCHIA M, MAINI MC, CROSIGNANI P, GHANEM K, GAUBATZ J, MORRISETT JD : Plasma Lp(a) concentration after estrogen and progestagen treatment in menopausal women. *Lancet* 1991 337: 612

406. CROOK D, SIDHU M, SEED M, O'DONNELL M, STEVENSON JC : Lipoprotein Lp(a) levels are reduced by danazol an anabolic steroid. *Atherosclerosis* 1992; 92: 41-47

407. SHEWMON DA, STOCK JL, ABUSAMRA LC, KRISTAN MA, BAKER S, HEINILUOMA KM : Tamoxifen decreases Lp(a) in patients with breast cancer. *Metabolism* 1994; 43: 531-532

408. SCANU AM, EDELSTEIN C, FLESS GM, EISENBART J, SITRIN M, KASAWA B, HINMAN J : Postprandial lipoprotein (a) response to a single meal containing either saturated or ω-3 polyunsaturated fatty acids in subjects with hypoalphalipoproteinemia. *Metabolism* 1992; 41: 1361-1366

409. GAVISH D, BRESLOW JL : Lipoprotein (a) reduction by N-acetylcysteine. *Lancet* 1991; 337: 203-204

410. SCANU AM : N-acetylcysteine and immunoreactivity of lipoprotein (a). *Lancet* 1991; 337: 1159

411. RADER DJ, CAIN W, ZECH LA, USHER D, BREWER HB : Variation in lipoprotein (a) concentrations among individuals with the same apolipoprotein (a) isoform is determined by the rate of lipoprotein (a) production. *J Clin Invest* 1993; 91: 443-447

412. RADER DJ, CAIN W, IKEWAKI K, TALLEY G, ZECH LA, USHER D, BREWER HB : The inverse association of plasma lipoprotein (a) concentrations with apolipoprotein (a) isoform size is not due to differences in Lp(a) catabolism but to differences in production rate. *J Clin Invest* 1994; 93: 2758-2763

413. HOFMANN SL, EATON DL, BROWN MS, Mc CONATHY WJ, GOLDSTEIN JL, HAMMER RE : Overexpression of human LDL receptors leads to accelerated catabolism of Lp(a) lipoprotein in transgenic mice. *J Clin Invest* 1990; 85: 1542-1547

414. KOSTNER GM, GAVISH D, LEOPOLD B, BOLZANO K, WEINTRAUB MS, BRESLOW JL : HMGCoA reductase inhibitors lower LDL cholesterol levels without reducing Lp(a) levels. *Circulation* 1989; 80: 1313-1319

415. NEVEN L, KHALIL A, PFAFFINGE D, FLESS GM, JACKSON E, SCANU AM : Rhesus monkey model for familial hypercholesterolemia: relation between plasma Lp(a) levels, apo(a) isoforms and LDL-receptor function. *J Lipid Res* 1990; 31: 633-643

416. SOUTAR AK, Mc CARTHY SN, SEED M, KNIGHT BL : Relationship between apolipoprotein (a) phenotype, lipoprotein (a) concentration in plasma, and low density lipoprotein receptor function in a large kindred with familial hypercholesterolemia due to Pro664→Leu mutation in the LDL receptor gene. *J Clin Invest* 1991; 88: 483-492

417. PEROMBELON NYF, GALLAGHER JJ, MYANT NB, SOUTAR AK, KNIGHT BL : Lipoprotein (a) in subjects with familial defective apolipoprotein B-100. *Atherosclerosis* 1992; 92: 203- 212

418. NIEMEIER A, WILLNOW T, DIEPLINGER H, JACOBSEN C, MEYER N, HILPERT J, BEISIEGEL U : Identification of megalin/gp330 as a receptor for lipoprotein(a) in vitro. *Arterioscler Thromb Vasc Biol* 1999; 19: 552-561

419. KRONENBERG F, TRENKWALDER E, LINGENHEL A, FRIEDRICH G, LHOTTA K, SCHOBER M, MOES N, KONIG P, UTERMANN G, DIEPLINGER H : Renovascular arteriovenous differences in Lp(a) concentrations suggest removal of Lp(a) from the renal circulation. *J Lipid Res* 1997; 38: 1755-1763

420. WILLIAMS KJ, FLESS GM, PETRIE K, SNYDER ML, BROCIA RW, SWENSON TL : Lipoprotein lipase enhances cellular catabolism of lipoprotein (a). *Circulation* 1991; 84: SII-566

421. JONES PH, GOTTO AM, POWNALL HJ, PATSCH W, HERD JA, FARMER JA, PAYTON-ROSS C, COCANOUGHER B, GHANEM KK, MORRISETT JD : Effect of Gemfibrozil on plasma lipoprotein (a) levels in type IIa hyperlipoproteinemic subjects. *Circulation* 1991; 84: SII-483

422. LAPLAUD PM, BEAUBATIE SJ, RALL SJ, LUC G, SABOURAUD M : Lipoprotein (a) is the major apoB-containing lipoprotein in plasma of a hibernator, the hedgehog. *J Lipid Res* 1988; 29: 1157-1170

423. LAWN RM, SCHWARTZ K, PATTHY L : Convergent evolution of apolipoprotein(a) in primates and hedgehog. *Proc Natl Acad Sci USA* 1997; 94: 11992-11997

424. LISCUM L, FINER-MOORE J, STROUD RM, LUSKEY KL, BROWN MS, GOLDSTEIN JL : Domain structure of 3-hydroxy-3-methylglutaryl coenzyme A reductase, a glycoprotein of the endoplasmic reticulum. *J Biol Chem* 1985; 260: 522-530

425. OLENDER EH and SIMONI RD : The intracellular targeting and membrane topology of 3-hydroxy-3-methylglutaryl coenzyme A reductase. *J Biol Chem* 1992; 267: 4223-4235

426. HUMPHRIES SE, TATA F, HENRY I, BARICHARD F, HOLM M, JUNIEN C, WILLIAMSON R : The isolation, characterization, and chromosomal assignment of the gene for human 3-hydroxy-3-methylglutaryl-coenzyme A reductase, (HMG-CoA reductase). *Hum Genet* 1985; 71: 254-258

427. LUSKEY KL : Conservation of promoter sequence but not complex intron splicing pattern in human and hamster genes for 3-hydroxy-3-methylglutaryl coenzyme A reductase. *Mol Cell Biol* 1987; 7: 1881-1893

428. REYNOLDS GA, BASU SK, OSBORNE TF, CHIN DJ, GIL G, BROWN MS, GOLDSTEIN JL, LUSKEY KL : HMGCoA reductase: a negatively regulated gene with unusual promoter and 5' untranslated regions. *Cell* 198; 38: 275-285

429. RAMHARACK R and DEELEY RG : The three discrete size classes of 3-hydroxy-3-methylglutaryl coenzyme A reductase mRNA have different intrinsic stabilities. *Circulation* 1991; 84: SII-373

430. LEITERSDORF E, HWANG M, LUSKEY KL : ScrFI polymorphism in the 2nd intron of the HMGCR gene. *Nucleic Acids Res* 1990; 18: 5584

431. LEITERSDORF E and LUSKEY KL : HgiAI polymorphism near the HMGCR promoter. *Nucleic Acids Res* 1990; 18: 5584

432. OSBORNE T : Single nucleotide resolution of sterol regulatory region in promoter for 3-hydroxy-3-methylglutaryl coenzyme A reductase. *J Biol Chem* 1991; 266: 13947-13951

433. WILKIN DJ and EDWARDS PA : Calcium ionophore treatment impairs sterol mediated suppression of 3-hydroxy-3-methylglutaryl coenzyme A reductase, 3-hydroxy-3-methylglutaryl coenzyme A synthase, and Farnesyl diphosphate synthetase. *J Biol Chem* 1992; 267: 2831-2836

434. CHUN KT and SIMONI RD : The role of the membrane domain in the regulated degradation of 3-hydroxy-3-methylglutaryl-coenzyme A reductase. *J Biol Chem* 1992; 267: 4236-4246

435. PANINI SR, SCHNITZER-POLOKOFF R, SPENCER TA, SINENSKY M : Sterol independent regulation of 3-hydroxy-3-methylglutaryl coenzyme A reductase by mevalonate in chinese hamster ovary cells, magnitude and specificity. *J Biol Chem* 1989; 264: 11044-11052

436. SATO R, GOLDSTEIN JL, BROWN MS: Replacement of serine-871 of hamster 3-hydroxy-methyl-glutaryl-CoA reductase prevents phosphorylation by AMP-activated kinase and blocks inhibition of sterol synthesis induced by ATP depletion. *J Biol Chem* 1993; 268: 9261-9265

437. OSBORNE T, GIL G, GOLDSTEIN JL, BROWN MS : Operator constitutive mutation of 3-hydroxy-3-methylglutaryl coenzyme A reductase promoter abolishes protein binding to sterol regulatory element. *J Biol Chem* 1988; 263: 3380-3387

438. DAWSON PA, METHERALL JE, RIDGWAY ND, BROWN MS, GOLDSTEIN JL : Genetic distinction between sterol-mediated transcriptional and posttranscriptional control of 3-hydroxy-3-methylglutaryl-coenzyme A reductase. *J Biol Chem* 1991; 266: 9128-9134

439. GOLDSTEIN JL and BROWN MS : Lipoprotein metabolism in the macrophage : Implication for cholesterol deposition in atherosclerosis. *Ann Rev Biochem* 1983; 52: 223-261

440. GOLDSTEIN JL, HO YK, BASU SK, BROWN MS : Binding site on macrophages that mediates uptake and degradation of acetylated low density lipoprotein, producing massive cholesterol deposition. *Proc Natl Acad Sci USA* 1979; 76: 333-337

441. KRIEGER M and HERZ J : Structures and functions of multiligand lipoprotein receptors : macrophage scavenger receptors and LDL receptor-related protein (LRP). *Annu Rev Biochem* 1994; 63: 601-637

442. DUNNE DW, RESNICK D, GREENBERG J, KRIEGER M, JOINER KA : The type I macrophage scavenger receptor binds to Gram-positive bacteria and recognizes lipoteichoic acid. *Proc Natl Acad Sci USA* 1994; 91: 1863-1867

443. PLATT N, Da SILVA RP, GORDON S : Class A scavenger receptors and the phagocytosis of apoptopic cells. *Immunol Lett* 1999; 65: 15-19

444. TERPSTRA V, Van BERKEL TJC : Scavenger receptors on liver Kupffer cells mediate the in vivo uptake of oxidatively damaged red blood cells in mice. *Blood* 2000; 95: 2157-2163

445. SANO H, HIGASHI T, MATSUMOTO K, MELKKO J, JINNOUCHI Y, IKEDA K, EBINA Y, MAKINO H, SMEDSRØD B, HORIUCHI S : Insulin enhances macrophage scavenger receptor-mediated endocytic uptake of advanced glycation end products. *J Biol Chem* 1998; 273: 8630-8637

446. FRASER I, HUGHES D, GORDON S : Divalent cation-independent macrophage adhesion inhibited by monoclonal antibody to murine scavenger receptor. *Nature* 1993; 364: 343-346

447. KRIEGER M, ACTON S, ASHKENAS J, PEARSON A, PENMAN M, RESNICK D : Molecular flypaper, host defence, and atherosclerosis. Structure, binding properties, and functions of macrophage scavenger receptors. *J Biol Chem* 1993; 268: 4569-4572

448. VAN BERKEL TJ, VAN ECK M, FLUITER K, NION S : Scavenger receptor class A and B. Their roles in atherogenesis and the metabolism of modified LDL and HDL. *Ann NY Acad Sci* 2000; 902: 113-126

449. DE WINTER MP, VAN DIJK KW, HAVEKES LM, HOFKER MH : Macrophage scavenger receptor class A: a multifunctional receptor in atherosclerosis. *Arterioscler Thromb Vasc Biol* 2000; 20: 290-297

450. GREAVES DR, GOUGH PJ, GORDON S : Recent progress in defining the role of scavenger receptors in lipid transport, atherosclerosis and host defence. *Curr Opin Lipidol* 1998; 9: 425-432

451. DE WINTER MP, GIJBELS MJ, VAN DIJK KW, HAVEKES LM, HOFKER MH : Transgenic mouse models to study the role of the macrophage scavenger receptor class A in atherosclerosis. *Int J Tissue React* 2000; 22: 85-91

452. SUZUKI H, KURIHARA Y, TAKEYA M, KAMADA N, KATAOKA M, JISHAGE K, SAKAGUCHI H, KRUIJT JK, HIGASHI T, VAN BERKEL TJC, HORIUCHI S, TAKAHASHI K, YAZAKI Y, KODAMA T : The multiple roles of macrophage scavenger receptors (MSR) in vivo: resistance to atherosclerosis and susceptibility to infection in MSR knockout mice. *J Atheroscler Thromb* 1997; 4: 1-11

453. EMI M, ASAOKA H, MATSUMOTO A, ITAKURA H, KURIHARA Y, WADA Y, KANAMORI H, YAZAKI Y, TAKAHASHI EI, LEPPERT M, LALOUEL JM, KODAMA T, MUKAI T: Structure, organization and chromosomal mapping of the human scavenger receptor gene. *J Biol Chem* 1993; 268: 2120-2125

454. KODAMA T, FREEMAN M, ROHRER L, ZABREWCKY J, MATSUDAIRA P, KRIEGER M : Type I macrophage scavenger receptor contain α-helical and collagen-like coiled coils. *Nature* 1990 343: 531-535

455. ROHRER L, FREEMAN M, KODAMA T, PENMAN M, KRIEGER M : Coiled-coil fibrous domains mediate ligand binding by macrophage scavenger receptor type II. *Nature* 1990; 343: 570- 572

456. GOUGH PJ, GREAVES DR, GORDON S: A naturally occurring isoform of the human macrophage scavenger receptor (SR-A) gene generated by alternative splicing blocks modified LDL uptake. *J Lipid Res* 1998; 39: 531-543

457. FREEMAN MW : Macrophage scavenger receptors. *Curr Opin Lipidol* 1994; 5: 143-148

458. PENMAN M, LUX A, FREEDMAN NJ, ROHRER L, EKKEL Y, Mc KINSTRY H, RESNICK D, KRIEGER M : The type I and type II bovine scavenger receptors expressed in chinese hamster ovary cells are trimeric proteins with collagenous triple helical domains comprising noncovalently associated monomers and CYS83-disulphide-linked dimers. *J Biol Chem* 1991; 266: 23985-23993

459. FRANK S, LUSTIG A, SCHULTHESS T, ENGEL J, KAMMERER RA : A distinct seven-residue trigger sequence is indispensable for proper coiled-coil formation of the

human macrophage scavenger receptor oligomerization domain. *J Biol Chem* 2000; 275: 11672-11677

460. ACTON S, RESNICK D, FREEMAN M, EKKEL Y, ASHKENAS J, KRIEGER M : The collagenous domains of macrophage scavenger receptors and complement component C1q mediate their similar, but not identical, binding specificities for polyanionic ligands. *J Biol Chem* 1993; 268: 3530-3537

461. DEJAGER S, MIETUS-SNYDER M, FRIERA A, PITAS RE : Dominant negative mutations of the scavenger receptor. Native receptor inactivation by expression of truncated variants. *J Clin Invest* 1993; 92: 894-902

462. DOI T, HIGASHINO K, KURIHARA Y, WADA Y, MIYAZAKI T, NAKAMURA H, UESUGI S, IMANISHI T, KAWABE Y, ITAKURA H, YAZAKI Y, MATSUMOTO A, KODAMA T : Charged collagen structure mediates the recognition of negatively charged macromolecules. *J Biol Chem* 1993; 268: 2126-2133

463. FREEMAN M, ASKHENAS J, REES KJG, KINGSLEY DM, COPELAND NG, JENKINS NA, KRIEGER M : An ancient, highly conserved family of cystein-rich protein domains revealed by cloning type I and type II murine macrophage scavenger receptors. *Proc Natl Acad Sci USA* 1990; 87: 8810-8814

464. MOULTON KS, WU H, BARNETT J, PARTHASARATHY S, GLASS C : Regulated expression of the human acetylated low density lipoprotein receptor gene and isolation of promoter sequences. *Proc Natl Acad Sci USA* 1992; 89: 8102-8106

465. NAITO M, KODAMA T, MATSUMOTO A, DOI T, TAKAHASHI K : Tissue distribution, intracellular localization, and in-vitro expression of bovine macrophage scavenger receptor. *Am J Pathol* 1991; 139: 1411-1423

466. BICKEL EP and FREEMAN MW : Rabbit aortic smooth muscle cells express inducible macrophage scavenger receptor messenger RNA that is absent from endothelial cells. *J Clin Invest* 1992; 90: 1450-1457

467. INABA T, GOTODA T, SHIMANO H, SHIMADA M, HARADA K, KOZAKI K, WATANABE Y, HOH E, MOTOYOSHI K, YAZAKI Y, YAMADA N : Platelet-derived growth factor induces c-fms and scavenger receptor genes in vascular smooth muscle cells. *J Biol Chem* 1992; 267: 13107-13112

468. PITAS RE, FRIERA A, Mc GUIRE J, DEJAGER S : Further characterization of the acetyl LDL (scavenger) receptor expressed by rabbit smooth muscle cells and fibroblasts. *Arterioscler Thromb* 1992; 12: 1235-1244

469. MIETUS-SNYDER M, GOWRI MS, PITAS RE : Class A scavenger receptor up-regulation in smooth muscle cells by oxidized low density lipoprotein. Enhancement by calcium flux and concurrent cyclooxygenase-2 up-regulation. *J Biol Chem* 2000; 275: 17661-17670

470. VIA DP, PONS L, DENNISON DK, FANSLOW AE, BERNINI B : Induction of acetyl-LDL receptor activity by phorbol ester in human monocyte cell line THP-1. *J Lipid Res* 1989; 30: 1515-1524

471. GENG YJ and HANSSON GK: Interferon-γ inhibits scavenger receptor expression and foam cell formation in human monocyte-derived macrophages. *J Clin Invest* 1992; 89: 1322-1330

472. TONTONOZ P, NAGY L : Regulation of macrophage gene expression by peroxisome-proliferator-activated receptor gamma: implications for cardiovascular disease. *Curr Opin Lipidol* 1999; 10: 485-490

473. RICOTE M, LI AC, WILLSON TM, KELLY CJ, GLASS CK : The peroxisome proliferator-activated receptor-γ is a negative regulator of macrophage activation. *Nature* 1998; 391: 79-82

474. TEUPSER D, THIERY J, SEIDEL D : Alpha-tocopherol down-regulates scavenger receptor activity in macrophages. *Atherosclerosis* 1999; 144: 109-115

475. SILVERSTEIN RL, FEBBRAIO M : CD36 and atherosclerosis. *Curr Opin Lipidol* 2000; 11: 483-491

476. AITMAN TJ, COOPER LD, NORSWORTHY PJ, WAHID FN, GRAY JK, CURTIS BR, McKEIGUE PM, KWIATKOWSKI D, GREENWOOD BM, SNOW RW, HILL AV, SCOTT J : Malaria susceptibility and CD36 deficiency. *Nature* 2000; 405: 1015-1016

477. ENDEMANN G, STANTON LW, MADDEN KS, BRYANT CM, WHITE RT, PROTTER AA : CD36 is a receptor for oxidized low density lipoprotein. *J Biol Chem* 1993; 268: 11811-11816

478. PODREZ EA, FEBBRAIO M, SHEIBANI N, SCHMITT D, SILVERSTEIN RL, HAJJAR DP, COHEN PA, FRAZIER WA, HOFF HF, HAZEN SL : Macrophage scavenger receptor CD36 is the major receptor for LDL modified by monocyte-generated reactive nitrogen species. *J Clin Invest* 2000; 105: 1095-1108

479. CONNELLY MA, KLEIN SM, AZHAR S, ABUMRAD NA, WILLIAMS DL : Comparison of class B scavenger receptors, CD36 and scavenger BI(SR-BI), shows that both receptors mediate high density lipoprotein-cholesteryl ester selective uptake but SR-BI exhibits a unique enhancement of cholesteryl ester uptake. *J Biol Chem* 1999; 274: 41-47

480. ARMESILLA AL , VEGA MA : Structural organization of the gene for human CD36 glycoprotein. *J Biol Chem* 1994; 269: 18985-18991

481. FERNANDEZ-RUIZ E, ARMESILLA AL, SANCHEZ-MADRID F, VEGA MA : Gene encoding the collagen type I and thrombospondin receptor CD36 is located on chromosome 7q11.2. *Genomics* 1993; 17: 759-761

482. CALVO D, DOPAZO J, VEGA MA : The CD36, CLA-1 (CD36L1), and LIMPII (CD36L2) gene family : cellular distribution, chromosomal location, and genetic evolution. *Genomics* 1995; 25: 100-106

483. ARMESILLA AL , VEGA MA : Structural organization of the gene for human CD36 glycoprotein. *J Biol Chem* 1994; 269: 18985-18991

484. NICHOLSON AC, FEBBRAIO M, HAN J, SIVERSTEIN RL, HAJJAR DP : CD36 in atherosclerosis. The role of a class B macrophage scavenger receptor. *Ann NY Acad Sci* 2000; 902: 128-131

485. NAGY L, TOTONOZ P, ALVAREZ JG, CHEN H, EVANS RM : Oxidized LDL regulates macrophage gene expression through ligand activation of PPARgamma. *Cell* 1998; 93: 229-240

486. FENG J, HAN J, PEARCE SF, SILVERSTEIN RL, GOTTO AM Jr, HAJJAR DP, NICHOLSON AC : Induction of CD36 expression by oxidized LDL and IL-4 by a

common signaling pathway dependent on protein kinase and PPAR-gamma. *J Lipid Res* 2000; 41: 688-696

487. RICCIARELLI R, ZINGG JM, AZZI A : Vitamin E reduces the uptake of oxidized LDL by inhibiting CD36 scavenger receptor expression in cultured aortic smooth muscle cells. *Circulation* 2000; 102: 82-87

488. IBRAHIMI A, BONEN A, BLINN WD, HAJRI T, XIN LI, ZHONG K, CAMERON R, ABUMRAD NA : Muscle-specific overexpression of FAT/CD36 enhances fatty acid oxidation by contracting muscle, reduces plasma triglycerides and fatty acids and increases plasma glucose and insulin. *J Biol Chem* 1999; 274: 26761-26766

489. FEBBRAIO M, ABUMRAD NA, HAJJAR DP, SHARMA IR, CHENG WL, PEARCE SFA, SILVERSTEIN RL : A null mutation in murine CD36 reveals an important role in fatty acid and lipoprotein metabolism. *J Biol Chem* 1999; 274: 19055-19062

490. FEBBRAIO M, PODREZ EA, SMITH JD, HAJJAR DP, HAZEN SL, HOFF HF, SHARMA K, SILVERSTEIN RL : Targeted disruption of the class B scavenger receptor CD36 protects against atherosclerosis lesion development in mice. *J Clin Invest* 2000; 105: 1049-1056

491. AITMAN TJ, GLAZIER AM, WALLACE CA, COOPER LD, NORSWORTHY PJ, WAHID FN, AL-MAJALI KM, TREMBLING PM, MANN CJ, SHOULDERS CC, GRAF D, St LEZIN E, KURTZ TW, KREN V, PRAVENEC M, IBRAHIMI A, ABUMRAD NA, STANTON LW, SCOTT J : Identification of CD36(fat) as an insulin-resistance gene causing defective fatty acid and glucose metabolism in hypertensive rats. *Nature Genet* 1999; 21: 76-83

492. GOTODA T, IIZUKA Y, KATON, OSUGA JI, BIHOREAU MT, MURAKAMI T, YAMORI Y, SHIMANO H, ISHIBASHI S, YAMADA N : Absence of CD36 mutation in the original spontaneously hypertensive rats with insulin resistance. *Nature Genet* 1999; 22: 226-228

493. PRAVENEC M, ZIDEK V, SIMAKOVA M, KREN V, KRENOVA D, HORKY K, JACHYMOVA M, MIKOVA B, KAZDOVA L, AITMAN TJ, CHURCHILL PC, WEBB RC, HINGARTH NH, YANG Y, WANG JM, LEZIN EM, KURTZ TW : Genetics of CD36 and the clustering of multiple cardiovascular risk factors in spontaneous hypertension. *J Clin Invest* 1999; 103: 1651-1657

494. KASHIWAGI H, TOMIYAMAY, HONDA S, KOSUGI S, SHIRAGA M, NAGAO N, SEKIGUCHI S, KANAYAMA Y, KURATA Y, MATSUZAWA Y : Molecular basis of CD36 deficiency. Evidence that a 478C→T substitution (proline90→serine) in CD36 cDNA accounts for CD36 deficiency. *J Clin Invest* 1995; 95: 1040-1046

495. NOZAKI S, TANAKA T, YAMASHITA S, SOHMIYA K, YOSHIZUMI T, OKAMOTO F, KITAURA Y, KOTAKE C, NISHIDA H, NAKATA A, NAKAGAWA T, MATSUMOTO K, KAMEDA-TAKEMURA K, TADOKORO S, KURATA Y,TOMIYAMA Y, KAWAMURA K, MATSUZAWA Y : CD36 mediates long-chain fatty acid transport in human myocardium: complete myocardial accumulation defect of radiolabelled long-chain fatty acid analog in subjects with CD36 deficiency. *Mol Cell Biochem* 1999; 192: 129-135

496. TANAKA T, SOHMIYA K, KAWAMURA K : Is CD36 deficiency an etiology of hereditary hypertrophic cardiomyopathy? *J Mol Cell Cardiol* 1997; 29: 121-127

497. NOZAKI S, KASHIWAGI H, YAMASHITA S, NAKAGAWA T, KOSTNER B, TOMIYAMA Y, ISHIGAMI M, MIYAGAWA J, KAMEDA-TAKEMURA K, et al.

Reduced uptake of oxidized low density lipoproteins in monocyte-macrophages from CD36-deficient subjects. *J Clin Invest* 1995; 96: 1859-1865

498. YAMADA Y, DOI T, HAMAKUBO T, KODAMA T : Scavenger receptor family of proteins: roles for atherosclerosis, host defence and disorders in the central nervous system. *Cell Mol Life Sci* 1998; 54: 628-640

499. KRAAL G, VAN DER LAAN LJ, ELOMAA O, TRYGGVASON K : The macrophage receptor MARCO. *Microbes Infect* 2000; 2: 313-316

500. SAWAMURA T, KUME N, AOYAMA T, MORIWAKI H, HOSHIKAWA H, AIBA Y, TANAKA T, MIWA S, KATSURA Y, KITA T, MASAKI T : An endothelial receptor for oxidized low-density lipoprotein. *Nature* 1997; 386: 73-77

501. NAGASE M, ABE J, TAKAHASHI K, ANDO J, HIROSE S, FUJITA T : Genomic organization and regulation of expression of the lectin-like oxidized low-density lipoprotein receptor (LOX-1) gene. *J Biol Chem* 1998; 273: 33702-33707

502. MORIWAKI H, KUME N, SAWAMURA T, AOYAMA T, HOSHIKAWA H, OCHI H, NISHI E, MASAKI T, KITA T : Ligand specificity of LOX-1 a novel endothelial receptor for oxidized low density lipoprotein. *Arterioscler Thromb Vasc Biol* 1998; 18: 1541-1547

503. MORAWIETZ H, RUECKSCHLOSS U, NIEMANN B, DUERRENSCHMIDT, GALLE J, HAKIM K, ZERKOWSKI HR, SAWAMURA T : Angiotensin-II induces LOX-1, the human endothelial receptor for oxidized lox-density-lipoprotein. *Circulation* 1999; 100: 899-902

504. LI DY, ZHANG YC, PHILIPS MI, SAWAMURA T, MEHTA JL : Upregulation of endothelial receptor for oxidized low-density lipoprotein (LOX-1) in cultured human coronary artery endothelial cells by angiotensin II type 1 receptor activation. *Circ Res* 1999; 84: 1043-1049

505. KORN ED : Clearing factor, a heparin–activated lipoprotein lipase. *J Biol Chem* 1955; 215: 1-14

506. BRUNZELL JD : Familial lipoprotein lipase deficiency and other causes of the chylomicronemia syndrome. In : Scriver CR, Beaudet AL, Sly WS, Valle D. *The metabolic and molecular bases of inherited disease.* 7th ed. Highstown : New Jersey. Mc Graw Hill 1995 chap 45: 1913-1932

507. ECKEL RH : Lipoprotein lipase A multifunctional enzyme relevant to common metabolic diseases. *N Engl J Med* 1989; 320: 1060-1068

508. SANTAMARINA-FOJO S, HAUDENSCHILD C : Role of hepatic and lipoprotein lipase in lipoprotein metabolism and atherosclerosis: studies in transgenic and knockout animal models and somatic gene transfer. *Int J Tissue React* 2000; 22: 39-47

509. GOLDBERG IJ : Lipoprotein lipase and lipolysis: central roles in lipoprotein metabolism and atherogenesis. *J Lipid Res* 1996; 37: 693-707

510. FOJO SS, DUGI KA : Structure, function and role of lipoprotein lipase in lipoprotein metabolism. *Curr Opin Lipidol* 1994; 5: 117-125

511. WION KL, KIRCHGESSNER TG, LUSIS AJ, SCHOTZ MC, LAWN RM : Human lipoprotein lipase complementary DNA sequence. *Science* 1987; 235: 1638-1641

512. KIRCHGESSNER TG, CHUAT JC, HEINZMANN C, ETIENNE J, GUILHOT S, SVENSON K, AMEIS D, PILON C, D'AURIOL L, ANDALIBI A, SCHOTZ M,

GALIBERT F, LUSIS AJ : Organization of the human lipoprotein lipase gene and evolution of the lipase gene family. *Proc Natl Acad Sci USA* 1989; 86: 9647-9651

513. SPARKES RS, ZOLLMAN S, KLISAK I, KIRCHGESSNER TG, KOMAROMY MC, MOHANDAS T, SCHOTZ MC, LUSIS AJ : Human genes involved in lipolysis of plasma lipoproteins: mapping of loci for lipoprotein lipase to 8p22 and hepatic lipase to 15q21. *Genomics* 1987; 1: 138-144

514. BRZORZOWSKI AM, DEREWENDA U, DEREWENDA ZS, DODSON GG, LAWSON DM, TURKENBURG JP, BJORKLING F, HUGE-JENSEN B, PATKAR SA, THIM L : A model for interfacial activation in lipases from the structure of a fungal lipase-inhibitor complex. *Nature* 1991; 351: 491-494

515. BOWNES M : Why is there sequence similarity between insect yolk proteins and vertebrate lipases. *J Lipid Res* 1992; 33: 777-790

516. HIDE WA, CHAN L, LI W-H : Structure and evolution of the lipase superfamily. *J Lipid Res* 1992; 33: 167-178

517. BENLIAN P, ÉTIENNE J, DE GENNES JL, NOE L, BRAULT D, RAISONNIER A, ARNAULT F, HAMELIN J, FOUBERT L, CHUAT JC, TSE C, GALIBERT F : A homozygous gene deletion of exon 9 causes lipoprotein lipase deficiency : possible intron-*Alu* recombination. *J Lipid Res* 1995; 36: 356-366

518. DEEB SS, PENG R : Structure of the human lipoprotein lipase gene. *Biochemistry* 1989 28: 4131-4135

519. OKA K, TKALCEVIC GT, NAKANO T, TUCKER H, ISHIMURA-OKA K, BROWN WV : Structure and polymorphic map of human lipoprotein lipase gene. *Biochem Biophys Acta* 1990; 1049: 21-26

520. WANG CS, HARTSUCK J, Mc CONATHY WJ : Structure and functional properties of lipoprotein lipase. *Biochem Biophys Acta* 1992; 1123: 1-17

521. WILSON DE, HATA A, KWONG LK, LINGAM A, SHUHUA J, RIDINGER DN, YEAGER C, KALTENBORN KC, IVERIUS PH, LALOUEL JM : Mutations in exon 3 of the lipoprotein lipase gene segregating in a family with hypertriglyceridemia, pancreatitis and non-insulin-dependent diabetes. *J Clin Invest* 1993; 92: 203-211

522. HENDERSON HE, MA Y, LIU MS, CLARK-LEWIS I, MAEDER DL, KASTELEIN JJP, BRUNZELL JD, HAYDEN MR : Structure-function relationships of lipoprotein lipase : mutation analysis and mutagenesis of the loop region. *J Lipid Res* 1993; 34: 1593-1602

523. DAVIS RC, WONG H, NIKAZY J, WANG K, HAN Q, SCHOTZ MC : Chimeras of hepatic lipase and lipoprotein lipase. Domain localization of enzyme-specific properties. *J Biol Chem* 1992; 267: 21499-21504

524. HATA A, RIDINGER DN, SUTHERLAND SD, EMI M, KWONG LK, SHUHUA J, MYERS RL, REN K, CHENG T, INOUE I, WILSON DE, IVERIUS PH, LALOUEL JM : Binding of lipoprotein lipase to heparin. Identification of five critical residues in two distinct segments of the amino-terminal domain. *J Biol Chem* 1993; 268: 8447-8457

525. LOOKENE A, NIELSEN MS, GLIEMANN J, OLIVECRONA G : Contribution of the carboxy-terminal domain of lipoprotein lipase to interaction with heparin and lipoproteins. *Biochem Biophys Res Commun* 2000; 27: 15-21

526. BUSCA R, MARTINEZ M, VILELLE E, PEINADO J, GELPI JL, DEEB S, AUWERX J, REINA M, VILARO S : The carboxy-terminal region of human lipoprotein lipase is necessary for its exit from the endoplasmic reticulum . *J Lipid Res* 1998; 39: 821-833

527. ENERBÄCK S, GIMBLE JM : Lipoprotein lipase gene expression: physiological regulators at the transcriptional and post-transcriptional level. *Biochem Biophys Acta* 1993; 1169: 107-125

528. YANG WS, DEEB SS : Sp1 and Sp3 trans activate the human lipoprotein lipase gene promoter through binding to a CT element: synergy with the sterol regulatory element binding protein and reduced trans activation of a naturally occurring promoter variant. *J Lipid Res* 1998; 39: 2054-2064

529. HOMMA H, KURACHI H, NISHIO Y, TAKEDA T, YAMAMOTO T, ADACHI K, MORISHIGE K, OHMICHI M, MATSUZAWA Y, MURATA Y : Estrogen suppresses transcription of lipoprotein lipase gene. *J Biol Chem* 2000; 275: 11404-11411

530. FORMAN BM, TONTONOZ P, CHEN J, BRUN RP, SPIEELMAN BM, EVANS RM : 15-deoxy-delta 12,14-prostaglandin J2 is a ligand for the adipocyte determination factor PPAR gamma. *Cell* 1995; 83: 803-812

531. SCHOONJANS K, PEINADO-ONSURBE J, LEFEBVRE AM, HEYMAN RA, BRIGGS M, DEEB S, STAELS B, AUWERX J : PPARα and PPARγ activators direct a distinct tissue specific transcriptional response via a PPRE in the lipoprotein lipase gene. *EMBO J* 1996; 15: 5336-5348

532. ZECHNER R, NEWMAN TC, SHERRY B, CERAMI A, BRESLOW JL : Recombinant human cachectin/Tumor necrosis factor but not Interleukin-1a downregulates lipoprotein lipase gene expression at the transcriptional level in mouse 3T3-L1 adipocytes. *Mol Cell Biol* 1988; 8: 2394-2401

533. POEHLMAN ET, DESPRES JP, MARCOTTE M, TREMBLAY A, THERIAULT G, BOUCHARD C : Genotype dependency of adaptation in adipose tissue metabolism after short-term overfeeding. *Am J Physiol* 1986; 250: E480

534. HERTTUALA SY, LIPTON BA, ROSENFELD ME, GOLDBERG IJ, STEINBERG D, WITZTUM JL : Macrophages and smooth muscle cells express lipoprotein lipase in human and rabbit atherosclerotic lesions. *Proc Natl Acad Sci USA* 1991; 88: 10143-10147

535. RENIER G, SKAMENE E, DE SANCTIS JB, RADZIOCH D : High macrophage lipoprotein lipase expression and secretion are associated in inbred murine strains with susceptibility to atherosclerosis. *Arterioscler Thromb* 1993; 13: 190-196

536. BEN-ZEEV O, DOOLITTLE MH, DAVIS RC, ELOVSON J, SCHOTZ MC : Maturation of lipoprotein lipase. Expression of full catalytic activity requires glucose trimming but not translocation to the cis-Golgi compartment. *J Biol Chem* 1992; 267: 6219-6227

537. BERNFIELD M, GOTTE M, PARK PW, REIZES O, FITZGERALD ML, LINCECUM J, ZAKO M : Functions of cell surface heparan sulfate proteoglycans. *Annu Rev Biochem* 1999; 68: 729-777

538. KALLUNKI P, TRYGGVASON K : Human basement membrane heparan sulfate proteoglycan core protein: a 467-kd protein containing multiple domains resembling elements of the low density lipoprotein receptor, laminin, neural cell adhesion molecules, and epidermal growth factor. *J Cell Biol* 1992; 116: 559-571

539. MURDOCH AD, DODGE GR, COHEN I, TUAN RS, IOZZO RV: Primary structure of the human heparan sulfate proteoglycan from basement membrane (HSPG2/Perlecan). A chimeric molecule with multiple domains homologous to the low density lipoprotein receptor, laminin, neural cell adhesion molecules, and epidermal growth factor. *J Biol Chem* 1992; 267: 8544-8557

540. CISAR LA, HOOGEWERF AJ, CUPP M, RAPPORT CA, BENSADOUN A : Secretion and degradation of lipoprotein lipase in cultured adipocytes. Binding of lipoprotein lipase to membrane heparan sulfate proteoglycans is necessary for degradation. *J Biol Chem* 1989; 264: 1767-1774

541. NICKERSON DA, TAYLOR SL, WEISS KM, CLARK AG, HUTCHINSON RG, STENGARD J, SALOMAA V, VARTIAINEN E, BOERWINKLE E, SING CF : DNA sequence diversity in a 9.7-kb region of the human lipoprotein lipase gene. *Nat Genet* 1998; 19: 233-240

542. TEMPELTON AR, CLARK AG, WEISS KM, NICKERSON DA, BOERWINKLE E, SING CF : Recombinational and mutational hotspots within the human lipoprotein lipase gene. *Am J Hum Genet* 2000; 66: 69-83

543. HENDERSON HE, HASSAN MF, BERGER GMB, HAYDEN MR : The lipoprotein lipase Gly188→Glu mutation in South Africans of Indian Descent : evidence suggesting common origins and an increased frequency. *J Med Genet* 1993; 29: 119-122

544. MA Y, HENDERSON HE, VEN MURTHY MR, ROEDERER G, MONSALVE MV, CLARKE LA, NORMAND T, JULIEN P, GAGNE C, LAMBERT M, DAVIGNON J, LUPIEN PJ, BRUNZELL J, HAYDEN MR : A mutation in the lipoprotein lipase gene as the most common cause of familial chylomicronemia in French Canadians. *N Engl J Med* 1991; 324: 1761-1766

545. FOUBERT L, DE GENNES JL, LAGARDE JP, EHRENBORG E, RAISONNIER A, GIRARDET JP, HAYDEN MR, BENLIAN P : Assessment of French patients with LPL deficiency for French Canadian mutations. *J Med Genet* 1997; 34: 672-675

546. FOUBERT L, BRUIN T, DE GENNES JL, EHRENBORG E, FURIOLI J, KASTELEIN JJP, BENLIAN P, HAYDEN MR : A single Ser259Arg mutation in the gene for lipoprotein lipase causes chylomicronemia in Moroccans of Berber ancestry. *Hum Mutat* 1997; 10: 179-185

547. HENDERSON HE, MA Y, HASSAN MF, MONSALVE MV, MARAIS AD, WINKLER F, GUBERNATOR K, PETERSON J, BRUNZELL JD, HAYDEN MR : Amino acid substitution (Ile194→Thr) in exon 5 of the lipoprotein lipase gene causes lipoprotein lipase deficiency in three unrelated probands. Support for multicentric origin. *J Clin Invest* 1991; 87: 2005-2011

548. SANTAMARINA-FOJO S : The familial chylomicronemia syndrome. *Endocrinol Metab Clin North Am* 1998; 27: 551-567

549. HAVEL RJ, GORDON RS : Idiopathic hyperlipidemia: metabolic studies in an affected family. *J Clin Invest* 1960; 39: 1777

550. AUWERX JH, BABIRAK SP, FUJIMOTO WY, IVERIUS PH, BRUNZELL JD : Defective enzyme protein in lipoprotein lipase deficiency. *Eur J Clin Invest* 1989; 19: 433-437

551. HATA A, RIDINGER DN, SUTHERLAND SD, EMI M, KWONG LK, SHUHUA J, LUBBERS A, GUY-GRAND B, BASDEVANT A, IVERIUS PH, WILSON DE,

LALOUEL JM : Missense mutations in exon 5 of the human lipoprotein lipase gene Inactivation correlates with loss of dimerization. *J Biol Chem* 1992; 267: 20132-20139

552. WEINSTOCK PH, BISGAIER CL, AALTO-SETÄLÄ K, RADNER H, RAMAKRISHNAN R, LEVA-FRANK S, ESSENBURG AD, ZECHNER R, BRESLOW JL : Severe hypertriglyceridemia, reduced high density lipoprotein and neonatal death in lipoprotein lipase knock out mice. *J Clin Invest* 1995; 96: 2555-2568

553. RESTA F, CHIMIENTI G, COLACICCO AM, LOVECCHIO M, DI PERNA V, PEPE G, CAPURSO A : Lipoprotein Lipase deficiency : a new case in an Apulian-Italian family. *Atherosclerosis* 1994; 109: 61

554. HÖLZL B, KRAFT HG, WIEBUSH H, SANDHOFER A, PATSH J, SANDHOFER F, PAULWEBER B : Two novel mutations in the lipoprotein lipase gene in a family with marked hypertriglyceridemia in heterozygous carriers: potential interaction with the polymorphic marker D1S104 on chromosome 1q21-q23. *J Lipid Res* 2000; 41: 734-741

555. REINA M, BRUNZELL JD, DEEB SS : Molecular basis of familial chylomicronemia : mutations in the lipoprotein lipase and apolipoprotein C-II genes. *J Lipid Res* 1992; 33: 1823-1832

556. FOUBERT L, DE GENNES JL, BENLIAN P, TRUFFERT J, MIAO L, HAYDEN MR : Compound heterozygosity for frameshift mutations in the gene for lipoprotein lipase in a patient with early-onset chylomicronemia. *Hum Mutat* 1998; S1: S141-S144

557. BENLIAN P, DE GENNES JL, FOUBERT L, ZHANG H, GAGNÉ E, HAYDEN MR. Premature atherosclerosis in familial chylomicronemia caused by mutations in the lipoprotein lipase gene. *N Engl J Med* 1996; 335: 848-854

558. DEEB SS, REINA M, PETERSON J, TAKATA K, KAJIYAMA G, BRUNZELL JD : Gene mutations in patients with lipoprotein lipase deficiency. *Circulation* 1991; 84 SII: 1815

559. MA Y, LIU MS, ZHANG H, FORSYTHE IJ, BRUNZELL JD, HAYDEN MR : A 4 basepair deletion in exon 4 of the human lipoprotein lipase gene results in type I hyperlipoproteinemia. *Hum Mol Genet* 1993; 2: 1049-1050

560. MAILLY F, PALMEN J, MULLER DP, GIBBS T, LLOYD J, BRUNZELL J, DURRINGTON P, MITROPOULOS K, BETTERIDGE J, WATTS G, LITHELL H, ANGELICO F, HUMPFRIES SE, TALMUD PJ : Familial lipoprotein lipase (LPL) deficiency: a catalogue of LPL gene mutations identified in 20 patients from the UK, Sweden and Italy. *Hum Mutat* 1997;10: 465-473

561. MA Y, BRUIN T, TUZGOL S, WILSON BI, ROEDERER G, LIU MS, DAVIGNON J, KASTELEIN JJP, BRUNZELL JD, HAYDEN MR : Two naturally occurring mutations at the first and second bases of codon aspartic acid 156 in the proposed catalytic triad of human lipoprotein lipase. In vivo evidence that aspartic acid is essential for catalysis. *J Biol Chem* 1992; 267: 1918-1923

562. HAUBENWALLNER S, HÖRL G, SHCHTER NS, PRESTA E, FRIED SK, HÖFLER G, KOSTNER GM, BRESLOW JL, ZECHNER R : A novel missense mutation in the gene for lipoprotein lipase resulting in a highly conservative amino acid substitution (Asp180→Glu) causes familial hyperchylomicronemia (type I hyperlipoproteinemia). *Genomics* 1993; 18: 392-396

563. TENKANEN H, TASKINEN MR, ANTIKAINEN M, ULMANEN I, KONTULA K, EHNHOLM C : A novel amino acid substitution (His183→Gln) in exon 5 of the

lipoprotein lipase gene results in loss of catalytic activity : phenotypic expression of the mutant gene in a heterozygous state. *J Lipid Res* 1994; 35: 220-228

564. TAKAGI A, IKEDA Y, MORI A, TSUTSUMI Z, OIDA K, NAKAI T, YAMAMOTO A : A newly identified heterozygous lipoprotein lipase gene mutation (Cys239→Stop/TGC972→TGA; LPLobama) in a patient with primary type IV hyperlipoproteinemia. *J Lipid Res* 1994; 35: 2008-2018

565. APPELMAN EEG, BIJVOET SM, WIEBUSCH H, MA Y, REYMER PWA, BRUIN T, HAYDEN MR : A de novo mutation in the lipoprotein lipase (LPL) gene causing LPL deficiency. *Atherosclerosis* 1994; 109: 63

566. KAO JT, HSIAO WH, CHIANG FT: Newly identified missense mutation reduces lipoprotein lipase activity in Taiwanese patients with hypertriglyceridemia. *J Formos Med Assoc* 1999; 98: 606-612

567. CHAN L, MAK Y, TOMLINSON B, BAUM L, WU X, MASAREI J, PANG C : Compound heterozygosity of Leu252Val and Leu252Arg causing lipoprotein lipase deficiency in a chinese patient with hypertriglyceridemia. *Eur J Clin Invest* 2000; 30: 33-40

568. MA Y, OOI TC, LIU MS, ZHANG H, MAC PHERSON R, EDWARDS AL, FORSYTHE I, FROHLICH J, BRUNZELL JD, HAYDEN MR : High frequency of mutations in the human lipoprotein lipase gene in pregnancy-induced chylomicronemia: possible association with apolipoprotein E2 isoform. *J Lipid Res* 1994; 35: 1066-1075

569. EVANS D, WENDT D, AHLE S, GUERRA A, BEISIEGEL U : Compound heterozygosity for a new (S259G) and a previously described (G188E) mutation in lipoprotein lipase (LpL) as a cause of chylomicronemia. *Hum Mutat* 1998; 12: 217

570. FUNKE H, WIEBUSCH H, PAULWEBER B, ASSMANN G : Identification of the molecular defect in a patient with type I hyperlipidemia. *Arteriosclerosis* 1990; 10: 830a

571. KOBAYASHI J, NAGASHIMA I, TAIRA K, HIKITA M, TAMURA K, BUJO H, MORISAKI N, SAITO Y : A novel frameshift mutation in exon 6 (the site of Asn291) of the lipoprotein lipase gene in type I hyperlipidemia. *Clin Chim Acta* 1999; 285: 173-182

572. BERTOLINI S, SIMONE ML, PES GM, GHISELLINI M, ROLLERI M, BELLOCHIO A, ELICIO N, MASTURZO P, CALANDRA S : Pseudodominance of lipoprotein lipase (LPL) deficiency due to nonsense mutation (Tyr302→Term) in exon 6 of LPL gene in an Italian family from Sardinia (LPL(Olbia)). *Clin Genet* 2000; 57: 140-147

573. HÖLZL B, HUBER R, PAULWEBER B, PATSCH JR, SANDHOFER F : Lipoprotein lipase deficiency due to a 3' splice site mutation in intron 6 of the lipoprotein lipase gene. *J Lipid Res* 1994; 35: 2161-2169

574. KOBAYASHI J, SASAKI N, TASHIRO J, INADERA H, SAITO Y, YOSHIDA S : A missense mutation (Ala334→ Thr) in exon 7 of the lipoprotein lipase gene in a case with type I hyperlipidemia. *Biochem Biophys Res Comm* 1993; 191: 1046-1054

575. PREVIATO L, GUARDAMAGNA O, DUGI KA, RONAN R, TALLEY GD, SANTAMARINA-FOJO S, BREWER HB : A novel missense mutation in the C-terminal domain of lipoprotein lipase (Glu410→Val) leads to enzyme inactivation and familial chylomicronemia. *J Lipid Res* 1994; 35: 1552-1560

576. HENDERSON H, LEISEGANG F, HASSAN F, HAYDEN M, MARAIS D : A novel Glu421Lys substitution in the lipoprotein lipase gene in pregnancy-induced hypertriglyceridemic pancreatitis. *Clin Chim Acta* 1998; 269: 1-12

577. KEILSON LM, VARY CPH, SPRECHER DL, RENFREW R : Hyperlipidemia and pancreatitis during pregnancy in two sisters with a mutation in the lipoprotein lipase gene. *Ann Intern Med* 1996; 124: 425-428

578. HUSSAIN MM, OBUNIKE JC, SHAHEEN A, HUSSAIN MJ, SHELNESS GS, GOLDBERG IJ: High affinity binding between lipoprotein lipase and lipoproteins involves multiple ionic and hydrophobic interactions, does not require enzyme activity, and is modulated by glycosaminoglycans. *J Biol Chem* 2000; 275: 29324-29330

579. HOEG JM, OSBORNE JC, GREGG RE, BREWER HB. Initial diagnosis of lipoprotein lipase deficiency in a 75-year-old man. *Am J Med* 1983; 75: 889-892

580. GAGNÉ C, BRUN DL, JULIEN P, MOORJANI S, LUPIEN P-J. Primary lipoprotein lipase activity deficiency: clinical investigation of a French Canadian population. *CMAJ* 1989; 140: 405-411.

581. HOKANSON JE : Functional variants of the lipoprotein lipase gene and risk of cardiovascular disease. *Curr Opin Lipidol* 1999; 10: 393-399

582. FISHER RM, HUMPHRIES SE, TALMUD PJ : Common variation in the lipoprotein lipase gene : effects on plasma lipids and risk of atherosclerosis. *Atherosclerosis* 1997; 135: 145-149

583. WILSON DE, EMI M, IVERIUS PH, HATA A, WU LL, HILLAS E, WILLIAMS RR, LALOUEL JM : Phenotypic expression of heterozygous lipoprotein lipase deficiency in the extended pedigree of a proband homozygous for a missense mutation. *J Clin Invest* 1990; 86: 735-750

584. MAILLY F, TUGRUL Y, REYMER PWA, BRUIN T, SEED M, GROENEMEYER BF, ASPLUND-CARLSON A, VALLANCE D, WINDER AF, MILLER GJ, KASTELEIN JJP, HAMSTEN A, OLIVECRONA G, HUMPHRIES SE, TALMUD PJ : A common variant in the gene for lipoprotein lipase (Asp9→Asn). Functional implications and prevalence in normal and hyperlipidemic subjects. *Arterioscler Thromb* 1995; 15: 468-478

585. REYMER PWA, GAGNÉ E, GROENEMEYER BE, ZHANG H, FORSYTHE I, JANSEN H, SEIDELL JC, KROMHOUT D, LIE KE, KASTELEIN JJP, HAYDEN MR : A lipoprotein lipase mutation (Asn291Ser) is associated with reduced HDL cholesterol levels in premature atherosclerosis. *Nature Genet* 1995; 10: 28-34

586. WU D-A, BU X, WARDEN CH, SHEN DDC, JENG C-Y, SHEU WHH, FUH MMT, KATSUYA T, DZAU VJ, REAVEN GM, LUSIS AJ, ROTTER JI, CHEN IYD : Quantitative trait locus mapping of human blood pressure to a genetic region at or near the lipoprotein lipase gene locus on chromosome 8p22. *J Clin Invest* 1996; 97: 2111-2118

587. SPRECHER DL, HARRIS BV, STEIN EA, BELLET PS, KEILSON LM, SIMBARTL LA : Higher triglycerides, lower high density lipoprotein cholesterol, and higher systolic blood pressure in lipoprotein lipase-deficient heterozygotes. *Circulation* 1996; 94: 3239-3245.

588. HUBEL CA, ROBERTS JM, FERRELL RE : Association of pre-eclampsia with common coding sequence variations in the lipoprotein lipase gene. *Clin Genet* 1999; 56: 289-296

589. HUMPHRIES SE, NICAUD V, MARGALEF J, TIRET L, TALMUD PJ: Lipoprotein lipase gene variation is associated with a paternal history of premature coronary artery

disease and fasting and postprandial plasma triglycerides. *Arterioscler Thromb Vasc Biol* 1998; 18: 526-534

590. GAGNE SE, LARSON MG, PIMSTONE SN, SCHAEFER EJ, KASTELEIN JJ, WILSON PW, ORDOVAS JM, HAYDEN MR : A common truncation variant of lipoprotein lipase (Ser447X) confers protection against coronary heart disease: the Framingham Offspring Study. *Clin Genet* 1999; 55: 450-454

591. OLIVECRONA G, OLIVECRONA T : Triglyceride lipases and atherosclerosis. *Curr Opin Lipidol* 1995; 6: 291-305

592. SANTAMARINA-FOJO S, HAUDENSCHILD C, AMAR M : The role of hepatic lipase in lipoprotein metabolism and atherosclerosis. *Curr Opin Lipidol* 1998; 9: 211-219

593. CONNELLY PW : The role of hepatic lipase in lipoprotein metabolism. *Clin Chim Acta* 1999; 286: 243-255

594. KRAPP A, AHLE S, KERSTING S, HUA Y, KNESER K, NIELSEN M, GLIEMANN J, BEISIEGEL U : Hepatic lipase mediates the uptake of chylomicrons and beta-VLDL into cells via the LDL receptor-related protein (LRP 1). *J Lipid Res* 1996; 37:926-936

595. BARRANS A, COLLET X, BARBARAS R, JASPARD B, MANENT J, VIEU C, CHAP H, PERRET B : Hepatic lipase induces the formation of pre-ß1 high density lipoprotein (HDL) from triacylglycerol-rich HDL2. A study comparing liver perfusion to in vitro incubation with lipases. *J Biol Chem* 1994; 269: 11572-11577

596. COLLET X, TALL AR, SERAJUDDIN H, GUENDOUZI K, ROYER L, OLIVEIRA H, BARABRAS R, JIANG X-C, FRANCONE O : Remodeling of HDL in CETP in vivo and in CETP and hepatic lipase in vitro results in enhanced uptake of HDL CE by cells expressing scavenger receptor B-1. *J Lipid Res* 1999; 40: 1185-1193

597. WANG N, WENG W, BRESLOW JL, TALL AR : Scavenger receptor BI (SR-BI) is up-regulated in adrenal gland in apolipoprotein A-I and hepatic lipase knock-out mice as a response to depletion of cholesterol stores. *J Biol Chem* 1996; 271: 21001-21004

598. DATTA S, LUO CC, LI WH, TUINENE PV, LEDBETTER DH, BROWN MA, CHEN SH, LIU SW, CHAN L : Human hepatic lipase. Cloned cDNA sequence, restriction fragment length polymorphisms, chromosomal localization, and evolutionary relationships with lipoprotein lipase and pancreatic lipase. *J Biol Chem* 1988; 263: 1107-1110

599. AMEIS D, STAHNKE G, KOBAYASHI J, Mc LEAN J, LEE G, BÜSCHER M, SCHOTZ MC, WILL H : Isolation and characterization of the human hepatic lipase gene. *J Biol Chem* 1990; 265: 6552-6558

600. KOMAROMY MC, SCHOTZ MC : Cloning of rat hepatic lipase cDNA: evidence for a lipase gene family. *Proc Natl Acad Sci USA* 1987; 84: 1526-1530

601. DUGI KA, AMAR MJ, HAUDENSCHILD CC, SHAMBUREK RD, BENSADOUN A, HOYT RF JR, FRUCHART-NAJIB J, MADJ Z, BREWER HB JR, SANTAMARINA-FOJO S : In vivo evidence for both lipolytic and nonlipolytic function of hepatic lipase in the metabolism of HDL. *Arterioscler Thromb Vasc Biol* 2000; 20: 793-800

602. COHEN JC, VEGA GL, GRUNDY SM : Hepatic lipase: new insights from genetic and metabolic studies. *Curr Opin Lipidol* 1999; 10: 259-267

603. KNUDSEN P, ANTIKAINEN M, EHNHOLM S, UNSI-OUKARI M, TENKANEN H, LAHDENPERÄ S, KAHRI J, TILLY-KIESI M, BENSADOUN A, TASKINEN M-R, EHNHOLM C : A compound heterozygote for hepatic lipase gene mutations

Leu334→Phe and Thr383→Met: Correlation between hepatic lipase activity and phenotypic expression. *J Lipid Res* 1996; 37: 825-834

604. BRAND K, DUGI KA, BRUNZELL JD, NEVIN DN, SANTAMARINA-FOJO S : A novel A→G mutation in intron I of the hepatic lipase gene leads to alternative splicing resulting in enzyme deficiency. *J Lipid Res* 1996; 37: 1213-1223

605. KNUDSEN P, ANTIKAINEN M, UNSI-OUKARI M, EHNHOLM S, LAHDENPERÄ S, BENSADOUN A, FUNKE H, WIEBUSCH H, ASSMANN G, TASKINEN M-R, EHNHOLM C: Heterozygous hepatic lipase deficiency, due to missense mutations of R186H and L334F, in the HL gene. *Atherosclerosis* 1997; 128: 165-174

606. DEREWENDA ZS and CAMBILLAU, C : Effects of gene mutations in lipoprotein and hepatic lipases as interpreted by a molecular model of the pancreatic triglyceride lipase. *J Biol Chem* 1991; 266: 23112-23119

607. DURSTENFELD A, BEN-ZEEV O, REUE K, STAHNKE G, DOOLITTLE MH : Molecular characterization of human hepatic lipase deficiency. In vitro expression of two naturally occurring mutations. *Arterioscler Thromb* 1994; 14: 381-385

608. GEHRISCH S, KOSTKA H, TIEBEL M, PATZAK A, PAETZOLD A, JULIUS U, SCHROEDER HE, HANEFELD M, JAROSS W : Mutations of the human hepatic lipase gene in patients with combined hypertriglyceridemia/hyperalphalipoproteinemia and in patients with familial combined hyperlipidemia. *J Mol Med* 1999; 77: 728-734

609. HOFFER MJV, SNIEDER H, BREDIE SJH, DEMACKER PNM, KASTELEIN JJP, FRANTS RR, STALENHOEF AFH : The V73M mutation in the hepatic lipase gene is associated with elevated cholesterol levels in four Dutch pedigrees with familial combined hyperlipidemia. *Atherosclerosis* 2000; 151: 443-450

610. COHEN JC, WANG Z, GRUNDY SM, STOESZ MR, GUERRA R : Variation at the hepatic lipase and apolipoprotein AI/CIII/AIV loci is a major cause of genetically determined variation in plasma HDL cholesterol levels. *J Clin Invest* 1994; 94: 2377-2384

611. GUERRA R, WANG J, GRUNDY SM, COHEN JC : A hepatic lipase (LIPC) allele associated with high plasma concentrations of high density lipoprotein cholesterol. *Proc Natl Acad Sci USA* 1997; 29: 4532-4537

612. VAN'T HOOFT FM, LUNDAHL B, RAGOGNA F, KARPE F, OLIVECRONA G, HAMSTEN A : Functional characterization of 4 polymorphisms in promoter region of hepatic lipase gene. *Arterioscler Thromb Vasc Biol* 2000; 20: 1335-1339

613. DEEB SS, PENG R : The C-514T polymorphism in the human hepatic lipase gene promoter diminishes its activity. *J Lipid Res* 2000; 41: 155-158

614. MURTOMÄKI S, TAHVANAINEN E, ANTIKAINEN M, TIRET L, NICAUD V, JANSEN H, EHNHOLM C, on behalf of the european atherosclerosis research study (EARS) group : Hepatic lipase gene polymorphisms influence plasma HDL levels. *Arterioscler Thromb Vasc Biol* 1997; 17: 1879-1884

615. JANSEN H, VERHOEVEN AJ, WEEKS L, KASTELEIN JJ, HALLEY DJ, VAN DEN OUWELAND A, JUKEMA JW, SEIDELL JC, BIRKENHAGER JC : Common C-to-T substitution at position -480 of the hepatic lipase promoter associated with a lowered lipase activity in coronary artery disease patients. *Arterioscler Thromb Vasc Biol* 1997; 17: 2837-2842

616. TAHVANAINEN E, SYVANNE M, FRICK H, MURTOMÄKI-REPO S, ANTIKAINEN M, KESANIEMI AY, KAUMA H, PASTERNAK A, TASKINEN M-R, EHNHOLM C for the LOCAT study investigators : Association of variation in hepatic lipase activity with promoter variation in the hepatic lipase gene. *J Clin Invest* 1998; 101: 956-960

617. JANSEN H, CHU G, EHNHOLM C, DALLONGEVILLE J, NICAUD V, TALMUD PJ : The allele of the hepatic lipase promoter variant C-480T is associated with increased fasting lipids and HDL and increased preprandial and postprandial LpCIII:B. *Arterioscler Thromb Vasc Biol* 1999; 19: 303-308

618. SHOHET RV, VEGA G, ANWAR A, CIGARROA JE, GRUNDY SM, COHEN JC : Hepatic lipase (LIPC) promoter polymorphism in men with coronary artery disease. *Arterioscler Thromb Vasc Biol* 1999; 19: 1975-1978

619. COUTURE P, OTVOS JD, CUPPLES A, LAHOZ C, WILSON PWF, SCHAEFER EJ, ORDOVAS JM : Association of the C-514T polymorphism in the hepatic lipase gene with variations in lipoprotein subclass profiles. *Arterioscler Thromb Vasc Biol* 2000; 20: 815-822

620. ZAMBON A, DEEB S, HOKANSON JE, BROWN G, BRUNZELL JD : Common variants in the promoter of the hepatic lipase gene are associated with lower levels of hepatic lipase activity, buoyant LDL, and higher LDL_2 Cholesterol. *Arterioscler Thromb Vasc Biol* 1998; 18: 1723-1729

621. PIHLAJAMAKI J, KARJALAINEN L, KARHAPAA P, VAUHKONEN I, TASKINEN MR, DEEB SS, LAAKSO M : G-250A substitution in promoter of hepatic lipase gene is associated with dyslipidemia and insulin resistance in healthy control subjects and in members of families with familial combined hyperlipidemia. *Arterioscler Thromb Vasc Biol* 2000; 20: 1789-1795

622. HOLM C, OSTERLUND T, LAURELL H, CONTRERAS JA : Molecular mechanisms regulating hormone-sensitive lipase and lipolysis. *Annu Rev Nutr* 2000; 20: 365-393

623. LANGIN D, LAURELL H, HOLST LS, BELFRAGE P, HOLM C : Gene organization and primary structure of human hormone-sensitive lipase : possible significance of a sequence homology with a lipase of Moraxella TA144, an Antarctic bacterium. *Proc Natl Acad Sci USA* 1993; 90: 4897-4901

624. HOLM C, KIRCHGESSNER TG, SVENSON K, FREDRICKSON G, NILSSON S, MILLER CG, SHIVELY JE, HEINZMANN C, SPARKES RS, MOHANDAS T, LUSIS AJ, BELFRAGE P, SCHOTZ MC : Hormone-sensitive lipase: sequence, expression, and chromosomal localization to 19 cent-q133. *Science* 1988; 241: 1502-1506

625. LI Z, SUMIDA M, BIRCHBAUER A, SCHOTZ MC, REUE K : Isolation and characterization of the gene for mouse hormone-sensitive lipase. *Genomics* 1994; 24: 259-265

626. HOLM C, DAVIS RC, OSTERLUND T, SCHOTZ MC, FREDRIKSON G : Identification of the active site serine of hormone-sensitive lipase by site directed mutagenesis. *FEBS Lett* 1994; 344: 234-238

627. SHEN WJ, PATEL S, NATU V, KRAEMER FB : Mutational analysis of structural features of rat hormone-sensitive lipase. *Biochemistry* 1998; 37: 8973-8979

628. ANTHONSEN MW, RONNSTRAND L, WERNSTEDT C, DEGERMAN E, HOLM C : Identification of novel phosphorylation sites in hormone-sensitive lipase that are

phosphorylated in response to isoproterenol and govern activation properties in vitro. *J Biol Chem* 1998; 273: 215-221

629. KRAEMER FB, TAVANGAR K, HOFFMAN AR : Developmental regulation of hormone-sensitive lipase mRNA in the rat : changes in steroidogenic tissues. *J Lipid Res* 1991; 32: 1303-1310

630. WILSON BE, DEEB S, FLORANT GL : Seasonal changes in hormone-sensitive and lipoprotein lipase mRNA concentrations in marmot white adipose tissue. *Am J Physiol* 1992; 262: R177-R181

631. OSUGA J-I, ISHIBASHI S, OKA T, YAGYU H, TOZAWA R, FUJIMOTO A, SHIONOIRI F, YAHAGI N, KRAEMER FB, TSUTSUMI O, YAMADA N : Targeted disruption of hormone-sensitive lipase results in male sterility and adipocyte hypertrophy, but not in obesity. *Proc Natl Acad Sci USA* 2000; 97: 787-792

632. ARNER P : Obesity - a genetic disease of adipose tissue? *Br J Nutr* 2000; 83 S1: S9-16

633. ESCARY JL, CHOY HA, REUE K, WANG XP, CASTELLANI LW, GLASS CK, LUSIS AJ, SCHOTZ MC : Paradoxical effect on atherosclerosis of hormone-sensitive lipase overexpression in macrophages. *J Lipid Res* 1999; 40: 397-404

634. LIDBERG U, NILSSON J, STRÖMBERG K, STENMAN G, SAHLIN P, ENERBÄCK S, BJURSELL G : Genomic organization, sequence analysis, and chromosomal localization of the human carboxyl ester lipase (CEL) gene and CEL-like (CELL) gene. *Genomics* 1992; 13: 630-640

635. SHAMIR R, JOHNSON WJ, MORLOCK-FITZPATRICK K, ZOLFAGHARI R, LI L, MAS E, LOMBARDO D, MOREL DW, FISHER EA : Pancreatic carboxyl ester lipase: A circulating enzyme that modifies normal and oxidized lipoproteins in vitro. *J Clin Invest* 1996; 97: 1696-1704

636. TAYLOR AK, ZAMBAUX JL, KLISAK I, MOHANDAS T, SPARKES RS, SCHOTZ MC, LUSIS AJ : Carboxyl ester lipase : a highly polymorphic locus on chromosome 9qter. *Genomics* 1991; 10: 425-431

637. LOWE ME : Structure and function of pancreatic lipase and colipase. *Annu Rev Nutr* 1997; 17:141-158

638. WINKLER FK, D'ARCY A, HUNZIKER W : Structure of human pancreatic lipase. *Nature* 1990; 343: 771-774

639. MICKEL FS, WEIDENBACH F, SWAROVSKY B, LAFORGE KS, SCHEELE GA : Structure of canine pancreatic lipase gene. *J Biol Chem* 1989; 264: 12895-12901

640. DAVIS RC, XIA Y, MOHANDAS T, SCHOTZ MC, LUSIS AJ : Assignment of human pancreatic colipase gene to chromosome 6p21 to pter. *Genomics* 1991; 10: 262-265

641. DAVIS RC, DIEP A, HUNZIKER W, KLISAK I, MOHANDAS T, SCHOTZ MC, SPARKES RS, LUSIS AJ : Assignment of human pancreatic lipase gene (PNLIP) to chromosome 10q24-q26. *Genomics* 1991; 11: 1164-1166

642. FIGARELLA C, DE CARO A, LEUPOLD D, POLEY JR : Congenital pancreatic lipase deficiency. *J Pediatr* 1980; 96: 412-416

643. UUSITUPA M : New aspects in the management of obesity: operation and the impact of lipase inhibitors. *Curr Opin Lipidol* 1999; 10: 3-7

644. RADER DJ, JAYE M : Endothelial lipase: a new member of the triglyceride lipase gene family. *Curr Opin Lipidol* 2000; 11: 141-147

645. BRIQUET-LAUGIER V, BEN-ZEEV O, WHITE A, DOOLITTLE MH : cld and lec 23 are disparate mutations that affect maturation of lipoprotein lipase in the endoplasmic reticulum. *J Lipid Res* 1999; 40: 2044-2058

646. JONG MC, HOFKER MH, HAVEKES LM : Role of apoCs in lipoprotein metabolism. Functional differences between ApoC1, ApoC2 and ApoC3. *Arterioscler Thromb Vasc Biol* 1999; 19: 472-484

647. MYKLEBOST O, ROGNE S : A physical map of the apolipoprotein gene cluster on human chromosome 19. *Hum Genet* 1988; 78: 244-247

648. DAS HK, Mc PHERSON J, BRUNS GAP, KARATHANASIS SK, BRESLOW JL : Isolation, characterization and mapping to chromosome 19 of the human apolipoprotein E gene. *J Biol Chem* 1985; 260: 6240-6247

649. LI WH, TANIMURA M, LUO CC, DATTA S, CHAN L : The apolipoprotein multigene family: biosynthesis, structure-function relationships, and evolution. *J Lipid Res* 1988; 29: 245-271

650. FORNAGE M, CHAN L, SIEST G, BOERWINKLE E : Allele frequency distribution of the (TG)n(AG)m microsatellite in the apolipoprotein C-II gene. *Genomics* 1992; 12: 63-68

651. HEGELE RA and TU L : Variation within intron 3 of the apolipoprotein CII gene. *Nucleic Acids Res* 1991; 19: 3162

652. MENZEL HJ, KANE JP, MALLOY MS, HAVEL RJ : A variant primary structure of apolipoprotein C-II in individuals of African descent. *J Clin Invest* 1986; 77: 595-601

653. STREICHER R, GEISEL J, WEISSHAAR C, AVCI H, OETTE K, MULLER-WIELAND D, KRONE W : A single nucleotide subsitution in the promoter region of the apolipoprotein C-II gene identified in individuals with chylomicronemia. *J Lipid Res* 1996; 37: 2599-2607

654. FOJO SS : Genetic dyslipoproteinemias : role of lipoprotein lipase and apolipoprotein C-II. *Curr Opin Lipidol* 1992 3: 186-195

655. SEPEHRNIA B, KAMBOH MI, ADAMS-CAMPBELL LL, BUNKER CH, NWANKWO MAJUMDER PP, FERRELL RE : Genetic studies of human apolipoproteins XI The effect of the apolipoprotein CII polymorphism on lipoprotein levels in Nigerian blacks. *J Lipid Res* 1989; 30: 1349-1355

656. PULLINGER CR, HENNESSY LK, LOVE JA, FROST PH, MALLOY MJ, KANE JP : Molecular cloning and characteristics of a new apolipoprotein C-II mutant identified in three unrelated individuals with hypercholesterolemia and hypertriglyceridemia. *Hum Mol Genet* 1993; 2: 69-74

657. BEIL FU, FOJO SS, BREWER HB, GRETEN H, BEISIEGEL U : Apolipoprotein C-II deficiency syndrome due to apoC-II Hamburg: clinical and biochemical features and HphI restriction enzyme polymorphism. *Eur J Clin Invest* 1992; 22: 88-95

658. ALLAN CM, WALKER D, SEGREST JP, TAYLOR JM : Identification and characterization of a new human gene (APOC4) in the apolipoproteins E, C-I, and C-II, gene locus. *Genomics* 1995; 28: 291-300

659. ALLAN CM, TAYLOR JM : Expression of a novel human apolipoprotein (apoC-IV) causes hypertriglyceridemia in transgenic mice. *J Lipid Res* 1996; 37: 1510-1518

660. TAYLOR JM, SIMONET WS, BUCAY N, LAUER SJ, DE SILVA HV : Expression of the human apolipoprotein E/apolipoprotein C-I gene locus in transgenic mice. *Curr Opin Lipidol* 1991; 2: 73-80

661. LAUER SJ, WALKER D, ELSHOURBAGY NA, REARDON CA, LEVY-WILSON B, TAYLOR JM : Two copies of the human apolipoprotein C-I gene are linked closely to the apolipoprotein E gene. *J Biol Chem* 1988; 263: 7277-7286

662. DANG Q, TAYLOR J : In vivo footprinting analysis of the hepatic control region of the human apolipoprotein E/CI/CIV/CII, gene locus. *J Biol Chem* 1996; 271: 28667-28676

663. FREITAS EM, GAUDIERI S, ZHANG WJ, KULSKI JK, vanBOCKXMEER FM, CHRISTIANSEN FT, DAWKINS RL: Duplication and diversification of the apolipoprotein CI (APOCI) genomic segment in association with retroelements. *J Mol Evol* 2000; 50: 391-396

664. FROSSARD PM, LIM RT, COLEMAN RT, FUNKE H, ASSMANN G, MALLOY MJ, KANE JP : Human apolipoprotein CI (ApoCI) gene locus : BglI dimorphic site. *Nucleic Acids Res* 1987; 15: 1344

665. FROSSARD PM, COLEMAN RT, MALLOY MJ, KANE JP, LEVY-WILSON B, APPLEBY VA : Human apolipoprotein CI (ApoCI) gene locus: DraI dimorphic site. *Nucleic Acids Res* 1987; 15: 1884

666. NILLESEN WM, SMEETS HJM, VAN OOST BA: Human ApoCI HpaI restriction site polymorphism revealed by the polymerase chain reaction. *Nucleic Acids Res* 1990; 18: 3428

667. BJÖRKEGREN J, BOQUIST S, SAMNEGÅRD A, LUNDMAN P, TORNVALL P, ERICSSON C-G, HAMSTEN A : Accumulation of apolipoprotein CI-rich and cholesterol-rich VLDL remnants during exaggerated postprandial triglyceridemia in normolipidemic patients with coronary artery disease. *Circulation* 2000; 101: 227-230

668. MAHLEY RW Apolipoprotein E : cholesterol transport protein with expanding role in cell biology. *Science* 1988; 240: 622-630

669. MAHLEY RW, RALL SC : Type III hyperlipoproteinemia (dysbetalipoproteinemia): the role of apolipoprotein E in normal and abnormal lipoprotein metabolism. In : Scriver CR, Beaudet AL, Sly WS, Valle D (eds). *The Metabolic and Molecular Bases of Inherited Disease*, 7th edn. Mac Graw Hill, New York, 1995; 1953-1980

670. DERGUNOV AD, ROSSENEU M : The significance of apolipoprotein E structure to the metabolism of plasma triglyceride-rich lipoproteins. *Biol Chem Hoppe Seyler* 1994; 375: 485-495

671. WILSON C, WARDELL MR, WEISGRABER KH, MAHLEY RW, AGARD DA : Three-dimensional structure of the LDL-receptor binding domain of apolipoprotein E. *Science* 1991; 252: 1817-1822

672. LALAZAR A, OU SHI, MAHLEY R : Human apolipoprotein E. Receptor binding activity of truncated variants with carboxyl-terminal deletions. *J Biol Chem* 1989; 264: 8447-8450

673. WILLEMS VAN DIJK K, HOFKER MH, HAVEKES LM : Use of transgenic mice to study the role of apolipoprotein E in lipid metabolism and atherosclerosis. *Int J Tissue React* 2000; 22: 49-58

674. BEISIEGEL U, KRAPP A, WEBER W, OLIVECRONA G : The role of alpha 2M receptor/LRP in chylomicron remnant metabolism. *Ann N Y Acad Sci* 1994; 737: 53-69

675. THUREN T, WEISGRABER KH, SISSON P, WAITE M : Role of apolipoprotein E in hepatic lipase catalyzed hydrolysis of phospholipid in high-density lipoproteins. *Biochemistry* 1992; 31: 2332-2338

676. PLUMP AS, SMITH JD, HAYEK T, AALTO-SETÄLÄ K, WALSH A, VERSTUYFT JG, RUBIN EM, BRESLOW JL : Severe hypercholesterolemia and atherosclerosis in apolipoprotein E-deficient mice created by homologous recombination in ES cells. *Cell* 1992; 71: 343-353

677. ZHANG SH, REDDICKRL, PIEDRAHITA JA, MAEDA N : Spontaneous hypercholesterolemia and arterial lesions in mice lacking apolipoprotein E. *Science* 1992; 258: 468-471

678. VAN DEN MAAGDENBERG AMJM, HOFKER MH, KRIMPENFORT PJA, DE BRUIJN IH, VAN VLIJMEN B, VAN DER BOOM H, HAVEKES LM, FRANTS RR : Transgenic mice carrying the apolipoprotein E3-Leiden gene exhibit hyperlipoproteinemia. *J Biol Chem* 1993; 268: 10540-10545

679. HO YY, AL-HAIDERI M, MAZZONE T, VOGEL T, PRESLEY JF, STURLEY SL, DECKELBAUM RJ : Endogenously expressed apolipoprotein E has different effects on cell lipid metabolism as compared to exogenous apolipoprotein E carried on triglyceride-rich particles. *Biochemistry* 2000; 39: 4746-4754

680. LINTON MF, HASTY AH, BABAEV VR, FAZIO S : Hepatic apo E expression is required for remnant lipoprotein clearance in the absence of the low density lipoprotein receptor. *J Clin Invest* 1998; 101: 1726-1736

681. CURTISS LK, BOISVERT WA : Apolipoprotein E and atherosclerosis. *Curr Opin Lipidol* 2000; 11: 243-251

682. SAXENA U, FERGUSON E, BISGAIER CL : Apolipoprotein E modulates low density lipoprotein retention by lipoprotein lipase anchored to the subendothelial matrix. *J Biol Chem* 1993; 268: 14812-14819

683. HUANG Y, VON ECKARDSTEIN A, WU S, MAEDA N, ASSMANN G : A plasma lipoprotein containing only apolipoprotein E and with γ mobility on electrophoresis releases cholesterol from cells. *Proc Natl Acad Sci USA* 1994; 91: 1834-1838

684. WEISGRABER KH, ROSES AD, STRITTMATTER WJ : The role of apolipoprotein E in the nervous system. *Curr Opin Lipidol* 1994; 5: 110-116

685. NATHAN BP, BELLOSTA S, SANAN DA, WEISGRABER KH, MAHLEY RW, PITAS RE : Differential effects of apolipoprotein E3 and E4 on neuronal growth in vitro. *Science* 1994; 264: 850-852

686. HOLTZMANDM, PITAS RE, KILBRIDGE J, NATHAN B, MAHLEY RW, BU G, SCHWARTZ AL : Low density lipoprotein receptor-related protein mediates apolipoprotein E-dependent neurite outgrowth in a central nervous system derived neuronal cell-line. *Proc Natl Acad Sci USA* 1995; 92: 9480-9484

687. FAGAN AM, BU G, SUN Y, DAUGHERTY A, HOLTZMAN DM : Apolipoprotein E-containing high density lipoprotein promotes neurite outgrowth and is a ligand for the low density lipoprotein receptor-related protein. *J Biol Chem* 1996; 271: 30121-30125

688. MICHIKAWA M, FAN QW, ISOBE I, YANAGISAWA K : Apolipoprotein E exhibits isoform-specific promotion of lipid efflux from astrocytes and neurons in culture. *J Neurochem* 2000; 74: 1008-1016

689. DYER CA, CURTISS LK : Apoprotein E-rich high density lipoproteins inhibit ovarian androgen synthesis. *J Biol Chem* 1988; 263: 10965-10973

690. DYER CA, SMITH RS, CURTISS LK : Only multimers of a synthetic peptide of human apolipoprotein E are biologically active. *J Biol Chem* 1991; 266: 15009-15015

691. LARSEN F, SOLHEIM J, PRYDZ H : A methylated CpG island 3' in the apolipoprotein-E gene does not repress its transcription. *Hum Mol Genet* 1993; 2: 775-780

692. KRAFT HG, MENZEL HJ, HOPPICHLER F, VOGEL W, UTERMANN G : Changes of genetic apolipoprotein phenotypes caused by liver transplantation Implications for apolipoprotein synthesis. *J Clin Invest* 1989; 83: 137-142

693. SMITH JD, MELIAN A, LEFF T, BRESLOW JL : Expression of human apolipoprotein E gene is regulated by multiple positive and negative elements. *J Biol Chem* 1988; 263: 8300-8308

694. WERNETTE-HAMMOND ME, LAUER SJ, CORSINI A, WALKER D, TAYLOR JM, RALL SC : Glycosylation of human apolipoprotein E. *J Biol Chem* 1989; 264: 9094-9101

695. ZANNI EE, KOUVASTI A, HADZOPOULOU-CLADARAS M, KRIEGER M, ZANNIS VI : Expression of apo E gene in chinese hamster cells with a reversible defect in O-Glycosylation Glycosylation is not required for apo E secretion. *J Biol Chem* 1989; 264: 9137-9140

696. STROBL W, GORDER NL, FIENUP GA, LIN-LEE YC, GOTTO AM, PATSCH W : Effect of sucrose diet on apolipoprotein biosynthesis in rat liver. Increase in apolipoprotein E gene transcription. *J Biol Chem* 1989; 264: 1190-1194

697. YE SQ, OLSON LM, REARDON CA, GETZ GS : Human plasma lipoproteins regulate apolipoprotein E secretion from a post-Golgi compartment. *J Biol Chem* 1992; 267: 21961-21966

698. WYNE KL, SCHREIBER JR, LARSEN AL, GETZ GS : Regulation of apolipoprotein E biosynthesis by cAMP and phorbol ester in rat ovarian granulosa cells. *J Biol Chem* 1989; 264: 981-989

699. SALOMON RN, UNDERWOOD R, DOYLE MV, WANG A, LIBBY P : Increased apolipoprotein E and c-fms gene expression without elevated interleukin 1 or 6 mRNA levels indicate selective activation of macrophage functions in advanced human atheroma. *Proc Natl Acad Sci USA* 1992; 89: 2814-2818

700. DAVIGNON J, GREGG RE, SING CF : Apolipoprotein E polymorphism and atherosclerosis. *Arteriosclerosis* 1988; 8: 1-21

701. CHAN L and LI WH : Apolipoprotein variation among different species. *Curr Opin Lipidol* 1991; 2: 96-103

702. MAHLEY RW, HUANG Y, RALL SCJr : Pathogenesis of type III hyperlipoproteinemia (dysbetalipoproteinemia): questions, quandaries, and paradoxes. *J Lipid Res* 1999; 40: 1933-1949

703. DE KNIJFF P, BOOMSMA DI, DE WIT E, KEMPEN HJM, GEVERS LEUVEN JA, FRANTS RR, HAVEKES LM : The effects of the apolipoprotein E phenotype on plasma lipids is not influenced by environmental variability : results of a Dutch twin study. *Hum Genet* 1993; 91: 268-272

704. BOERWINKLE E and UTERMANN G : Simultaneous effects of the apolipoprotein E polymorphism on apolipoprotein E, apolipoprotein B, and cholesterol metabolism. *Am J Hum Genet* 1988; 42: 104-112

705. DALLONGEVILLE J, ROY M, LEBOEUF N, XHIGNESSE M, DAVIGNON J, LUSSIER-CACAN S : Apolipoprotein E polymorphism association with lipoprotein profile in endogenous hypertriglyceridemia and familial hypercholesterolemia. *Arterioscler Thromb* 1991; 11: 272-278

706. MAEDA H, NAKAMURA H, KOBORI S : Molecular cloning of human apolipoprotein E variant: E5 (Glu3→Lys3). *Biochem J* 1989; 105: 491

707. LOSHE P, MANN WA, STEIN EA, BREWER HB : Apolipoprotein E-4 Philadelphia (Glu13→Lys, Arg145→Cys). Homozygosity for two rare point mutations in the apolipoprotein E gene combined with severe type III hyperlipoproteinemia. *J Biol Chem* 1991; 266: 10479-10484

708. FEUSSNER G, FEUSSNER V, HOFFMANN MM, LOHRMANN J, WIELAND H, MARZ W : Molecular basis of type III hyperlipoproteinemia in Germany. *Hum Mutat* 1998; 11: 417-423.

709. ORTH M, WENG W, FUNKE H, STEINMETZ A, ASSMANN G, NAUCK M, DIERKES J, AMBROSCH A, WEISGRABER KH, MAHLEY RW, WIELAND H, LULEY C : Effects of a frequent apolipoprotein E isoform, apo E4$_{Freiburg}$ (Leu28→Pro), on lipoproteins and the prevalence of coronary artery disease in whites. *Arterioscler Thromb Vasc Biol* 1999; 19: 1306-1315

710. FEUSSNER G, FUNKE H, WENG W, ASSMANN G, LACKNER KJ, ZIEGLER R : Severe type III hyperlipoproteinemia associated with unusual apolipoprotein E1 phenotype and e1/'null' genotype. *Eur J Clin Invest* 1992; 22: 599-608

711. CLADARAS C, HADZOPOULOU-CLADARAS M, FELBERG BK, PAVLAKIS G, ZANNIS VI : The molecular basis of familial apo E deficiency. An acceptor splice site mutation in the third intron of the deficient apo E gene. *J Biol Chem* 1987; 262: 2310-2315

712. DE KNIJFF P, VAN DEN MAAGDENBERG AMJM, STALENHOEF AFH, GEVERS LEUVEN AFH, DEMACKER PNM, KUYT LP, HAVEKES LM : Familial dysbetalipoproteinemia associated with apolipoprotein E3-Leiden in an extended multigeneration pedigree. *J Clin Invest* 1991; 88: 643-655

713. STEINMETZ A, ASSEFBARKHI N, ELTZE C, EHLENZ K, FUNKE H, PIES A, ASSMANN G, KAFFARNI K : Normolipidemic dysbetalipoproteinemia and hyperlipoproteinemia type III in subjects homozygous for a rare genetic apolipoprotein E variant (ApoE1). *J Lipid Res* 1990; 31: 1005-1013

714. WENHAM PR, MAC DOWELL IFW, HODGES VM, MAC ENENY J, O'KANE MJ, DAVIES RJH, NICHOLLS DP, TRIMBLE ER, BLUNDELL G : Rare apolipoprotein E variant identified in a patient with type III hyperlipidemia. *Atherosclerosis* 1993; 99: 261-271

715. MINNICH A, WEIGRABERN KH, NEWHOUSE Y, DONG LM, FORTIN LJ, TREMBLAY M, DAVIGNON J : Identification and characterization of a novel apolipoprotein E variant, apolipoprotein E3' (Arg136→His) : association with mild dyslipidemia and double pre-ß very low density lipoproteins. *J Lipid Res* 1995 36: 57-66

716. WARDELL MR, BRENNAN SO, JANUS ED, FRASER R, CARRELL RW : Apolipoprotein E2-Christchurch (136 Arg→Ser). New variant of human apolipoprotein E in a patient with type III hyperlipoproteinemia. *J Clin Invest* 1987; 80: 483-490

717. RALL SC, NEWHOUSE YM, CLARKE HRG, WEISGRABER KH, MAC CARTHY BJ, MAHLEY RW, BERSOT TP : Type III hyperlipoproteinemia associated with apolipoprotein E phenotype E3/3 : structure and genetics of an apolipoprotein E3 variant. *J Clin Invest* 1989; 83: 1095-1101

718. RICHARD P, PASCUAL DE ZUELETA M, BEUCLER I, DE GENNES JL, CASSAIGNE A, IRON A : Identification of a new apolipoprotein E variant (E2 Arg142→Leu) in type III hyperlipidemia. *Atherosclerosis* 1995; 112: 19-28

719. RALL SC, WEISGRABER KH, INNERARITY TL, MAHLEY RW : Structural basis for receptor binding heterogeneity of apolipoprotein E from type III hyperlipoproteinemic subjects. *Proc Natl Acad Sci USA* 1982; 79: 4696-4700

720. SMIT M, DE KNIJFF P, VAN DER KOOIJ-MEIJS E, GROENENDIJK C, VAN DEN MAAGDENBERG AMJM, GEVERS LEUVEN AFH, STALENHOEF AFH, STUYT PM, FRANTS RR, HAVEKES LM : Genetic heterogeneity in familial dysbetalipoproteinemia The E2 (Lys146→Gln) variant results in a dominant mode of inheritance. *J Lipid Res* 1990; 31: 45-53

721. MANN WA, GREGG RE, RONAN R, THOMAS F, ZECH LA, BREWER HB : Apolipoprotein E1-Harrisburg: a new variant of apolipoprotein E dominantly associated with type III hyperlipoproteinemia. *Biochim Biophys Acta* 1989; 1005: 239-244

722. MORIYAMA K, SASAKI J, MATSUNAGA A, ARAKAWA F, TAKADA Y, ARAKI K, KANEKO S, ARAKAWA K : Apolipoprotein E1 Lys-146→Glu with type III hyperlipoproteinemia. *Biochim Biophys Acta* 1992; 1128: 58-64

723. LOSHE P, BREWER HBIII, MENG MS, SKARLATOS JC, LA ROSA J, BREWER HB : Familial apolipoprotein E deficiency and type III hyperlipoproteinemia due to a premature stop codon in the apolipoprotein E gene. *J Lipid Res* 1992; 33: 1583-1590

724. WARDELL MR, RALL SC, BRENNAN SO, NYE ER, GEORGE PM, JANUS ED, WEISGRABER KH : Apolipoprotein E2-Dunedin (228 Arg→Cys): an apolipoprotein E2 variant with normal receptor binding activity. *J Lipid Res* 1990; 31: 535-543

725. MAEDA H, NAKAMURA H, KOBORI S : Identification of human apolipoprotein E variant gene: Apolipoprotein E7 (Glu244,245→Lys244,245). *Biochem J* 1989; 105: 51

726. HAZZARD WR, WARNICK GR, UTERMANN G, ALBERS JJ : Genetic transmission of isoapolipoprotein E phenotypes in a large kindred : relationship to dysbetalipoproteinemia and hyperlipidemia. *Metabolism* 1981; 30: 79-88

727. HOPKINS PN, WU LL, SCHUMACHER MC, EMI M, HEGELE RM, HUNT SC, LALOUEL JM, WILLIAMS RR : Type III dyslipoproteinemia in patients heterozygous for familial hypercholesterolemia and apolipoprotein E2. Evidence for gene-gene interaction. *Arterioscler Thromb* 1991; 11: 1137-1146

728. ELBAZ A, AMARENCO P : Genetic susceptibility and ischaemic stroke. *Curr Opin Neurol* 1999; 12: 47-55

729. WILSON PWF, SCHAEFER EJ, LARSON MG, ORDOVAS JM : Apolipoprotein E alleles and risk of coronary disease. *Arterioscler Thromb Vasc Biol* 1996; 16: 1250-1255

730. FRIKKE-SCHMIDT R, TYBJAERG-HANSEN A, STEFFENSEN R, JENSEN G, NORDESTGAARD BG : Apolipoprotein E genotype: epsilon32 women are protected while epsilon 43 and epsilon 44 men are susceptible to ischemic heart disease: the Copenhagen city heart study. *J Am Coll Cardiol* 2000; 35: 1192-1199

731. SNYDER SM, TERDIMAN JF, CAAN B, FEINGOLD KR, HUBL ST, SMITH RS, YOUNG SG : Relationship of apolipoprotein E phenotypes to hypocholesterolemia. *Am J Med* 1993; 95: 480-488

732. SCHÄCHTER F, FAURE-DELANEF L, GUENOT F, ROUGER H, FROGUEL P, LESUEUR-GINOT L, COHEN D : Genetic associations with human longevity at the APOE and ACE loci. *Nature Genet* 1994; 6: 29-32

733. VUORISTO M, FÄRKKILÄ M, GYLLING H, KARVONEN A-L, LEINO R, LEHTOLA J, MAKINEN J, MATTILA J, TILVIS R, MIETTINEN TA : Expression and therapeutic response related to apolipoprotein E polymorphism in primary biliary cirrhosis. *J Hepatol* 1997; 27: 136-142

734. ROSES AD: Apolipoprotein E and Alzheimer's disease. The tip of the susceptibility iceberg. *Ann NY Acad Sci* 1998; 855: 738-743

735. HORSBURG K, McCARRIN MO, WHITE F, NICOLL JA : The role of apolipoprotein E in Alzheimer's disease, acute brain injury and cerebrovascular disease: evidence of common mechanisms and utility of animal models. *Neurobiol Aging* 2000; 21: 245-255

736. TETER B, XU PT, GILBERT JR, ROSES AD, GALASKO D, COLE GM : Human apolipoprotein E isoform-specific differences in neuronal sprouting in organotypic hippocampal culture. *J Neurochem* 1999; 73: 2613-2616

737. MAYEUX R, SAUNDERS AM, SHEA S, MIRRA S, EVANS D, ROSES AD, HYMAN BT, CRAIN B, TANG MX, PHELPS CH : Utility of the apolipoprotein E genotype in the diagnosis of Alzheimer's disease. Alzheimer's disease centers consortium on apolipoprotein E and Alzheimer's disease. *N Engl J Med* 1998; 338: 506-511

738. SAINT GEORGE-HYSLOP P, CRAPPER MAC LACHLAN D, TUDA T, ROGAEV E, KARLINSKY H, LIPPA CF, POLLEN D : Alzheimer's disease and possible gene interaction. *Science* 1994; 263: 537

739. CORDER EH, ROBERTSON K, LANNFELT L, BOGDANOVIC N, EGGERTSEN G, WILKINS J, HALL C : HIV-infected subjects with E4 allele for APOE have excess dementia and peripheral neuropathy. *Nature Med* 1998; 4: 1182-1184

740. SOUIED EH, BENLIAN P, AMOUYEL P, FEINGOLD J, LAGARDE J-P, MUNNICH A, KAPLAN J, COSCAS G, SOUBRANE G : The ε4 allele of the apolipoprotein E gene as a potential protective factor for age-related macular degeneration. *Am J Ophtalmol* 1998; 125: 353-359

741. KLAVER CC, KLIFFEN M, VAN DUIJN CM, HOFMAN A, CRUTS M, GEOBBEE DE, VAN BROECKHOVEN C, DE JONG PT : Genetic association of apolipoprotein E with age-related macular degeneration. *Am J Hum Genet* 1998; 63: 200-206

742. BEISIEGEL U, WEBER W, IHRKE G, HERZ J, STANLEY KK : The LDL-receptor-related protein, LRP, is an apolipoprotein E-binding protein. *Nature* 1989; 341: 162-164

743. MYKLEBOST O, ARHEDEN K, ROGNE S, VAN KESSEL AG, MANDAHL N, HERZ J, STANLEY K, HEIM S, MITELMAN F : The gene for the human putative apo E receptor is on chromosome 12 in the segment q13-14. *Genomics* 1989; 5: 65-69

744. NIMPF J, STIFANI S, BILOUS PT, SCHNEIDER WJ : The somatic cell-specific low density lipoprotein receptor-related protein of the chicken. Close kinship to mammalian low density lipoprotein receptor gene family members. *J Biol Chem* 1994; 269: 212-219

745. VAN LEUVEN F, STAS L, HILLIKER C, LORENT K, UMANS L, SERNEELS L, OVERBERG L, TORREKENS S, MOECHARS D, DE STROOPER B, VAN DEN BERGHE H : Structure of the gene (LRP1) coding for the human α2-macroglobulin receptor lipoprotein receptor-related protein. *Genomics* 1994; 24: 78-89

746. ZULIANI G, HOBBS HH : Tetranucleotide length polymorphism 5' of the α2-macroglobulin receptor (A2MR)/LDL receptor-related protein (LRP) gene. *Hum Mol Genet* 1994; 3: 215

747. HUSSAIN MM, STRICKLAND DK, BAKILLAH A : The mammalian low-density lipoprotein receptor family. *Annu Rev Nutr* 1999; 19: 141-172

748. NEELS JG, VAN DEN BERG BMM, LOOKENE A, LOOKENE A, OLIVECRONA G, PANNEKOEK H, VAN ZONNEVELD A-J : The second and fourth cluster of class A cysteine-rich repeats of the low-density lipoprotein receptor-related protein share ligand-binding properties. *J Biol Chem* 1999; 274: 31305-31311

749. SPRINGER TA : An extracellular beta-propeller module predicted in lipoprotein and scavenger receptors, tyrosine kinases, epidermal growth factor precursor, and extracellular matrix components. *J Mol Biol* 1998; 283: 837-862

750. TROMMSDORFF M, BORG J-P, MARGOLIS B, HERZ J : Interaction of cytosolic adaptor proteins with neuronal apolipoprotein E receptors and the amyloid precursor protein. *J Biol Chem* 1998; 273: 33556-33560

751. MOESTRUP SK, GLIEMANN J, PALLESEN G : Distribution of the alpha 2-macroglobulin receptor/low density lipoprotein receptor-related protein in human tissues. *Cell Tissue Res* 1992; 269: 375-82

752. SCHMOELZL S, BENN SJ, LAITHWAITE JE, GREENWOOD SJ, MARSHALL WS, MUNDAY NA, FITZGERALD DJ, LAMARRE J: Expression of hepatocyte low-density lipoprotein receptor-related protein is post-transcriptionally regulated by extracellular matrix. *Lab Invest* 1998; 78: 1405-1413

753. KOUNNAS MZ, ARGRAVES WS, STRICKLAND DK : The 39-kDa receptor-associated protein interacts with two members of the low density lipoprotein receptor family, α-2 macroglobulin receptor and glycoprotein 330. *J Biol Chem* 1992 267: 21162-21166

754. VAN LEUVEN F, HILLIKER C, SERNEELS L, UMANS L, OVERBERGH L, DE STROOPER B, FRYNS JP, VAN DEN BERGHE : Cloning, characterization, and chromosomal localization to 4p16 of the human gene (LRPAP1) coding for the α₂-macroglobulin receptor-associated protein and structural comparison with the murine gene coding for the 44kDa heparin-binding protein. *Genomics* 1995; 25: 492-500

755. JOU Y-S, GOOD RD, MYERS RM : Localization of the α_2-macroglobulin receptor associated-protein 1 gene (LRPAP1) and other gene fragments to human chromosome 4p16.3 by direct cDNA selection. *Genomics* 1994; 24: 410-413

756. STRIEKLAND DK, ASHCOM JD, WILLIAMS S, BATTEY F, BEHRE E, McTIGUE K, BATTEY JF, ARGRAVES WS : Primary structure of alpha 2-macroglobulin receptor-associated protein. Human homologue of a Heymann nephritis antigen. *J Biol Chem* 1991; 266: 13364-13369

757. ORLANDO RA, FARQUHAR MG : Functional domains of the receptor-associated protein (RAP). *Proc Natl Acad Sci USA* 1994; 91: 3161-3165

758. WILNOW TE, ROHLMANN A, HORTON J, OTANI H, BRAUN JR, HAMMER RE, HERTZ J : RAP a specialized chaperone, prevents ligand-induced ER retention and degradation of LDL receptor-related endocytic receptors. *EMBO J* 1996; 15: 2632-2639

759. WILLNOW TE, SHENG Z, ISHIBASHI S, HERZ J : Inhibition of hepatic chylomicron remnant uptake by gene transfer of a receptor antagonist. *Science* 1994; 264: 1471-1474

760. HERTZ J, WILLNOW TE : Lipoprotein and receptor interactions in vivo. *Curr Opin Lipidol* 1995; 6: 97-103

761. MAHLEY RW, JI Z-S: Remnant lipoprotein metabolism: key pathways involving cell-surface heparan sulfate proteoglycans and apolipoprotein E. *J Lipid Res* 1999; 40: 1-16

762. MELLINGER M, GSCHWENTNER C, BURGER I, HAUMER M, WAHRMANN M, SZOLLAR L, NIMPF J, HUETTINGER M : Metabolism of activated complement component C3 is mediated by the low-density lipoprotein receptor-related protein /α2-macroglobulin receptor. *J Biol Chem* 1999; 274: 38091-38096

763. HYMAN BT, STRICKLAND D, REBECK GW : Role of the low-density lipoprotein receptor elated protein in beta-amyloid metabolism and Alzheimer's disease. *Arch Neurol* 2000; 57: 646-650

764. BLACKER D, CRYSTAL AS, WILCOX MA, LAIRD NM, TANZI RE : α2-macroglobulin gene and Alzheimer disease. Reply. *Nature Genet* 1999; 22: 21-22

765. WIJNBERG MJ, QUAX PH, NIEUWENBROECK NM, VERHEIJEN JH : The migration of human smooth muscle cells in vitro is mediated by plasminogen activation and can be inhibited by alpha2-macroglobulin receptor associated protein. *Thromb Haemost* 1997; 78: 880-886

766. YOCHEM J, GREENWALD I : A gene for low density lipoprotein receptor related protein in the nematode *Caenorhabditis* elegans. *Proc Natl Acad Sci USA* 1993; 90: 4572-4576

767. CHRISTENSEN EI, WILLNOW TE : Essential role of megalin in renal proximal tubule for vitamin homeostasis. *J Am Soc Nephrol* 1999; 10: 2224-2236

768. FARQUHAR MG, SAITO A, KERJASCHKI D, ORLANDO RA : The Heyman nephritis antigenic complex; megalin (gp330) and RAP. *J Am Soc Nephrol* 1995; 6: 35-47

769. WILLNOW TE , HILPERT J, ARMSTRONG SA, ROHLMANN A, HAMMER RE, BURNS DK, HERZ J : Defective forebrain development in mice lacking gp330/megalin. *Proc Natl Acad Sci USA* 1996; 93: 8460-8464

770. HAMMAD SM, RANGANATHAN S, LOUKINOVA E, TWAL WO, ARGRAVES WS : Interaction of apolipoprotein J-amyloid beta-peptide complex with low density

lipoprotein receptor-related protein 2/megalin. A mechanism to prevent pathological accumulation of amyloid beta-peptide. *J Biol Chem* 1997; 272: 18644-18649

771. SAITO A, PIETROMONACO S, LOO AK, FARQUHAR MG : Complete cloning and sequencing of rat gp330/"megalin", a distinctive member of the low-density lipoprotein receptor gene family. *Proc Natl Acad Sci USA* 1994; 91: 9725-9729

772. HJALM G, MURRAY E, CRUMLEY G, HARAZIM W, LUNDGREN HW, ONYANGO I, EK B, LARSSON M, JUHLINC, HELLMAN P, DAVIS H, AKERSTROM G, RASK L, MORSE B : Cloning and sequencing of human gp330, a Ca(2+)-binding receptor with potential intracellular signaling properties. *Eur J Biochem* 1996; 239: 132-137

773. ORLANDO RA, EXNER M, CZERKAY RP, YAMAZAKI H, SAITO A, ULLRICH R, KERJASCHKI D, FARQUHAR MG : Identification of the second cluster of ligand-binding repeats in megalin as a site for receptor-ligand interactions. *Proc Natl Acad Sci USA* 1997; 94: 2368-2373

774. JINGAMI H, YAMAMOTO T : The VLDL Receptor: wayward brother of the LDL Receptor. *Curr Opin Lipidol* 1995; 6: 104-108

775. NIMPF J, SCHNEIDER WJ : The VLDL Receptor: an LDL Receptor relative with eight ligand binding repeats, LR8. *Atherosclerosis* 1998; 141: 191-202

776. SAKAI J, HOSHINO A, TAKAHASHI S, MIURA Y, ISHII H, SUZUKI H, KAWARABAYASI Y, YAMAMOTO T : Structure, chromosome location, and expression of the human very low density lipoprotein receptor gene. *J Biol Chem* 1994; 269: 2173-2182

777. RETTENBERGER PM, OKA K, ELLGAARD L, PETERSEN HH, CHRISTENSEN A, MARTENSEN PM, MONARD D, ETZERODT M, CHAN L, ANDREASEN PA : Absence of the third complement-type repeat encoded by exon 4 is associated with reduced binding of Mr 40,000 receptor-associated protein. *J Biol Chem* 1999; 274: 8973-8980

778. WEBB JC, SUN X-M, PATEL DD, Mc CARTHY SN, KNIGHT BL, SOUTAR AK : Characterization and tissue-specific expression of the human "very low density lipoprotein (VLDL) receptor" mRNA. *Hum Mol Genet* 1994; 3: 531-537

779. KOBAYASHI K, OKA K, FORTE T, ISHIDA B, TENG B, ISHIMURA-OKA K, NAKAMUTA M, CHAN L : Reversal of hypercholesterolemia in low density lipoprotein receptor knockout mice by adenovirus-mediated gene transfer of the very low density lipoprotein receptor. *J Biol Chem* 1996; 271: 6852-6860

780. VAN DIJK KW, VAN VLIJMEN BJ, VAN DER ZEE A, VAN'T HOF B, VAN DER BOOM H, KOBAYASHI K, CHAN L, HAVEKES LM, HOFKER MH : Reversal of hypercholesterolemia in apolipoprotein E2 and apolipoprotein E3-Leiden transgenic mice by adenovirus-mediated gene transfer of the VLDL receptor. *Arterioscler Thromb Vasc Biol* 1998; 18: 7-12

781. BUJO H, HERMANN M, KADERLI MO, JACOBSEN L, SUGAWARA S, NIMPF J, YAMAMOTO T, SCHNEIDER WJ : Chicken oocyte growth is mediated by an eight ligand binding repeat member of the LDL receptor family. *EMBO J* 1994; 13: 5165-5175

782. BUJO H, YAMAMOTO T, HAYASHI K, HERMANNM, NIMPF J, SCHNEIDER WJ : Mutant oocytic low density lipoprotein receptor gene family member causes atherosclerosis and female sterility. *Proc Natl Acad Sci USA* 1995; 92: 9905-9909

783. HILTUNEN TP, LUOMA JS, NIKKARI T, YLÄ-HERTTUALA S : Expression of LDL receptor, VLDL receptor, LDL receptor-related protein, and scavenger receptor in rabbit atherosclerotic lesions. *Circulation* 1998; 97: 1079-1086

784. FRYKMAN PK, BROWN MS, YAMAMOTO T, GOLDSTEIN JL, HERZ J : Normal plasma lipoproteins and fertility in gene-targeted mice homozygous for a disruption in the gene encoding very low density lipoprotein receptor . *Proc Natl Acad Sci USA* 1995; 92: 8453-8457

785. HERZ J, GOTTHARDT M, WILLNOW TE : Cellular signalling by lipoprotein receptors. *Curr Opin Lipidol* 2000;11:161-166

786. KIM DH, MAGOORI K, INOUE TR, MAO CC, KIM H-J, SUZUKI H, FUJITA T, ENDO Y, SAEKI S, YAMAMOTO TT: Exon/intron organization, chromosome localization, alternative splicing and transcription units of human apolipoprotein E receptor 2 gene. *J Biol Chem* 1997; 272: 8498-8504

787. JACOBSEN L, MADSEN P, MOESTRUP SK, LUND AH, TOMMERUP N, NYKJAER A, SOTTRUP-JENSEN L, GLIEMANN J, PETERSEN CM : Molecular characterization of a novel hybrid-type receptor that binds the alpha2-macroglobulin receptor-associated protein. *J Biol Chem* 1996; 271: 31379-31383

788. POSSE DE CHAVES EI, VANCE DE, CAMPENOT RB, KISS RS, VANCE JE : Uptake of lipoproteins for axonal growth of sympathetic neurons. *J Biol Chem* 2000; 275: 19883-19890

789. KANAKI T, BUJO H, HIRAYAMA S, ISHII I, MORISAKI N, SCHNEIDER WJ, SAITO Y : Expression of LR11, a mosaic LDL receptor family member, is markedly increased in atherosclerotic lesions. *Arterioscler Thromb Vasc Biol* 1999; 19: 2687-2695

790. YEN FT, MASSON M, CLOSSAIS-BERNARD N, ANDRÉ P, GROSSET J-M, BOUGLERET L, DUMAS J-B, GUERASSIMENKO O, BIHAIN BE. Molecular cloning of a lipolysis-stimulated remnant receptor expressed in the liver. *J Biol Chem* 1999; 274: 13390-13398

791. BRESLOW JL : Familial disorders of high density lipoprotein metabolism. In: Scriver CR, Beaudet AL, Sly WS, Valle D (eds). *The metabolic and molecular bases of inherited disease*, 7th edn. Mac Graw Hill, New York. 1995.

792. ROTHBLAT GH, MAHLBERG FH, JOHNSON WJ, PHILLIPS MC : Apolipoproteins, membrane cholesterol domains, and the regulation of cholesterol efflux. *J Lipid Res* 1992; 33: 1091-1097

793. KALOPISSIS AD, CHAMBAZ J : Transgenic animals with altered high-density lipoprotein composition and functions. *Curr Opin Lipidol* 2000; 11: 149-153

794. KARATHANASIS SK : Apolipoprotein multigene family: tandem organization of human apolipoprotein AI, CIII, AIV genes. *Proc Natl Acad Sci USA* 1985; 82: 6374-6378

795. KARATHANASIS SK, ZANNIS VI, BRESLOW JL : Isolation and characterization of the human apolipoprotein A-I gene. *Proc Natl Acad Sci USA* 1983; 80: 6147-6151

796. SEGREST JP, LI L, ANATHARAMAIAH GM, HARVEY SC, LIADAKI KN, ZANNIS V : Structure and function of apolipoprotein A-I and high density lipoprotein. *Curr Opin Lipidol* 2000; 11: 105-115

797. FRANK PG, MARCEL YL : Apolipoprotein A-I: structure function relationships. *J Lipid Res* 2000; 41: 853-872

798. HAYEK T, OIKNINE J, DRANKNER G, BROOK JG, AVIRAM M : HDL apolipoprotein A-I attenuates oxidative modification of low density lipoproteins: studies in transgenic mice. *Eur J Clin Chem Clin Biochem* 1995; 33: 721-725

799. OWENS RJ, ANANTHARAMAIAH GM, KALHON RV, SRINIVAS RV, COMPANS RW, SEGREST JP : Apolipoprotein A-I and its amphipathic peptide analogues inhibit human immunodeficiency virus-induced syncytium formation. *J Clin Invest* 1990; 86: 1142-1150

800. MARTIN I, DUBOIS MC, SAERMARK T, RUYSSCHAERT JM : Apolipoprotein A-1 interacts with the N-terminal fusogenic domains of SIV (simian immunodeficiency virus) GP32 and HIV (human immunodeficiency virus) GP41: implications in viral entry. *Biochem Biophys Res Comm* 1992; 186: 95-101

801. FOLZ RJ, GORDON JI : The effect of deleting the propeptide from human preproapolipoprotein A-I on co-translational translocation and signal peptidase processing. *J Biol Chem* 1987; 262: 17221-17230

802. BEG ZH, STONIK JA, HOEG JM, DEMOSKY SJ, FAIRWELL T, BREWER HB : Human apolipoprotein A-I. Post-translational modification by covalent phosphorylation. *J Biol Chem* 1989; 264: 6913-6921

803. HADDAD IA, ORDOVAS JM, FITZPATRICK T, KARATHANASIS SK : Linkage, evolution and expression of the rat apolipoprotein A-I, C-III, A-IV genes. *J Biol Chem* 1986; 261: 13268-13277

804. LAMON-FAVA S, SASTRY R, FERRARI S, RAJAVASHISTH TB, LUSIS AJ, KARATHANASIS SK : Evolutionary distinct mechanisms regulate apolipoprotein A-I gene expression: differences between avian and mammalian apo A-I gene transcription control regions. *J Lipid Res* 1992; 33: 831-842

805. PAPAZAFIRI PK, OGAMI DP, RAMJI DP, NICOSIA A, MONACI P, CLADARAS C, ZANNIS VI: Promoter elements and factors involved in hepatic transcription of the human apo A-I gene positive and negative regulators bind to overlapping sites. *J Biol Chem* 1991; 266: 5790-5797

806. VU-DAC N, SCHOONJANS K, LAINE B, FRUCHART JC, AUWERX J, STAELS B : Negative regulation of the human apolipoprotein A-I promoter by fibrates can be attenuated by the interaction of the peroxisome proliferator-activated receptor with its response element. *J Biol Chem* 1994; 269: 31012-31018

807. KAPTEIN A, DE WIT ECM, PRINCEN HMG : Retinoids stimulate apo A-I synthesis by induction of gene transcription in primary hepatocytes cultures from cynomolgus monkey (Macaca fascicularis). *Arterioscler Thromb* 1993; 13: 1505-1514

808. ELSHOURBAGY N, BOGUSKI MS, LIAO WS, JEFFERSON LS, GORDON JI, TAYLOR JM: Expression of rat apolipoprotein A-IV and A-I genes: mRNA induction during development and in response to glucocorticoids and insulin. *Proc Natl Acad Sci USA* 1985; 82: 8242-8246

809. SRIVASTAVA RAK, TANG J, KRUL ES, PFLEGER B, KITCHENS RT, SCHONFELD G : Dietary fatty acids and dietary cholesterol differ in their effect on the in vivo regulation of apolipoprotein A-I and A-II gene expression in inbred strains of mice. *Biochim Biophys Acta* 1992; 1125: 251-261

810. WALSH A, ITO Y, BRESLOW JL: High levels of apolipoprotein A-I in transgenic mice result in increased plasma levels of small high density lipoprotein (HDL) particles comparable to human HDL3. *J Biol Chem* 1989; 264: 16488-6494

811. RUBIN EM, ISHIDA BY, CLIFT SM, KRAUSS RM : Expression of human apolipoprotein A-I in transgenic mice results in reduced plasma levels of murine apolipoprotein A-I and the appearance of two new high density lipoprotein size subclasses. *Proc Natl Acad Sci USA* 1991; 88: 434-438

812. HAYEK T, ITO Y, VERDERY RB, AALTO-SETÄLÄ K, WALSH A, BRESLOW JL : Dietary fat increases high density lipoprotein HDL levels both by increasing the transport rates and decreasing the fractional catabolic rates of HDL cholesterol ester and apolipoprotein (Apo) A-I. *J Clin Invest* 1993; 91: 1665-1671

813. WALSH A, AZROLAN N, MARCIGLIANO A, O'CONNELL A, BRESLOW JL: Intestinal expression of the human apoA-I gene in transgenic mice is controlled by a DNA region 3' to the gene in the promoter of the adjacent convergently transcribed apoC-III gene. *J Lipid Res* 1993; 34: 617-623

814. SHEMER R, KAFRI T, O'CONNELL A, EISENBERG S, BRESLOW JL, RAZIN A : Methylation changes in the apolipoprotein AI gene during embryonic development of the mouse. *J Biol Chem* 1991; 266: 11300-11304

815. BENLIAN P, BOILEAU C, LOUX N, PASTIER D, MASLIAH J, COULON M, NIGOU M, RAGAB A, GUIMARD J, RUIDAVETS JB, BONAITI-PELIE C, FRUCHART JC, DOUSTE-BLAZY P, BEREZIAT G, JUNIEN C : Extended haplotypes and linkage disequilibrium between 11 markers at the APOAI-CIII-AIV gene cluster on chromosome 11. *Am J Hum Genet* 1991; 48: 903-910

816. HAVILAND MB, KESSLING AM, DAVIGNON J, SING CF : Estimation of hardy-weinberg and pairwise disequilibrium in the apolipoprotein AI-CIII-AIV gene cluster. *Am J Hum Genet* 1991; 49: 350-365

817. LAMBERT-PRENGER V, BEATY TH, KWITEROVITCH PO : Genetic determination of high-density lipoprotein-cholesterol and apolipoprotein A-I levels in a family study of catheterization patients. *Am J Hum Genet* 1992; 51: 1047-1057

818. TALMUD PJ, WATERWORTH DM : In-vivo and in-vitro nutrient-gene interactions. *Curr Opin Lipidol 2000*; 11: 31-36

819. ORDOVAS JM, SCHAEFER EJ : Genetic determinants of plasma lipid response to dietary intervention: the role of the APOA1/C3/A4 gene cluster and APOE gene. *Br J Nutr* 2000; 83: S127-S136

820. SMITH JD, BRINTON EA, BRESLOW JL : Polymorphism of the human apolipoprotein A-I gene promoter region. Association of the minor allele with decreased production rate in vivo and promoter activity in vitro. *J Clin Invest* 1992; 89: 1796-1800

821. JUO SH, WYSZYNSKI DF, BEATY TH, HUANG NY, BAILEY-WILSON JE : Mild association between the A/G polymorphism in the promoter of the apolipoproteins A-I gene and apolipoproteins A-I levels: a meta-analysis. *Am J Med Genet* 1999; 82: 235-241

822. NAGANAWA S, GINSBERG HN, GLICKMAN RM, GINSBURG GS : Intestinal transcription and synthesis of Apolipoproteins A-I is regulated by five natural polymorphisms upstream of the Apolipoproteins CIII gene. *J Clin Invest* 1997; 99: 1958-1965

823. NG DS, LEITER LA, VEZINA C, CONNELLY PW, HEGELE RA : Apolipoprotein A-I Q[-2]X causing isolated apolipoprotein A-I deficiency in a family with analphalipoproteinemia. *J Clin Invest* 1994; 93: 223-229

824. LACKNER KJ, DIEPLINGER H, NOWICKA G, SCHMITZ G : High density lipoprotein deficiency with xanthomas. A defect in reverse cholesterol transport caused by a point mutation in the apolipoprotein A-I gene. *J Clin Invest* 1993; 92: 2262-2273

825. BOOTH DR, TAN S-Y, BOOTH SE, TENNENT GA, HUTCHINSON WL, HSUAN JJ, TOTTY NF, TRUONG O, SOUTAR AK, HAWKINS PN, BRUGUERA M, CABALLERIA J, SOLE M, CAMPISTOL JM, PEPYS MB : Hereditary hepatic and systemic amyloidosis caused by a new deletion/insertion mutation in the apolipoprotein AI gene. *J Clin Invest* 1996; 97: 2714-2721

826. UTERMANN G, HAAS J, STEINMETZ A, PAETZOLD R, WILK J, FEUSSNER G, KAFFARNIK C, MULLER-ECKHARDT C, SEIDEL D, VOGELBERG KH, ZIMMER F : Apolipoprotein A-I Marburg. Studies on two kindreds with a mutant of human apolipoprotein A-I. *Hum Genet* 1982; 62: 329-337

827. UTERMANN G, STEINMETZ A, PAETZOLD R, RALL SC, WEISGRABER KH, MAHLEY RW : Apo A-I Giessen (Pro143→Arg). *Eur J Biochem* 1984; 144: 325-331

828. MORIYAMA K, SASAKI J, TAKADA Y, MATSUNAGA A, KUKUI J, ALBERS JJ, ARAKAWA K : A cysteine-containing truncated apo A-I variant associated with HDL deficiency. *Arterioscler Thromb Vasc Biol* 1996; 16: 1416-1423

829. FUNKE H, VON ECKARDTSEIN A, PRITCHARD PH, KARAS M, ALBERS JJ, ASSMANN G: A frameshift mutation in the human apolipoprotein A-I gene causes high density lipoprotein deficiency, partial lecithin:cholesterol-acyltransferase deficiency, and corneal opacities. *J Clin Invest* 1991; 87: 371-376

830. HAN H, SASAKI J, MATSUNAGA A, HAKAMATA H, HUANG W, AGETA M, TAGUCHI T, KOGA T, KUGI M, HORIUCHI S, ARAKAWA K : A novel mutant, apo A-I Nichinan (Glu235→0), is associated with low HDL cholesterol levels and decreased cholesterol efflux from cells. *Arterioscler Thromb Vasc Biol* 1999; 19: 1447-1455

831. NG DS, VEEZINA C, WOLEVER TS, KUKSIS A, HEGELE RA, CONNELLY PW : Apolipoproteins A-I deficiency. *Arterioscler Thromb Vasc Biol* 1995; 15: 2157-2164

832. DEEB SS, CHEUNG MC, PENG R, WOLF AC, STERN R, ALBERS JJ, KNOPP RH : A mutation in the human apolipoprotein A-I gene. Dominant effect on the level and characteristics of plasma high density lipoproteins. *J Biol Chem* 1991; 266: 13654-13660

833. KASTELEIN JJP, HAINES JL, HAYDEN MR : The gene causing familial hypoalphalipoproteinemia is not caused by a defect in the apo AI-CIII-AIV gene cluster in a Spanish family. *Hum Genet* 1990; 84: 396-400

834. LADIAS JAA, KWITEROVITCH PO, SMITH HH, KARATHANASIS SK, ANTONARAKIS SE : Apolipoprotein A1 Baltimore (Arg10→Leu), a new ApoA1 variant. *Hum Genet* 1990; 84: 439-445

835. LI H, REDDICK RL, MAEDA N : Lack of apo A-I is not associated with increased susceptibility to atherosclerosis in mice. *Arterioscler Thromb* 1993; 13: 1814-1821

836. BROWN VW, BAGINSKY ML : Inhibition of lipoprotein lipase by an apolipoprotein of human very low density lipoprotein. *Biochem Biophys Res Comm* 1972; 46: 375-382

837. KINNUNEN PKJ, EHNHOLM C : Effect of serum and C apolipoproteins from very low density lipoproteins on human post heparin plasma hepatic lipase. *FEBS Lett* 1976; 86: 9647-9651

838. AALTO-SETÄLÄ K, FISHER EA, CHEN X, CHAJEK-SHAUL T, HAYEK T, ZECHNER R, WALSH A, RAMAKRISHNAN R, GINSBERG HN, BRESLOW JL : Mechanisms of hypertriglyceridemia in human apolipoprotein (Apo) CIII transgenic mice. Diminished very low density lipoprotein fractional catabolic rate associated with increased Apo CIII and reduced Apo E on the particles. *J Clin Invest* 1992; 90: 1889-1900

839. DE SILVA HV, LAUER SJ, WANG J, SIMONET WS, WEISGRABER KH, MAHLEY RW, TAYLOR JM : Overexpression of human apolipoprotein C-III in transgenic mice results in an accumulation of apolipoprotein B-48 remnants that is corrected by excess apolipoprotein E. *J Biol Chem* 1994; 269: 2324-2335

840. THOMPSON GR : Angiographic evidence for the role of triglyceride-rich lipoproteins in progression of coronary artery disease. *Eur heart J* 1998; 19 suppH: H31-36

841. KARATHANASIS SK : Apolipoprotein multigene family: tandem organization of human apolipoprotein AI, CIII, AIV genes. *Proc Natl Acad Sci USA* 1985; 82: 6374-6378

842. SEGREST JP, JONES MK, DE LOOF H, BROUILLETTE CG, VENKATACHALAPATHI YV, ANANTHARAMAIAH GM : The amphipathic helix in the exchangeable apolipoproteins: a review of secondary structure and function. *J Lipid Res* 1992; 33: 141-166

843. ZULIANI G, HOBBS HH : Tetranucleotide repeat polymorphism in the apolipoprotein C-III gene. *Nucleic Acids Res* 1990; 18: 4299

844. TALMUD PJ, HUMPHRIES SE: Apolipoprotein C-III gene variation and dyslipidemia. *Curr Opin Lipidol* 1997; 8: 154-158

845. REUE K, LEFF T, BRESLOW JL : Human apolipoprotein CIII gene expression is regulated by positive and negative cis-acting elements and tissue-specific protein factors. *J Biol Chem* 1988; 263: 6857-6864

846. MIETUS-SNYDER M, SLADEK FM, GINSBURG GS, KUO CF, LADIAS JAA, DARNELL JE, KARATHANASIS SK : Antagonism between apolipoprotein AI regulatory protein 1, Ear3/COUP-TF, and hepatocyte nuclear factor 4 modulates apolipoprotein CIII gene expression in liver and intestinal cells. *Mol Cell Biol* 1992; 12: 1708-1718

847. AUWERX J, SCHOONJANS K, FRUCHART JC, STAELS B: Transcriptional control of triglyceride metabolism: fibrates and fatty acids change the expression of the LPL and apo C-III genes by activating the nuclear receptor PPAR. *Atherosclerosis* 1996; 124 supp: S29-37

848. DAMMERMAN M, SANDKUIJL LA, HALAAS J, CHUNG W, BRESLOW JL : An apolipoprotein CIII haplotype protective against hypertriglyceridemia is specified by promoter and 3' untranslated region polymorphisms. *Proc Natl Acad Sci USA* 1993; 90: 4562-4566

849. LI WW, DAMMERMAN M, SMITH JD, METZGER S, BRESLOW JL, LEFF T: Common genetic variation in the promoter of the human apo CIII gene abolishes regulation by insulin and may contribute to hypertriglyceridemia. *J Clin Invest* 1995; 96: 2601-2605

850. ROGHANI A, ZANNIS V : Mutagenesis of the glycosylation site of apo CIII O-Linked glycosylation is not required for apo CIII secretion and lipid binding. *J Biol Chem* 1988; 263: 17925-17932

851. MAEDA H, HASHIMOTO R, OGURA T, HIRAGA S, UZAWA H : Molecular cloning of human apo C-III variant : Thr 74→ Ala 74 mutation prevents O-glycosylation. *J Lipid Res* 1987; 28: 1405-1409

852. KARATHANASIS SK, FERRIS E, HADDAD IA : DNA inversion within the apolipoproteins AI/CIII/AIV-encoding gene cluster of certain patients with premature atherosclerosis. *Proc Natl Acad Sci USA* 1984; 84: 7198-7202

853. NORUM RA, LACKIER JB, GOLDSTEIN S, ANGEL A, GOLDBERG RB, BLOCK DK, NOFZE DK, DOLPHIN PJ, EDELGLASS J, BOGORAD D, ALAUPOVIC P : Familial deficiency of apolipoproteins A-I and C-III and precocious coronary heart disease. *N Engl J Med* 1982; 306: 1513-1519

854. LÜTTMANN S, VON ECKARDSTEIN A, WEI W, FUNKE H, KÖHLER E, MAHLEY RW, ASSMANN G : Electrophoretic screening for genetic variation in apolipoprotein C-III : identification of a novel apo C-III variant, apo C-III (Asp45→Asn), in a Turkish patient. *J Lipid Res* 1994; 35: 1431-1440

855. VON ECKARDSTEIN A, HOLZ H, SANDKAMP M, WENG W, FUNKE H, ASSMANN G : Apolipoprotein C-III (Lys58 →Glu) Identification of an apolipoprotein C-III variant in a family with hyperalphalipoproteinemia. *J Clin Invest* 1991; 87: 1724-1731

856. PULLINGER CR, MALLOY MJ, SHAHIDI AK, GHASSEMZADEH M, DUCHATEAU P, VILLAGOMEZ J, ALLAART J, KANE JP : A novel apolipoprotein C-III variant, apoC-III (Gln38→Lys), associated with moderate hypertriglyceridemia in a large kindred of Mexican origin. *J Lipid Res* 1997; 38: 1833-1840

857. FUJIMOTO K, FUKAGAWA K, SAKATA T, TSO P: Suppression of food intake by apolipoprotein A-IV is mediated through the central nervous system in rats. *J Clin Invest* 1993; 91: 1830-1833

858. LAGROST L, GAMBERT P, BOQUILLON M, LALLEMANT C : Evidence for high density lipoproteins as the major apolipoprotein A-IV-containing fraction in normal human serum. *J Lipid Res* 1989; 30: 1525-1534

859. DIEPLINGER H, LOBENTANZ EM, KÖNIG P, GRAF H, SANDHOLZER C, MATTHYS E, ROSSENEU M, UTERMANN G : Plasma apolipoprotein A-IV metabolism in patients with chronic renal disease. *Eur J Clin Invest* 1992; 22: 166-174

860. GOLDBERG IJ, SCHERALDI CA, YACOUB LK, SAXENA U, BISGAIER CL : Lipoprotein apo C-II activation of lipoprotein lipase. Modulation by apolipoprotein A-IV. *J Biol Chem* 1990; 265: 4266-4272

861. BOYLES JK, NOTTERPEK LM, ANDERSON LJ : Accumulation of apolipoproteins in the regenerating and remyelinating mammalian peripheral nerve. Identification of apolipoprotein D, apolipoprotein A-IV, apolipoprotein E, and apolipoprotein A-I. *J Biol Chem* 1990; 265: 17805-17815

862. QIN XF, SWERTFEGER DK, ZHENG SQ, HUI DY, TSO P : Apolipoprotein A-IV : a potent endogenous inhibitor of lipid oxidation. *Am J Physiol* 1998; 274: H1836-H1840

863. COHEN RD, CASTELLANI LW, QIAO JH, VAN LENTEN BJ, LUSIS AJ, REUE K : Reduced aortic lesions and elevated high density lipoprotein levels in transgenic mice overexpressing mouse apolipoprotein A-IV. *J Clin Invest* 1997; 99: 1906-1916

864. DUVERGER N, TREMP G, CAILLAUD JM, EMMANUEL F, CASTRO G, FRUCHARD JC, STEINMETZ A, DENEFLE P : Protection against atherogenesis in mice mediated by human apolipoprotein A-IV. *Science* 1996; 273: 966-968

865. STEINMETZ A, CZEKELIUS P, THIEMANN E, MOTZNY S, KAFFARNI K : Changes of apolipoprotein A-IV in human neonate : evidence for different inductions of apolipoproteins A-IV and A-I in the post partum period. *Atherosclerosis* 1988; 69: 21-27

866. STROBL W, KNERER B, GRATZL R, ARBEITER K, LIN-LEE YC, PATSCH W : Altered regulation of apolipoprotein A-IV gene expression in the liver of the genetically obese Zucker rat. *J Clin Invest* 1994; 92: 1766-1773

867. VERGES B, RADER D, SCHAEFER J, ZECH L, KINDT M, FAIRWELL T, GAMBERT P, BREWER HB : In vivo metabolism of apolipoprotein A-IV in severe hypertriglyceridemia : a combined radiotracer and stable isotope kinetic study. *J Lipid Res* 1994; 35: 2280-2291

868. ELSHOURBAGY N, WALKER DW, PAIK YK, BOGUSKI MS, FREEMAN M, GORDON JL, TAYLOR JM : Structure and expression of the human apolipoprotein A-IV gene. *J Biol Chem* 1987; 262: 7973-7981

869. VON ECKARDSTEIN A, FUNKE H, SCHULTE M, ERREN M, SCHULTE H, ASSMANN G : Nonsynonymous polymorphic sites in the apolipoprotein (apo) A-IV gene are associated with changes in the concentration of apo B- and apo A-I-containing lipoproteins in a normal population. *Am J Hum Genet* 1992; 50: 1115-1128

870. TENKANEN H, KOSKINEN P, METSO J, BAUMANN M, LUKKA M, KAUPPINEN-MÄKELIN R, KONTULA K, TASKINEN MR, MÄNTTÄRI M, MANNINEN V, EHNHOLM C : A novel polymorphism of apolipoprotein A-IV is the result of an asparagine to serine substitution at residue 127. *Biochim Biophys Acta* 1992; 1138: 27-33

871. DEEBS SS, NEVIN DN, IWASAKI L, BRUNZELL JD : Two novel apolipoprotein A-IV variants in individuals with familial combined hyperlipidemia and diminished levels of lipoprotein lipase activity. *Hum Mutat* 1996; 8: 319-325

872. LOSHE P and BREWER HB : Genetic polymorphism of apolipoprotein A-IV. *Curr Opin Lipidol* 1991; 2: 90-95

873. EHNHOLM C, TENKANEN H, DE KNIJFF P, HAVEKES L, ROSSENEU M, MENZEL HJ, TIRET L and the EARS Group : Genetic polymorphism of apolipoprotein A-IV in five different regions of Europe relations to plasma lipoproteins and to history of myocardial infarction : the EARS study. *Atherosclerosis* 1994; 107: 229-238

874. ZAIOU M, VISVIKIS S, GUEGUEN R, PARRA HJ, FRUCHART JC, SIEST G : DNA polymorphisms of human apolipoprotein A-IV gene: frequency and effects on lipid, lipoprotein and apolipoprotein levels in a French population. *Clin Genet* 1994; 46: 248-254

875. MATA P, ORDOVAS JM, LOPEZ-MIRANDA J, LICHTENSTEIN AH, CLEVIDENCE B, JUDD JT, SCHAEFER EJ : Apo A-IV phenotype affects diet-induced plasma LDL cholesterol lowering. *Arterioscler Thromb* 1994; 14: 884-891

876. McCOMBS RJ, MARCADIS DE, ELLIS J, WEINBERG RB : Attenuated hypercholesterolemic response to a high-cholesterol diet in subjects heterozygous for the apolipoprotein A-IV-2 allele. *N Engl J Med* 1994; 331: 706-710

877. OSTOS MA, LOPEZ-MIRANDA J, ORDOVAS JM, MARIN C, BLANCO A, CASTRO P, LOPEZ-SEGURA F, JIMENEZ-PEREPEREZ J, PEREZ-JIMENEZ F : Dietary fat clearance is modulated by genetic variation in apolipoprotein A-IV gene locus. *J Lipid Res* 1998; 39: 2493-2500

878. HIXSON JE, KAMMMERER CM, MOTT GE, BRITTEN ML, BIRNBAUM S, POWERS PK, VANDEBERG JL : Baboon apolipoprotein A-IV Identification of Lys76→Glu that distinguishes two common isoforms and detection of length polymorphisms at the carboxyl terminus. *J Biol Chem* 1993; 268: 15667-15673

879. HIXSON JE, POWERS PK: Restriction isotyping of human apolipoprotein AIV: rapid typing of known isoforms and detection of a new isoform that deletes a conserved repeat. *J Lipid Res* 1991; 32: 1529-1535

880. OSADA J, POCOVI M, NICOLOSI RJ, SCHAEFER EJ, ORDOVAS JM : Nucleotide sequences of the Macaca fascicularis apolipoprotein C-III and A-IV genes. *Biochim Biophys Acta* 1993; 1172: 335-339

881. REUE K, LEETE TH : Genetic variation in mouse apolipoprotein AIV due to insertion and deletion in a region of tandem repeat. *J Biol Chem* 1991; 266: 12715-12721

882. WEIGRABER KH, MAHLEY RW : Apoprotein (E-AII) complex of human plasma lipoproteins. I. Characterization of this mixed disulfide and its identification in a high density lipoprotein subfraction. *J Biol Chem* 1978; 253: 6281-6288

883. OSADA J, GARCES C, SASTRE J, SCHAEFER EJ, ORDOVAS JM : Molecular cloning and sequence of the cynomolgus monkey apolipoprotein A-II gene. *Biochim Biophys Acta* 199; 1172: 340-342

884. GONG EL, STOLTZFUS LJ, BRION CM, MURUGESH D, RUBIN EM : Contrasting in vivo effects of murine and human apolipoprotein A-II. *J Biol Chem* 1996; 271: 5984-5987

885. TSAO TK, WEI CF, ROBBERSON DL, GOTTO AM, CHAN L : Isolation and characterization of the human apolipoprotein A-II gene. *J Biol Chem* 1985; 260: 15222-15231

886. MIDDLETON-PRICE HR, VAN DEN BERGHE JA, SCOTT J, KNOTT TJ, MALCOLM S : Regional chromosomal localisation of APOA2 to 1q21-1q23. *Hum Genet* 1988; 79: 283-285

887. WEBER JL, MAY PE : Abundant class of human DNA polymorphisms which can be typed using the polymerase chain reaction. *Am J Hum Genet* 1989; 44: 388-396

888. SCOTT J, PRIESTLEY LM, KNOTT TJ, ROBERTSON ME, MANN DV, KOSTNER G, MILLER NE : High density lipoprotein composition is altered by a common DNA polymorphism adjacent to apoprotein AII gene in man. *Lancet* 1985; I: 771-773

889. CHAMBAZ J, CARDOT P, PASTIER D, ZANNIS VI, CLADARAS C: Promoter elements and factors required for hepatic transcription of the human apoA-II gene. *J Biol Chem* 1991 266: 11676-11685

890. CARDOT P, CHAMBAZ J, CLADARAS C, ZANNIS VI : Regulation of the human apo A-II gene by the synergistic action of factor binding to the proximal and distal regulatory elements. *J Biol Chem* 1991; 266: 24460-24470

891. CHIESA G, PAROLINI C, CANAVESI M, COLOMBO N, SIRTORI CR, FUMAGALLI R, FRANCESCHINI G, BERNINI F : Human apolipoprotein A-I and A-

II in cell cholesterol efflux. Studies with transgenic mice. *Arterioscler Thromb Vasc Biol* 1998; 18: 1417-1423

892. WENG W, BRANDENBURG N, ZHONG S, HALKIAS J, WU L, JIANG XC, TALL A, BRESLOW JL : ApoA-II maintains HDL levels in part by inhibition of hepatic lipase: studies in ApoA-II and hepatic lipase double knockout mice. *J Lipid Res* 1999; 40: 1064-1070

893. BOISFER E, LAMBERT G, ATGER V, TRAN NQ, PASTIER D, BENETOLLO C, TROTTIER J-F, BEAUCAMPS I, ANTONUCCI M, LAPLAUD M, GRIGLIO S, CHAMBAZ J, KALOPISSIS A-D : Overexpression of human apolipoprotein A-II in mice induces hypertriglyceridemia due to defective very low density lipoprotein hydrolysis. *J Biol Chem* 1999; 274: 11564-11572

894. WENG W, BRESLOW JL : Dramatically decreased high density lipoprotein cholesterol, increased remnant clearance, and insulin hypersensitivity in apolipoproteins A-II knockout mice suggest a complex role for apolipoprotein A-II in atherosclerosis susceptibility. *Proc Natl Acad Sci USA* 1996; 93: 1488-1494

895. CASTELLANI LW, NAVAB M, VAN LENTEN BJ, HEDRICK CC, HAMA SY, GOTO AM, FOGELMAN AM, LUSIS AJ : Overexpression of apolipoprotein A-II in transgenic mice converts high density lipoproteins to proinflammatory particles. *J Clin Invest* 1997; 100: 464-474

896. WARDEN CH, HEDRICK CC, QIAO JH, CASTELLANI LW, LUSIS AJ : Atherosclerosis in transgenic mice overexpressing apolipoprotein A-II. *Science* 1993 261: 469-472

897. GERVAIS A, AYRAULT-JARRIER M, REBOUL J, SCHULLER E, MEILLET D, APARTIS E, LYON-CAEN O, BRICAIRE F, GENTILLINI M : Apolipoprotein A-II in HIV-1 infection. *Lancet* 1992; 340: 730-731

898. NAKAMURA M, TANAKA Y, YAMASHITA T, SALVI F, FERLINI A, GOBBI P, PATROSSO C, UCHINO M, ANDO M : Decreased affinity of apolipoproteins A-II to high-density lipoprotein in patients with transthyretin-related amyloidosis (Met30, Gln 89, Pro36, and Thr34). *Biochem Biophys Res Commun* 1996; 219: 316-321

899. DEEB SS, TAKATA K, PENG RL, KAJIYAMA G, ALBERS JJ : A splice junction mutation responsible for familial apolipoprotein A-II deficiency. *Am J Hum Genet* 1990; 46: 822-827

900. KUNISADA T, HIGUCHI K, AOTA SI, TAKEDA T, YAMAGISHI : Molecular cloning and nucleotide sequence of cDNA for senile amyloid protein : nucleotide substitutions found in apolipoprotein A-II cDNA of senescence accelerated mouse (SAM). *Nucleic Acids Res* 1986; 14: 5729-5740

901. WARDEN CH, DALUISKI A, BU X, PURCELL-HUYNH DA, DE MEESTER C, SHIEH BH, PUPPIONE D, GRAY RM, REAVEN GM, CHEN YDI, ROTTER JI, LUSIS AJ : Evidence for linkage of the apolipoprotein A-II locus to plasma apolipoprotein A-II and free fatty acid levels in mice and humans. *Proc Natl Acad Sci USA* 1993; 90: 10886-10890

902. BU X, WARDEN CH, XIA YR, DE MEESTER C, PUPPIONE DL, TERUYA S, LOKENSGARD B, DANESHMAND S, BROWN J, GRAY RJ, ROTTER J, LUSIS AJ : Linkage analysis of the genetic determinants of high density lipoprotein concentrations and composition : evidence for involvement of the apolipoprotein A-II and cholesteryl ester transfer protein loci. *Hum Genet* 1994; 93: 639-648

903. HEDRICK CC, CASTELLANI LW, WARDEN CH, PUPPIONE DL, LUSIS AJ : Influence of mouse apolipoprotein A-II on plasma lipoproteins in transgenic mice. *N Engl J Med* 1986; 315: 1509-1515

904. SANTAMARINA-FOJO S, LAMBERT G, HOEG JM, BREWER HB Jr : Lecithin-cholesterol acyltransferase: role in lipoprotein metabolism, reverse cholesterol transport and atherosclerosis. *Curr Opin Lipidol* 2000; 11: 267-275

905. PEELMAN F, VANDEKERKHOVE J, ROSSENEU M : Structure and function of lecithin cholesterol acyl transferase: new insights from structural predictions and animal models. *Curr Opin Lipidol* 2000; 11: 155-160

906. JONAS A : Regulation of lecithin cholesterol acyl transferase activity. *Prog Lipid Res* 1998; 37: 209-234

907. GLOMSET JA, ASSMANN G, GJONE E, NORUM KR : Lecithin:cholesterol acyl transferase deficiency and fish-eye disease. *In* The Metabolic and Molecular Bases of Inherited Disease. Scriver CR, Beaudet AL, Sly WS and Valle D, eds. *Mac Graw Hill,* New York. 1995; 2: 1933-1951

908. VOHL MC, NEVILLE TA, KUMARATHASAN R, BRASCHI S, SPARKS DL : A novel lecithin-cholesterol acyl transferase antioxidant activity prevents the formation of oxidized lipids during lipoprotein oxidation. *Biochem* 1999; 38: 5976-5981

909. FORTE TM, ODA MN, KNOFF L, FREI B, SUH J, HARMONY JAK, STUART WD, RUBIN EM, NG DS : Targeted disruption of the murine lecithin:cholesterol acyltransferase gene is associated with reductions in plasma paraoxonase and platelet-activating factor acetylhydrolase activities but not in apo J concentration. *J Lipid Res* 1999; 40: 1276-1283

910. JIMI S, UESUGI N, SAKU K, ITABE H, ZHANG B, ARAKAWA K, TAKEBAYASHI S : Possible induction of renal dysfunction in patients with lecithin:cholesterol acyltransferase deficiency by oxidized phosphatidylcholine in glomeruli. *Arterioscler Thromb Vasc Biol* 1999; 19: 794-801

911. COLLET X, FRANCONE O, BESNARD F, FIELDING CJ : Secretion of lecithin:cholesterol acyltransferase by brain neuroglial cell lines. *Biochem Biophys Res Comm* 1999; 258: 73-76

912. AZOULAY M, HENRY I, TATA F, WEIL D, GRZESCHIK H, CHAVES ME, Mc INTYRE N, WILLIAMSON R, HUMPHRIES SE, JUNIEN C : The structural gene for lecithin:cholesterol acyl transferase (LCAT) maps to 16q22. *Ann Hum Genet* 1987; 51: 129-136

913. BROWN ML, HESLER C, TALL AR : Plasma enzymes and transfer proteins in cholesterol metabolism. *Curr Opin Lipidol* 1990; 1: 122-127

914. Mac LEAN J, WION K, DRAYNA D, FIELDING C, LAWN R : Human lecithin-cholesterol acyltransferase gene : complete gene sequence and sites of expression. *Nucleic Acids Res* 1986; 14: 9397-9406

915. PEELMAN F, VERSCHELDE J-L, VANLOO B, AMPE C, LABEUR C, TAVERNIER J, VANDEKERKHOVE J, ROSSENEU M : Effects of natural mutations in lecithin:cholesterol acyltransferase on the enzyme structure and activity. *J Lipid Res* 1998; 40: 59-69

916. KUIVENHOVEN JA, PRITCHARD H, HILL J, FROHLICH J, ASSMANN G, KASTELEIN JJ : The molecular pathology of lecithin:cholesterol acyltransferase (LCAT) deficiency syndromes. *J Lipid Res* 1997; 38: 191-205

917. ARGYROPOULOS G, JENKINS A, KLEIN RL, LYONS T, WAGENHORST B, ST ARMAND J, MARCOVINA SM, ALBERS JJ, PRITCHARD HP, GARVEY TW : Transmission of two novel mutations in a pedigree with familial lecithin:cholesterol acyltransferase deficiency: structure-function relationships and studies in a compound heterozygous proband. *J Lipid Res* 1998; 39: 1870-1876

918. TEH EM, CHISHOLM JW, DOLPHIN PJ, POULIQUEN Y, SAVOLDELLI M, DE GENNES JL, BENLIAN P : Classical LCAT deficiency resulting from a novel homozygous dinucleotide deletion in exon 4 of the human lecithin:cholesterol acyltransferase gene causing a frameshift and stop codon at residue 144. *Atherosclerosis* 1999; 146: 141-151

919. GUERIN M, DACHET C, GOULINET S, CHEVET D, DOLPHIN PJ, CHAPMAN MJ, ROUIS M : Familial lecithin:cholesterol acyltransferase deficiency: molecular analysis of a compound heterozygote: LCAT (Arg147→Trp) and LCAT (Tyr171→Stop). *Atherosclerosis* 1997; 131: 85-95

920. MIETTINEN HE, GYLLING H, TENHUNEN J, VIRTAMO J, JAUHIAINEN M, HUTTUNEN JK, KANTOLA I, MIETTINEN TA, KONTULA K : Molecular genetic study of Finns with hypoalphalipoproteinemia and hyperalphalipoproteinemia. A novel $Gly^{230}Arg$ mutation ($LCAT_{FIN}$) of lecithin:cholesterol acyltransferase (LCAT) accounts for 5% of cases with very low serum HDL cholesterol levels. *Arterioscler Thromb Vasc Biol* 1998; 18: 591-598

921. : KUIVENHOVEN JA, WEIBUSCH H, PRITCHARD HP, FUNKE H, BENNE R, ASSMANN G, KASTELEIN JJP : An intronic mutation in a lariat branchpoint sequence is a direct cause of inherited human disorder (fish eye disease). *J Clin Invest* 1996; 98: 358-364

922. WEBB JC, PATEL DD, SHOULDERS CC, KNIGHT BL, SOUTAR AK : Genetic variation at a splicing branch point in intron 9 of the low density lipoprotein (LDL)-receptor gene: a rare mutation that disrupts mRNA splicing in a patient with familial hypercholesterolemia and a common polymorphism. *Hum Mol Genet* 1996; 5: 1325-1331

923. STEYRER E, DUROVIC S, FRANK S, GRIEßAUF W, BURGER A, DIEPLINGER H, ZECHNER R, KOSTNER G : The role of lecithin:cholesterol acyltransferase for lipoprotein (a) assembly. Structural integrity of low density lipoprotein is a prerequisite for Lp(a) formation in human plasma. *J Clin Invest* 1994; 94: 2330-2340

924. AOUIZERAT BE, ALLAYEE H, CANTOR RM, DALLINGA-THIE GM, LANNING CD, DE BRUIN TWA, LUSIS AJ, ROTTER JI : Linkage of a candidate gene locus to familial combined hyperlipidemia. Lecithin:cholesterol acyltransferase on 16q. *Arterioscler Thromb Vasc Biol* 1999; 19: 2730-2736

925. OELKERS P, TINKELENBERG A, ERDENIZ N, CROMLEY D, BILHEIMER JT, STURLEY SL : A lecithin:cholesterol acyltransferase-like gene mediates diacylglycerol esterification in yeast. *J Biol Chem* 2000; 275: 15609-15612

926. BRUCE C, CHOUINARD RA, Jr, TALL AR : Plasma lipid transfer proteins, high density lipoproteins and reverse cholesterol transport. *Ann Rev Nutr* 1998; 18: 297-330

927. YAMASHITA S, SAKAI N, HIRANO K, ARAI T, ISHIGAMI M, MARUYAMA T, MATSUZAWA Y : Molecular genetics of plasma cholesteryl ester transfer protein. *Curr Opin Lipidol* 1997; 8: 101-110

928. LAGROST L, PERSÉGOL L, LALLEMANT C, GAMBERT P : Influence of apolipoprotein composition of high density lipoprotein particles on cholesteryl ester transfer protein activity. Particles containing various proportions of apolipoproteins AI and AII. *J Biol Chem* 1994; 269: 3189-3197

929. JIANG XC, MASUCCI-MAGOULAS L, MAR J, LIN M, WALSH A, BRESLOW JL, TALL AR : Down-regulation of mRNA for the low density lipoprotein receptor in transgenic mice containing the gene for human cholesteryl ester transfer protein. Mechanism to explain accumulation of lipoprotein particles. *J Biol Chem* 1993; 268: 27406-27412

930. HAYEK T, MASUCCI-MAGOULAS L, JIANG X, WALSH A, RUBIN E, BRESLOW JL, TALL AR : Decreased early atherosclerotic lesions in hypertriglyceridemic mice expressing cholesteryl ester transfer protein transgene. *J Clin Invest* 1995; 96: 2071-2074

931. DRAYNA DT, JARNAGIN AS, Mac LEAN J, HENZEL W, KOHR W, FIELDING C, LAWN R : Cloning and sequencing of human cholesteryl ester transfer protein cDNA. *Nature* 1987; 327: 632-634

932. AGELLON LB, QUINET EM, GILLETTE TG, DRAYNA DT, BROWN ML, TALL AR : Organization of the human cholesteryl ester transfer protein gene. *Biochemistry* 1990; 29: 1373-1376

933. LUSIS AJ, ZOLLMAN S, SPARKES RS, KLISAK I, MOHANDAS T, DRAYNA D, LAWN RM : Assignment of the human gene for cholesteryl ester transfer protein to chromosome 16q12-16q21. *Genomics* 1987; 1: 232-235

934. INAZU A, BROWN ML, HESLER CB, AGELLON LB, KOIZUMI J, TAKATA K, MARUHAMA Y, MABUCHI H, TALL AR : Increased high density lipoprotein levels caused by a common cholesterol-ester transfer protein gene mutation. *N Engl J Med* 1990; 323: 1234-1238

935. ZULIANI G, HOBBS HH : EcoN I polymorphism in the human cholesteryl ester transfer protein (CETP) gene. *Nucleic Acids Res* 1990; 18: 2834

936. ORDOVAS JM, CUPPLES AL, CORELLA D, OTVOS JD, OSGOOD D, MARTINEZ A, LAHOZ C, COLTELL, WILSON PWF, SCHAEFER EJ : Association of cholesteryl ester transfer protein-TaqIB polymorphism with variations in lipoprotein subclasses and coronary heart disease risk. The Framingham study. *Arterioscler Thromb Vasc Biol* 2000; 20: 1323-1329

937. KUIVENHOVEN JA, JUKEMA LW, ZWINDERMAN AH, DE KNIJFF P, McPHERSON R, BRUSCHKE AVG, LIE KI, KASTELEIN JJP : The role of a common variant of the cholesteryl ester transfer protein gene in the progression of coronary atherosclerosis. *N Engl J Med* 1998; 338: 86-93

938. LUO Y, TALL AR : Sterol upregulation of human CETP expression in vitro and in transgenic mice by an LXR element. *J Clin Invest* 2000; 105: 513-520

939. MASUCCI-MAGOULAS L, MOULIN P, JIANG XC, RICHARDSON H, WALSH A, BRESLOW JL, TALL A : Decreased cholesteryl ester transfer protein (CETP) mRNA and protein and increased high density lipoprotein following lipopolysaccharide administration in human CETP transgenic mice. *J Clin Invest* 1995; 95: 1587-1594

940. RITSCH A, DREXEL H, AMANN FW, PFEIFHOFER C, PATSCH JR : Deficiency of cholesteryl ester transfer protein. Description of the molecular defect and the dissociation of cholesteryl ester and triglyceride transport in plasma. *Arterioscler Thromb Vasc Biol* 1997; 17: 3433-3441

941. TEH EM, DOLPHIN PJ, BRECKENRIDGE WC, TAN M-H : Human plasma CETP deficiency: identification of a novel mutation in exon 9 of the CETP gene in a caucasian subject from north america. *J Lipid Res* 1998; 39: 442-456

942. KAKKO S, TAMMINEN M, KESÄNIEMI YA, SAVOLAINEN MJ : R451Q mutation in the cholesteryl ester transfer protein (CETP) gene is associated with high plasma CETP activity. *Atherosclerosis* 1998; 136: 233-240

943. ZHONG S, SHARP DS, GROVE JS, BRUCE C, YANO K, CURB DJ, TALL AR: Increased coronary heart disease in japanese-american men with mutations in the cholesteryl ester transfer protein gene despite increased HDL levels. *J Clin Invest* 1996; 12: 2917-2923

944. HIRANO K-I, YAMASHITA S, KUGA Y, SAKAI N, NOZAKI S, KIHARA S, ARAI T, YANAGI K, TAKAMI S, MENJU M, ISHIGAMI M, YOSHIDA Y, KAMEDA-TAKEMURA K, HAYASHI K, MATSUZAWA Y : Atherosclerotic disease in marked hyperalphalipoproteinemia. Combined reduction of cholesteryl ester transfer protein and hepatic triglyceride lipase. *Arterioscler Thromb Vasc Biol* 1995; 15: 1849-1856

945. AGERHOLM-LARSEN B, NORDESTGAARD BG, STEFFENSEN R, JENSEN G, TYBJAERG-HANSEN A : Elevated HDL cholesterol is a risk factor for ischemic heart disease in white women when caused by a common mutation in the cholesteryl ester transfer protein gene. *Circulation* 2000; 101: 1907-1912

946. BRUCE C, SHARP DS, TALL AR : Relationship of HDL and coronary heart disease to a common amino acid polymorphism in the cholesteryl ester transfer protein in men with and without hypertriglyceridemia. *J Lipid Res* 1998; 39: 1071-1078

947. DACHET C, POIRIER O, CAMBIEN F, CHAPMAN J, ROUIS M : New functional promoter polymorphism, CETP/-629, in cholesteryl ester transfer protein (CETP) gene related to CETP mass and high density lipoprotein cholesterol levels: role of Sp1/Sp3 in transcriptional regulation. *Arterioscler Thromb Vasc Biol* 2000; 20: 507-515

948. HUUSKONEN J, ENHOLM C : Phospholipid transfer protein in lipid metabolism. *Curr Opin Lipidol* 2000; 11: 285-289

949. DAY JR, ALBERS JJ, LOFTON-DAY CE, GILBERT TL, CHING AFT, GRANT FJ, O'HARA PJ, MARCOVINA SM, ADOLPHSON JL : Complete cDNA encoding human phospholipid transfer protein from human endothelial cells. *J Biol Chem* 1994; 269: 9388-9391

950. KIRSCHNING CJ, AU-YOUNG J, LAMPING N, REUTER D, PFEIL D, SEILHAMER JJ, SCHUMANN RR : Similar organization of the lipopolysaccharide-binding protein (LBP) and phospholipid transfer protein (PLTP) genes suggest a common gene family of lipid-binding proteins. *Genomics* 1997; 46: 416-425

951. WHITMORE TE, DAY JR, ALBERS JJ : Localization of the human phospholipid transfer protein gene to human chromosome 20q12-q13.1. *Genomics* 1995; 28: 599-600

952. HUUSKONEN J, WOHLFAHRT G, JAUHIAINEN M, ENHOLM C, TELEMAN O, OLKKONEN VM : Structure and phospholipid transfer activity of human PLTP: analysis by molecular modeling and site-directed mutagenesis. *J Lipid Res* 1999; 40: 1123-1130

953. GUYARD-DANGREMONT G, DESRUMAUX C, GAMBERT P, LALLEMANT C, LAGROST L : Phospholpid and cholesteryl ester transfer activities in plasma from 14 vertebrate species. Relation to atherogenesis susceptibility. *Comp Biochem Physiol Biochem Mol Biol* 1998; 120: 517-525

954. LAGROST L, MENSINK RP, GUYARD-DANGREMONT V, TEMME EH, DESRUMAUX C, ATHIAS A, HORNSTRA G, GAMBERT P : Variation in serum cholesteryl ester transfer and phospholipid transfer activities in healthy women and men consuming diets enriched in lauric, palmitic or oleic acids. *Atherosclerosis* 1999; 142: 395-402

955. URIZAR NL, DOWHAN DH, MOORE DD : The farnesoid X-activated receptor mediates bile acid activation of phospholipid transfer protein gene expression. *J Biol Chem* 2000; in press.

956. TU AY, ALBERS JJ : DNA sequences responsible for reduced promoter activity of human phospholipid transfer protein by fibrate. *Biochem Biophys Res Comm* 1999; 264: 802-807

957. TAHVANAINEN E, JAUHIAINEN M, FUNKE H, VARTIAINEN E, SUNDVALL J, EHNHOLM C : Serum phospholipid transfer protein activity and genetic variation of the PLTP gene. *Atherosclerosis* 1999; 146: 107-115

958. DESRUMAUX C, ATHIAS A, BESSEDE G, VERGES B, FARNIER M, PERSEGOL L, GAMBERT P, LAGROST L : Mass conservation of plasma phospholipid transfer protein in normolipidemic, type IIa hyperlipidemic, type IIb hyperlipidemic, and non-insulin-dependent diabetic subjects as measured by specific ELISA. *Arterioscler Thromb Vasc Biol* 1999; 19: 266-275

959. FREDRICKSON DS, ALTROCCHI PH, AVIOLI LV, GOODMAN DS, GOODMAN HC : Tangier Disease. *Ann Intern Med* 1961; 55: 1016-1031

960. ASSMANN GA, VON ECKARDSTEIN A, BREWER HB Jr: Familial high density lipoprotein deficiency: Tangier disease. In *The Metabolic and Molecular Bases of Inherited Disease*. Scriver CR, Beaudet AL, Sly WS and Valle D, eds. Mac Graw Hill, New York. 1995; 2: 2053-2072

961. SERFATY-LACRONIERE C, CIVEIRA F, LANZBERG A, ISAIA P, BERG J, JANUS ED, SMITH MP Jr, PRITCHARD HP, FROHLICH J, LEES RS, BARNARD GF, ORDOVAS JM, SCHAEFER EJ : Homozygous Tangier disease and cardiovascular disease. *Atherosclerosis* 1994; 107: 85-98

962. SCHMITZ G, FISCHER H, BEUCK M, HOECKER KP, ROBENEK H : Dysregulation of lipid metabolism in Tangier monocyte-derived macrophages. *Arteriosclerosis* 1990 10: 1010-1019

963. ROEGLER G, TRUMBACH B, KLIMA B, LACKNER KJ, SCHMITZ G : HDL-mediated efflux of intracellular cholesterol is impaired in fibroblasts from Tangier disease patients. *Arterioscler Thromb Vasc Biol* 1995; 15: 683-690

964. SCHAEFER EJ, ZECH LA, SCHWARTZ DE, BREWER HB Jr : Coronary heart disease prevalence and other clinical features in familial high-density lipoprotein deficiency (Tangier Disease). *Ann Intern Med* 1980; 93: 261-266

965. LAWN RM, WADE DP, GARVIN MR, WANG XB, SCHWARTZ K, PORTER JG, SEILHAMER JJ, VAUGHAN AM, ORAM JF : The Tangier disease gene product ABC1 controls the cellular apolipoprotein-mediated lipid removal pathway. *J Clin Invest* 1999; 104: R25-R31

966. RUST S, WALTER M, FUNKE H, VON ECKARDSTEIN A, CULLEN P, KROES HY, HORDIJK R, GEISEL J, KASTELEIN J, MOLHUIZEN HO, SCHREINER M, MISCHKE A, HAHMANN HW, ASSMANN G : Assignment of Tangier disease to chomosome 9q31 by graphical linkage exclusion strategy. *Nature Genet* 1998; 20: 96-98

967. LANGMANN T, KLUCKEN J, REIL M, LIEBISH G, LUCIANI MF, CHIMINI G, KAMINSKI WE, SCHMITZ G : Molecular cloning of the human ATP-binding cassette transporter 1 (hABC1): evidence for sterol-dependent regulation in macrophages. *Biochem Biophys Res Commun* 1999; 257: 29-33

968. LUCIANI MF, DENIZOT F, SAVARY S, MATTEI MG, CHIMINI G : Cloning of two novel ABC transporters mapping on human chromosome 9. *Genomics* 1994; 21: 150-159

969. DEAN M, ALLIKMETS R : Evolution of ATP binding cassette transporter genes. *Curr Opin Genet Dev* 1995; 5: 779-785

970. BROUSSEAU ME, SCHAEFER EJ, DUPUIS J, EUSTACHE B, VAN EERDEWGH P, GOLDKAMP AL, THURSTON LM, FITZGERALD MG, YASEK-McKENNA D, O'NEILL G, EBERHART G, WEIFFENBACH B, ORDOVAS JM, FREEMAN MW, BROWN RH Jr, GU JZ : Novel mutations in the gene encoding ATP-binding cassette 1 in four Tangier disease kindreds. *J Lipid Res* 2000; 41: 433-441

971. REMALEY AT, RUST S, ROSIER M, KNAPPER C, NAUDIN L, BROCCARDO C, PETERSON KM, KOCH C, ARNOULD I, PRADES C, DUVERGER N, FUNKE H, ASSMANN G, DINGER M, DEAN M, CHIMINI G, SANTAMARINA-FOJO S, FREDRICKSON DS, DENEFLE P, BREWER HB Jr : Human ATP-binding cassette transporter 1 (ABC1): genomic organization and identification of the genetic defect in the original Tangier disease kindred. *Proc Natl Acad Sci USA* 1999; 96: 12685-12690

972. SANTAMARINA-FOJO S, PETERSON KM, KNAPPER C, QIU Y, FREEMAN L, CHENG JF, OSORIO J, REMALEY A, YANG XP, HAUDENSCHILD C, PRADES C, CHIMINI G, BLACKMON E, FRANCOIS T, DUVERGER N, RUBIN EM, ROSIER M, DENEFLE P, FREDRICKSON DS, BREWER HB Jr : Complete genomic sequence of the human ABCA1 gene: analysis of the human and mouse ATP-binding cassette A promoter. *Proc Natl Acad Sci USA* 2000; 97: 7987-7992

973. McNEISH J, AIELLO RJ, GUYOT D, TURI T, GABEL C, ALDINGER C, HOPPE KL, ROACH ML, ROYER LJ, DE WET J, BROCCARDO C, CHIMINI G, FRANCONE OL : High density lipoprotein deficiency and foam cell accumulation in mice with targeted disruption of ATP-binding cassette transporter-1. *Proc Natl Acad Sci USA* 2000; 97: 4245-4250

974. PANOUSIS CG, ZUKERMAN SH : Interferon-γ induces downregulation of Tangier disease gene (ATP-binding cassette transporter 1) in macrophage-derived foam cells. *Arterioscler Thromb Vasc Biol* 2000; 20: 1565-1571

975. COSTET P, LUO Y, WANG N, TALL AR : Sterol-dependent transactivation of the ABC1 promoter by liver X receptor/retinoid X receptor. *J Biol Chem* 2000; 275: 28240-28245

976. REPA JJ, TURLEY SD, LOBACCARO J-M, MEDINA J, LI L, LUSTIG K, SHAN B, HEYMAN RA, DIETSCHY JM, MANGELSDORF DJ : Regulation of absorption and ABC1-mediated efflux of cholesterol by RXR heterodimers. *Science* 2000; 289: 1524-1529

977. BROTNICK AE, ROTHBLAT GH, STOUDT G, HOPPE KL, ROYER LJ, McNEISH J, FRANCONE OL : The correlation of ATP-binding cassette 1 mRNA levels with cholesterol efflux from various cell lines. *J Biol Chem* 2000; 275: 28634-28640

978. MARCIL M, BROOKS-WILSON A, CLEE SM, ROOMP K, ZHANG LH, YU L, COLLINS JA, VAN DAM M, MOLHUIZEN HOF, LOUBSTER O, OUELETTE FBF, SENSEN CW, FICHTER K, MOTT S, DENIS M, BOUCHER B, PIMSTONE S, GENEST J Jr, KASTELEIN JJP, HAYDEN MR : Mutations in the ABC1 gene in familial HDL deficiency with defective cholesterol efflux. *Lancet* 1999; 354: 1341-1346

979. GLASS C, PITTMAN RC, WEINSTEIN DB, STEINBERG D : Dissociation of tissue uptake of cholesterol ester from that of apoprotein A-I of rat plasma high density lipoprotein: selective delivery of cholesterol ester to the liver. *Proc Natl Acad Sci USA* 1983; 80: 5435-5439

980. STEIN Y, DABACH Y, HOLLANDER G, HALPERN G, STEIN O : Metabolism of HDL-cholesteryl ester in the rat, studied with a nonhydrolyzable analog, cholesteryl linoleyl ether. *Biochim Biophys Acta* 1983; 752: 98-105

981. PLUMP AS, ERICKSON SK, WENG W, PARTIN JS, BRESLOW JL, WILLIAMS DL : Apolipoprotein A-I is required for cholesteryl ester accumulation in steroidogenic cells and for normal adrenal stgeroid production. *J Clin Invest* 1996; 97: 2660-2671

982. CALVO D, VEGA MA : Identification, primary structure, and distribution of CLA-1, a novel member of the CD36/LIMPII gene family. *J Biol Chem* 1993; 268: 18929-18935

983. ACTON SL, SCHERER PE, LODISH HF, KRIEGER M : Expression cloning of SR-BI, a CD36-related class B scavenger receptor. *J Biol Chem* 1994; 269: 21003-21009

984. ACTON S, RIGOTTI A, LANDSCHULZ KT, XU S, HOBBS HH, KRIEGER M: Identification of scavenger SR-BI as a high density lipoprotein receptor. *Science* 1996; 271: 518-520

985. TRIGATTI B, RIGOTTI A, KRIEGER M : The role of high-density lipoprotein receptor SR-BI in cholesterol metabolism. *Curr Opin Lipidol* 2000; 11: 123-131

986. ACTON S, OSGOOD D, DONOGHUE M, CORELLA D, POCOVI M, CENARRO A, MOZAS P, KEITLY J, SQUAZZO S, WOOLF EA, ORDOVAS JM : Association of polymorphisms at the SR-BI gene locus with plasma lipid levels and body mass index in a white population. *Arterioscler Thromb Vasc Biol* 1999; 19: 1734-1743

987. MATVEEV S, VAN DER WESTHUYEZEN DR, SMART EJ : Co-expression of scavenger receptor-BI and caveolin-1 is associated with enhanced selective cholesteryl ester uptake in THP-1 macrophages. *J Lipid Res* 1999; 40: 1647-1654

988. IKEMOTO M, ARAI H, FENG DD, TANAKA K, AOKI J, DOHMAE N, TAKIO K, ADACHI H, TSUJIMOTO M, INOUE K : Identification of a PDZ-domain-containing protein that interacts with the scavenger receptor class B type I. *Proc Natl Acad Sci USA* 2000; 97: 6538-6543

989. GU XJ, KOZARSKY K, KRIEGER M : Scavenger receptor class B type I-mediated [^{3}H]cholesterol efflux to high density and low density lipoproteins is dependent on lipoprotein binding to the receptor. *J Biol Chem* 2000; 275: 29993-30001

990. LOPEZ D, McLEAN MP : Sterol regulatory element-binding protein-1a binds to cis elements in the promoter of the rat high density lipoprotein receptor SR-BI gene. *Endocrinology* 1999; 140: 5669-5681

991. KOZARSKY K, DONAHEE MH, RIGOTTI A, IQBAL SN, EDELMAN ER, KRIEGER M : Overexpression of the HDL receptor SR-BI alters plasma HDL and bile cholesterol levels. *Nature* 1997; 387: 414-417

992. JI Y, WANG N, RAMAKRISHNAN R, SEHAYEK E, HUSZAR D, BRESLOW JL, TALL AR : Scavenger receptor BI promotes rapid clearance of high density lipoprotein free cholesterol and its transport into bile. *J Biol Chem* 1999; 274: 33398-33402

993. CHINETTI G, GBAGUIDI FG, GRIGLIO S, MALLAT Z, ANTONUCCI M, POULAIN P, CHAPMAN J, FRUCHART JC, TEDGUI A, NAJIB-FRUCHART J, STAELS B : CLA-1/SR-B1 is expressed in atherosclerotic lesion macrophages and regulated bby activators of peroxisome proliferator-activated receptors. *Circulation* 2000; 101: 24211-2417

994. JIAN B, de la LLERA MOYA M, JI Y, WANG N, PHILLIPS MC, SWANEY JB, TALL AR, ROTHBLAT GH: Scavenger receptor class B type I as a mediator of cellular cholesterol efflux to lipoproteins and phospholipid acceptors *J Biol Chem* 1998; 273: 5599-5606

995. STANGL H, HYATT M, HOBBS HH : Transport of lipids from high and low density lipoproteins via scavenger receptor-BI. *J Biol Chem* 1999; 274: 32692-32698

996. GLASS C, PITTMAN RC, KELLER GA, STEINBERG D : Tissue sites of degradation of apolipoprotein A-I in the rat. *J Biol Chem* 1983; 258: 7161-7167

997. BORK P, BECKMANN G : The CUB domain. A widespread module in developmentally regulated proteins. *J Mol Biol* 1993; 231: 539-545

998. BIRN H, VERROUST P, NEXO E, HAGER H, JACOBSEN C, CHRISTENSEN EI, MOESTRUP SK : Characterization of an epithelial approximately 460-kDa protein that faciliates endocytosis of intrinsic factor-vitamin B12 and binds receptor-associated protein. *J Biol Chem* 1997; 272: 26497-26504

999. HAMMAD SM, BARTH JL, KNAAK C, ARGRAVES WS : Megalin acts in concert with cubilin to mediate endocytosis of high density lipoproteins. *J Biol Chem* 2000; 275: 12003-12008

1000. BIRN H, FYFE JC, JACOBSEN C, MOUNIER F, VERROUST PJ, ORSKOV H, WILLNOW TE, MOESTRUP SK, CHRISTENSEN EI : Cubilin is an albumin binding protein important for renal tubular albumin reabsorption. *J Clin Invest* 2000; 105: 1353-1361

1001. KOZYRAKI R, FYFE J, KRISTIANSEN M, GERDES C, JACOBSEN C, CUI S, CHRISTENSEN EI, AMINOFF M, DE LA CHAPELLE A, KRAHE R, VERROUST PJ, MOESTRUP : The intrinsic factor-vitamin B12 receptor, cubilin, is a high-affinity apolipoprotein-A-I receptor facilitating endocytosis of high-density lipoprotein. *Nature Med* 1999; 5: 656-661

1002. HAMMAD SM, STEFANSSON S, TWAL WO, DRAKE CJ, FLEMING P, REMALEY A, BREWER HB Jr, ARGRAVES WS : Cubilin the endocytic receptor for intrinsic factor-vitamin B12 complex, mediates high-density lipoprotein holoparticle endocytosis. *Proc Natl Acad Sci USA* 1999; 96: 10158-10163

1003. BRASCHI S, NEVILLE TA, VOHL MC, SPARKS DL : Apolipoprotein-A-I charge and conformation regulate the clearance of reconstituted high density lipoprotein in vivo. *J Lipid Res* 1999; 40: 522-532

1004. KOZYRAKI R, KRISTAINSEN M, SILAHTAROGLU A, HANSEN C, JACOBSEN C, TOMMERUP N, VERROUST PJ, MOESTRUP SK : The human intrinsic factor-vitamin B12 receptor, cubilin: molecular characterization and chromosomal mapping of the gene to 10p within the autosomal recessive megaloblastic anemia (MGA1) region. *Blood* 1998; 91: 3593-3600

1005. AMINOFF M, CARTER JE, CHADWICK RB, JOHNSON C, GRASBECK R, ABDELAAL MA, BROCH H, JENNER LB, VERROUST PJ, MOESTRUP SK, DE LA CHAPELLE A, KRAHE R : Mutations in CUBN, encoding the intrinsic factor-vitamin B12 receptor, cubilin, cause of the hereditary megaloblastic anemia 1. *Nat Genet* 1999; 21: 309-313

1006. KRISTIANSEN M, AMINOFF M, JACOBSEN C, DE LA CHAPELLE A, KRAHE R, VERROUST PJ, MOESTRUP SK : Cublin P1297L mutation associated with hereditary megaloblastic anemia 1 causes impaired recognition of intrinsic factor-vitamin B12 by cubilin. *Blood* 2000; 96: 405-409

1007. MUKHERJEE S, MAXFIELD FR : Cholesterol stuck in traffic. *Nat Cell Biol* 1999; 1: E37-E38

1008. LISCUM L, KLANSEK JJ : Niemann-Pick disease type C. *Curr Opin Lipidol* 1998; 9: 131-135

1009. PENTCHEV PG, VANIER MT, SUZUKI K, PATTERSON MC : Niemann-Pick disease type C : cellular cholesterol lipidosis. In: Scriver CR, Beaudet AL, Sly WS, Valle D (eds). *The metabolic and molecular bases of inherited disease*, 7th edn. Mac Graw Hill. New York, 1995; 1655-1676

1010. CARSTEA ED, POLYMEROPOULOS MH, PARKER CC, DETERA-WADLEIGH SD, O'NEILL RR, PATTERSON MC, GOLDIN E, XIAO H, STRAUB RE, VANIER MT, BRADY RO, PENTCHEV PG : Linkage of Niemann-Pick disease type C to human chromosome 18. *Proc Natl Acad Sci USA* 1993; 90: 2002-2004

1011. CARSTEA ED, MORRIS JA, COLEMAN KG, LOFTUS SK, ZHANG D, CUMMINGS C, GU J, ROSENFELD MA, PAVAN WJ, KRIZMAN DB, NAGLE J, POLYMEROPOULOS MH, STURLEY SL, IOANNOU YA, HIGGINS ME, COMLY M, COONEY A, BROWN A, KANESKI CR, BLANCHETTE-MACKIE J, DWYER NK, NEUFELD EB, CHANG T-Y, LISCUM L, STRAUSS JF, OHNO K, ZIEGLER M, CARMI R, SOKOL J, MARKIE D, O'NEILL RR, VAN DIGGELEN OP, ELLEDER M, PATTERSON MC, BRADY RO, VANIER MT, PENTCHEV PG, TAGLE DA : Niemann-Pick C1 disease gene: homology to mediators of cholesterol homeostasis. *Science* 1997; 277: 228-231

1012. LOFTUS SK, MORRIS JA, CARSTEA ED, GU JZ, CUMMINGS C, BROWN A, ELLISON J, OHNO K, ROSENFELD MA, TAGLE DA, PENTCHEV PG, PAVAN WJ : Murine model of Niemann-Pick type C disease: mutation in a cholesterol homeostasis gene. *Science* 1997; 277: 232-235

1013. MORRIS JA, ZHANG D, COLEMAN KG, NAGLE J, PENTCHEV PG, CARSTEA ED : The genomic organization and polymorphism analysis of the human Niemann-Pick C1 gene. *Biochem Biophys Res Commun* 1999; 261: 493-498

1014. WATARI H, BLANCHETTE-MACKIE EJ, DWYER NK, WATARI M, NEUFELD EB, PATE S, PENTCHEV PG, STRAUSS III JF : Mutations in the leucine zipper motif and sterol-sensing domain inactivate the Niemann-Pick C1 glycoprotein. *J Biol Chem* 1999; 274: 21861-21866

1015. MILLAT G, MARCAIS C, RAFI MA, YAMAMOTO T, MORRIS JA, PENTCHEV PG, OHNO K, WENGER DA, VANIER MT : Niemann-Pick C1 disease: the I1061T substitution is a frequent mutant allele in patients of western european descent and correlates with a classic juvenile phenotype. *Am J Hum Genet* 1999; 65: 1321-1329

1016. NAURECKIENE S, SLEAT DE, LACKLAND H, FENSOM A, VANIER MT, WATTIAUX R, JADOT M, LOBEL P: Identification of HE1 as the second gene of Niemann-Pick C disease. *Science* 2000; 290: 2298-22301

1017. SCHMITZ G, ASSMANN G : Acid lipase deficiency : Wolman disease and cholesteryl ester storage disease In: Scriver CR, Beaudet AL, Sly WS, Valle D (eds). *The metabolic basis of inherited disease*, 6th edn. Mac Graw Hill, New York, 1989 1623-1644

1018. KOCH GA, Mc AVOY M, SHOWS TB : Assignment of LIPA, associated with human acid lipase deficiency, to human chromosome 10 and comparative assignment to mouse chromosome 19. *Somatic Cell Genet* 1981; 7: 345-358

1019. ANDERSON RA, SANDO GN : Cloning and expression of cDNA encoding human lysosomal acid lipase/cholesteryl ester hydrolase. Similarities to gastric and lingual lipases. *J Biol Chem* 1990; 265: 22479-22484

1020. DU H, WITTE DP, GRABOWSKI GA : Tissue and cellular specific expression of murine lysosomal acid lipase mRNA and protein. *J Lipid Res* 1996; 37: 937-949

1021. ANDERSON RA, BRYSON GM, PARKS JS : Lysosomal acid lipase mutations that determine phenotype in Wolman and cholesterol ester storage disease. *Mol Genet Metab* 1999; 68: 333-345

1022. DU H, DUNAMU M, WITTE DP , GRABOWSKI GA : Targeted disruption of the mouse lysosomal acid lipase gene: long-term survival with massive cholesteryl ester and triglyceride storage. *Hum Mol Genet* 1998; 7: 1347-1354

1023. STURLEY SL : Molecular aspects of intracellular sterol esterification: the Acyl CoA:cholesterol Acyltransferase reaction. *Curr Opin Lipidol* 1997; 8: 167-173

1024. BELL FP, GAMMILL RB, ST JOHN LC : U-73482 : a novel ACAT inhibitor that elevates HDL-cholesterol, lowers plasma triglyceride and facilitates hepatic cholesterol mobilization in the rat. *Atherosclerosis* 1992; 92: 115-122

1025. FARESE RV Jr : Acyl CoA:cholesterol acyltransferase genes and knockout mice. *Curr Opin Lipidol* 1998; 9: 119-123

1026. LIN S, CHENG D, LIU MS, CHEN J, CHANG TY : Human Acyl CoA:cholesterol Acyltransferase-1 in the endoplasmic reticulum contains seven transmembrane domains. *J Biol Chem* 1999; 274: 23276-23285

1027. LI BL, LI XL, DUAN ZJ, LEE O, LIN S, MA ZM, CHANG CCY, YANG XY, PARK JP, MOHANDAS TK, NOLL W, CHAN L, CHANG TY : Human Acyl CoA:cholesterol acyltransferase-1 (ACAT-1) gene organization and evidence that the 4.3kb ACAT-1 m-RNA is produced from two different chromosomes. *J Biol Chem* 1999; 274: 11060-11071

1028. SAKASHITA N, MIYAZAKI A, TAKEYA M, HORIUCHI S, CHANG CCY, CHANG TY, TAKAHASHI K : Localization of human Acyl Coenzyme A:cholesterol acyltransferase-1 (ACAT-1) in macrophages and in various tissues. *Am J Pathol* 2000; 156: 227-236

1029. CHANG CCY, SAKASHITA N, ORNVOLD K, LEE O, CHANG ET, DONG R, LIN S, LEE CYG, STROM S, KASHYAP R, FUNG J, FARESE RV Jr , PATOISEAU JF,

DELHON A, CHANG TY : Immunological quantitation and localization of ACAT-1 and ACAT-2 in human liver and small intestine. *J Biol Chem* 2000; 275: 28083-28092

1030. MEINER VL, WELCH CL, CASES S, MYERS HM, SANDE E, LUSIS AJ, FARESE RV Jr : Adrenocortical lipid depletion gene (ald) in AKR mice is associated with an Acyl CoA:cholesterol acyltransferase (ACAT) mutation. *J Biol Chem* 1998; 273: 1064-1069

1031. ACCAD M, SMITH SJ, NEWLAND DL, SANAN DA, KING LE Jr, LINTON MF, FAZIO S, FARESE RV Jr : Massive xanthomatosis and aletered composition of atherosclerotic lesions in hyperlipidemic mice lacking acyl CoA:cholesterol acyltransferase 1. *J Clin Invest* 2000; 105: 711-719

1032. YAGYU H, KITAMINE T, OSUGA JI, TOZAWA RI, CHEN Z, KAJI Y, OKA T, PERREY S, TAMURA Y, OHASHI K, OKAZAKI H, YAHAGI N, SHIONOIRI F, IIZUKA Y, HARADA K, SHIMANO H, YAMASHITA H, GOTODA T, YAMADA N, ISHIBASHI S : Absence of ACAT-1 attenuates atherosclerosis but causes dry eye and cutaneous xanthomatosis in mice with congenital hyperlipidemia. *J Biol Chem* 2000; 275: 21324-21330

1033. JOYCE C, SKINNER K, ANDERSON RA, RUDEL LL : Acyl CoenzymeA:cholesteryl acyltransferase 2. *Curr Opin Lipidol* 1999; 10: 89-95

1034. HAMILTON JA, KAMP F : How are free fatty acids transported in membranes? Is it by proteins or by free diffusion through the lipids? *Diabetes* 1999; 48: 2255-2269

1035. SWEETSER DA, BIRKENMEIER EH, KLISAK IJ, ZOLLMAN S, SPARKES RS, MOHANDAS T, LUSIS AJ, GORDON JI : The human and rodent intestinal fatty acid binding protein genes: a comparative analysis of their structure, expression, and linkage relationships. *J Biol Chem* 1987; 262: 16060-16071

1036. SWEETSER DA, BIRKENMEIER EH, KLISAK IJ, ZOLLMAN S, SPARKES RS, MOHANDAS T, LUSIS AJ, GORDON JI : The human and rodent intestinal fatty acid binding protein genes: a comparative analysis of their structure, expression, and linkage relationships. *J Biol Chem* 1987; 262: 16060-16071

1037. SPARKES RS, MOHANDAS T, HEINZMANN C, GORDON JI, KLISAK I, ZOLLMAN S, SWEETSER DA, RAGUNATHAN L, WINOKUR S, LUSIS AJ : Human fatty acid binding protein assignments: intestinal to 4q28-4q31 and liver to 2p11. *Cytogenet Cell Genet* 1987; 46: 697

1038. HEGELE RA : A review of intestinal fatty acid binding protein gene variation and the plasma lipoprotein response to dietary components. *Clin Biochem* 1998; 31: 609-612

1039. WAKIL SJ : Fatty acid synthase, a proficient multifunctional enzyme. *Biochemistry* 1989; 28: 4523-4530

1040. JAYAKUMAR A, TAI MH, HUANG WY, AL-FEEL W, HSU M, ABU-ELHEIGA L, CHIRALA SS, WAKIL SJ : Human fatty acid synthase: properties and molecular cloning. *Proc Natl Acad Sci USA* 1995; 92: 8695-8699

1041. JAYAKUMAR A, CHIRALA SS, CHINAULT AC, BALDINI A, ABU-ELHEIGA L, WAKIL SJ : Isolation and chromosomal mapping of genomic clones encoding the human fatty acid synthase. *Genomics* 1994; 23: 420-424

1042. SUL HS, WANG D : Nutritional and hormonal regulation of enzymes in fat synthesis: studies of fatty acid synthase and mitochondrial glycerol-3-phosphate acyltransferase gene transcription. *Ann Rev Nutr* 1998; 18: 331-351

1043. FORETZ M, GUICHARD C, FERRÉ P, FOUFELLE F : Sterol regulatory element binding protein-1c is a major mediator of insulin action on the hepatic expression of glucokinase and lipogenesis-related genes. *Proc Natl Acad Sci USA* 1999; 96: 12737-12742

1044. LOFTUS TM, JAWORSKY DE, FREHYWOT GL, TOWNSEND CA, RONNETT GV, LANE MD, KUHAJDA FP : Reduced food intake and body weight in mice treated with fatty acid synthase inhibitors. *Science* 2000; 288: 2379-2381

1045. FARESE RV, CASES S, SMITH SJ : Triglyceride synthesis: insights from the cloning of diacylglycerol acyltransferase. *Curr Opin Lipidol* 2000; 11: 229-234

1046. NTAMBI JM : Regulation of steatroyl-CoA desaturase by polyunsturated fatty acids and cholesterol. *J Lipid Res* 1999; 40: 1549-1558

1047. ZHANG L, GE L, PARIMOO S, STENN K, PROUTY SM : Human stearoyl-CoA desaturase: alternative transcripts generated from a single gene by usage of tandem polyadenylation sites. *Biochem J* 1999; 340: 255-264

1048. ZHENG Y, EILERTSEN KJ, ZHANG L, SUNDBERG JP, PROUTY SM, STENN KS, PARIMOO S : Scd1 is expressed in sebaceous glands and is disrupted in the asebia mouse. *Nat Genet* 1999; 23: 268-270

1049. MIYAZAKI M, KIM YC, KELLER MP, ATTIE AD, NTAMBI JM : The biosynthesis of hepatic cholesterol esters and triglycerides is impaired in mice with a disruption of the gene for stearoyl-CoA desaturase 1. *J Biol Chem* 2000; 275: 30132-30138

1050. BONNEFONT JP, DEMAUGRE F, PRIP-BUUS C, SAUDUBRAY JM, BRIVET M, ABADI N, THUILLIER L : Carnitine palmitoyltransferase deficiencies. *Mol Genet Metab* 1999; 68: 424-440

1051. TAMAI I, OHASHI R, NEZU J, YABUUCHI H, OKU A, SHIMANE M, SAI Y, TSUJI A : Molecular and functional identification of sodium ion-dependent, high affinity human carnitine transporter OCTN2. *J Biol Chem* 1998; 273: 20378-20382

1052. WANG Y, YE J, GANAPATHY V, LONGO N : Mutations in the organic cation/carnitine transporter OCTN2 in primary carnitine deficiency. *Proc Natl Acad Sci USA* 1999; 96: 2356-2360

1053. NEZU J, TAMAI I, OKU A, OHASHI R, YABUUCHI H, HASHIMOTO N, NIKAIDO H, SAI Y, KOIZUMI A, SHOJI Y, TAKADA G, MATSUISHI T, YOSHINO M, KATO H, OHURA T, TSUJIMOTO G, HAYAKAWA J, SHIMANE M, TSUJI A : Primary systemic carnitine deficiency is caused by mutations in a gene encoding sodium ion-dependent carnitine transporter. *Nat Genet* 1999; 21: 91-94

1054. ZHU Y, JONG MC, FRAZER KA, GONG E, KRAUSS RM, CHENG J-F, BOFFELLI D, RUBIN EM : Genomic interval engineering of mice identifies a novel modulator of triglyceride production. *Proc Natl Acad Sci USA* 2000; 97: 1137-1142

1055. BROWN MS, GOLDSTEIN JL : The SREBP pathway: regulation of cholesterol metabolism by proteolysis of a membrane-bound transcription factor. *Cell* 1997; 89: 331-340

1056. HUA X, WU J, GOLDSTEIN JL, BROWN MS, HOBBS HH : Structure of human gene encoding sterol regulatory element binding protein-1 (SREBF1) and localization of

SREBF1 and SREBF2 to chromosomes 17p11.2 and 22q13. *Genomics* 1995; 25: 667-673

1057. DE BOSE-BOYD RA, BROWN MS, LI WP, NOHTURFFT A, GOLDSTEIN JL, ESPENSHADE PJ : Transport-dependent proteolysis of SREBP: relocation of site-1 protease from Golgi to ER obviates the need for SREBP transport to Golgi. *Cell* 1999; 99: 703-712

1058. BIST A, FIELDING PE, FIELDING CJ : Two sterol regulatory element-like sequences mediate up-regulation of caveolin gene transcription in response to low density lipoprotein free cholesterol. *Proc Natl Acad Sci USA* 1997; 94: 10693-10698

1059. SATO R, MIYAMOTO W, INOUE J, TERADA T, IMANAKA T, MAEDA M : Sterol regulatory element-binding protein negatively regulates microsomal triglyceride transfer protein gene transcription. *J Biol Chem* 1999; 274: 24714-24720

1060. SATO R, OKAMOTO A, INOUE J, MIYAMOTO W, SAKAI Y, EMOTO N, SHIMANO H, MAEDA M : Transcriptional regulation of the APT citrate-lyase gene by sterol regulatory element-binding proteins. *J Biol Chem* 2000; 275: 12497-12502

1061. SHIMANO H, SHIMOMURA I, HAMMER RE, HERZ J, GOLDSTEIN JL, BROWN MS, HORTON JD : Elevated levels of SREBP-2 and cholesterol synthesis in livers of mice homozygous for targeted disruption of the SREBP-1 gene. *J Clin Invest* 1997; 100: 2115-2124

1062. TONTONOZ P, KIM JB, GRAVES RA, SPIEGELMAN BM : ADD1: a novel helix-loop-helix transcription factor associated with adipocyte determination and differentiation. *Mol Cell Biol* 1993; 13: 4753-4759

1063. SHIMOMURA I, HAMMER RE, RICHARDSON JA, IKEMOTO S, BASHMAKOV Y, GOLDSTEIN JL, BROWN MS : Insulin resistance and diabetes mellitus in transgenic mice expressing nuclear SREBP-1c in adipose tissue: model for congenital generalized lipodystrophy. *Genes Dev* 1998; 12: 3182-3194

1064. KORN BS, SHIMOMURA I, BASHMAKOV Y, HAMMER RE, HORTON JD, GOLDSTEIN JL : Blunted feedback suppression of SREBP processing by dietary cholesterol in transgenic mice expressing sterol-resistant SCAP(D443N). *J Clin Invest* 1998; 102: 2050-2060

1065. CLARKE SD, THUILLIER P, BAILLIE RA, SHA XM : Peroxisome proliferator-activated receptors: a familly of lipid-activated transcription factors. *Am J Clin Nutr* 1999; 70: 566-571

1066. ISSEMANN I, GREEN S : Activation of a member of the steroid hormone receptor superfamily by peroxisome proliferators. *Nature* 1990; 347: 645-650

1067. LAZAROW PB, MOSER HW : Disorders of peroxisome biogenesis. In SCRIVER CR, BEAUDET AL, SLY W, VALLE D eds: *The Metabolic and Molecular Bases of Inherited Disease*, 7th edn. *Mac Graw Hill,* New York. 1995.

1068. STAELS B, DALLONGEVILLE J, AUWERX J, SCHOONJANS K, LEITERSDORF E, FRUCHART JC : Mechanism of action of fibrates on lipid and lipoprotein metabolism. *Circulation* 1998; 98: 2088-2093

1069. KERSTEN S, DESVERGNE B, WAHLI W : Roles of PPARs in health and disease. *Nature* 2000; 405: 421-424

1070. STAELS B, KOENIG W, HABIB A, MERVAL, LEBRET M, PINEDA TORRA I, DELERIVE P, FADEL A, CHINETTI G, FRUCHART JC, NAJIB J, MACLOUF J,

TEDGUI A : Activation of human aortic smooth-muscle cells is inhibited by PPARα but not by PPARγ activators. *Nature* 1998; 393: 790-793

1071. COSTET P, LEGENDRE C, MORE J, EDGAR A, GALTIER P, PINEAU T : Peroxisome proliferator-activated receptor α-isoform deficiency leads to progressive dyslipidemia with sexual dimorphic obesity and steatosis. *J Biol Chem* 1998; 273: 29577-29585

1072. VOHL M-C, LEPAGE P, GAUDET D, BREWER CG, BETARD C, PERRON P, HOUDE G, CELLIER C, FAITH JM, DESPRES J-P, MORGAN K, HUDSON TJ : Molecular scanning of the human PPARα gene: association of the L162V mutation with hyperapobetalipoproteinemia. *J Lipid Res* 2000; 41: 945-952

1073. SAPONE A, PETERS JM, SAKAI S, TOMITA S, PAPILA SS, DAI R, FRIEDMAN FK, GONZALEZ FJ : The human peroxisome proliferator-activated receptor alpha gene: identification and functional characterization of two natural allelic variants. *Pharmacogenetics* 2000; 10: 321-333

1074. VAMECQ J, LATRUFFE N : Medical significance of peroxisome proliferator-activated receptors. *Lancet* 1999; 354: 141-148

1075. GREENE ME, BLUMBERG B, Mc BRIDE OW, YI HF, KRONQUIST K, KWAN K, HSIEH L, GREENE G, NIMER SD : Isolation of the human peroxisome proliferator activated receptor gamma cDNA : expression in hematopoietic cells and chromosomal mapping. *Gene Expr* 1995; 4: 281-299

1076. FAJAS L, AUBOEUF D, RASPE E, SCHOONJANS K, LEFEBVRE AM, SALADIN R, NAJIB J, LAVILLE M, FRUCHART JC, DEEB S, VIDAL-PUIG A, FLIER J, BRIGGS MR, STAELS B, VIDAL H, AUWERX J : The organization, promoter analysis, and expression of the human PPARgamma gene. *J Biol Chem* 1997; 272:18779-18789

1077. MILES PD, BARAK Y, HE W, EVANS RM, OLEFSKY JM : Improved insulin-sensitivity in mice heterozygous for PPAR-gamma deficiency. *J Clin Invest* 2000; 105: 287-292

1078. BARROSO I, GURNELL M, CROWLEY VE, AGOSTINI M, SCHWABE JW, SOOS MA, MASLEN GL, WILLIAMS TD, LEWIS H, SCHAFER AJ, CHATTERJEE VK, O'RAHILLY S : Dominant negative mutations in human PPARgamma associated with severe insulin resistance, diabetes mellitus and hypertension. *Nature* 1999; 402: 880-883

1079. ALTSHULER D, HIRSCHHORN JN, KLANNEMARK M, LINDGREN CM, VOHL MC, NEMESH J, LANE CR, SCHAFFNER SF, BOLK S, BREWER C, TUOMI T, GAUDET D, HUDSON TJ, DALY M, GROOP L, LANDER ES : The common PPARγ Pro12Ala polymorphism is associated with decreased risk of type 2 diabetes. *Nature Genet* 2000; 26: 76-80

1080. SARRAF P, MUELLER E, SMITH WM, WRIGHT HM, KUM JB, AALTONEN LA, DE LA CHAPELLE A, SPIEGELMAN BM, ENG C : Loss-of-function mutations in PPAR-gamma associated with human colon cancer. *Mol Cell* 1999; 3: 799-804

1081. MUKHERJEE R, STRASSER J, JOW L, HOENER P, PATERNITI JR, Jr, HEYMAN RA : RXR agonists activate PPARα–inducible genes, lower triglycerides and raise HDL levels in vivo. *Arterioscler Thromb Vasc Biol* 1998; 18: 272-276

1082. BLUMBERG B, EVANS RM : Orphan nuclear receptors-new ligands and new possibilities. *Genes Dev* 1998; 12: 3149-3155

1083. PEET DJ, JANOWSKI BA, MANGELSDORF DJ : The LXRs: a new class of oxysterol receptors. *Curr Opin Genet Dev* 1998; 8: 571-575

1084. MAKISHIMA M, OKAMOTO AY, REPA JJ, TU H, LEARNED RM, LUK A, HULL MV, LUSTIG KD, MANGELSDORF DJ, SHAN B : Identification of a nuclear receptor for bile acids. *Science* 1999; 284: 1362-1368

1085. PARKS DJ, BLANCHARD SG, BLEDSOE RK, CHANDRA G, CONSLER TG, KLIEWER SA, STIMMEL JB, WILLSON TM, ZAVACKI AM, MOORE DD, LEHMANN JM : Bile acids: natural ligands for an orphan nuclear receptor. *Science* 1999; 284: 1365-1368

1086. FATKIN D, MAC RAE C, SASAKI T, WOLFF MR, PORCU M, FRENNEAUX M, ATHERTON J, VIDAILLET HJ, SPUDICH S, DE GIROLAMI U, SEIDMAN JG, SEIDMAN CE : Missense mutations in the rod domain of the lamin A/C gene as causes of dilated cardiomyopathy and conduction-system disease. *N Engl J Med* 1999; 341: 1715-1724

1087. BONNE G, DI BARLETTA MR, VARNOUS S, BECANE HM, HAMMOUDA E, MERLINI L, MUNTONI F, GREENBERG CR, GARY F, URTIZBEREA JA, DUBOC D, FARDEAU M, TONIOLO D, SCHWARTZ K : Mutations in the gene encoding lamin A/C cause autosomal dominant Emery-Dreifuss muscular dystrophy. *Nature Genet* 1999; 21: 285-288

1088. CAO H, HEGELE RA : Nuclear lamin A/C R482Q mutation in canadian kindreds with Dunnigan-type familial partial lipodystrophy. *Hum Mol Genet* 2000; 9: 109-112

1089. CLEMENTS L, MANILAL S, LOVE DR, MORRIS GE : Direct interaction between emerin and lamin A. *Biochem Biophys Res Commun* 2000; 27: 709-714

1090. FAIRLEY EA, KENDRICK-JONES J, ELLIS JA : The Emery-Dreifuss muscular dystrophy phenotype arises from aberrant targeting of emerin at the inner nuclear membrane. *J Cell Science* 1999; 112: 2571-2582

1091. SHACKELTON S, LLOYD DJ, JACKSON SNJ, EVANS R, NIEMEIJER MF, SINGH BM, SCHMIDT H, BRABANT G, KUMAR S, DURRINGTON PN, GREGORY S, O'RAHILLY S, TREMBATH RC : LMNA, encoding lamin A/C, is mutated in partial lipodystrophy. *Nature Genet* 2000; 24: 153-156

1092. FARRELLY D, BROWN KS, TIEMAN A, REN J, LIRA SA, HAGAN D, GREGG R, MOOKHTIAR KA, HARIHARAN N : Mice mutant for glucokinase regulatory protein exhibit decreased liver glucokinase: a sequestration mechanism in metabolic regulation. *Proc Natl Acad Sci USA* 1999; 96: 14511-14516

1093. RUSSELL DW, SETCHELL KDR : Bile acid biosynthesis. *Biochemistry* 1992; 31: 4737-4749

1094. HIROHASHI T, SUZUKI H, TAKIKAWA H, SUGIYAMA Y : ATP-dependent transport of bile salts by rat multidrug resistance-associated protein 3 (Mrp-3). *J Biol Chem* 2000; 275: 2905-2910

1095. SCHWARZ M, LUND EG, RUSSEL DW : Two 7 α-hydroxylase enzymes in bile acid biosynthesis. *Curr Opin Lipidol* 1998; 9: 113-118

1096. COHEN JC, CALI JJ, JELINEK DF, MEHRABIAN M, SPARKES RS, LUSIS AJ, RUSSELL DW, HOBBS HH : Cloning of the human cholesterol 7α–hydroxylase gene (CYP7) and localization to chromosome 8q11-q12. *Genomics* 1992; 14: 153-161

1097. SPADY DK, CUTHBERT JA, WILLARD MN, MEIDELL RS : Adenovirus-mediated transfer of a gene encoding cholesterol 7α-hydroxylase into hamsters increases hepatic enzyme activity and reduces plasma total and low density lipoprotein cholesterol. *J Clin Invest* 1995; 96: 700-709

1098. WANG JP, FREEMAN DJ, GRUNDY SM, LEVINE DM, GUERRA R, COHEN JC : Linkage between cholesterol 7α-hydroxylase and high plasma low-density lipoprotein cholesterol concentrations. *J Clin Invest* 1998; 101: 1283-1361

1099. HAWKINS JL, LUNDT EG, BRONSON AD, RUSSELL DW : Expression cloning of an oxysterol 7α-hydroxylase selective for 24-hydroxycholesterol. *J Biol Chem* 2000; 276: 16543-16549

1100. SETCHELL KDR, SCHWARZ M, O'CONNELL NC, LUNDT EG, DAVIS DL, LATHE R, THOMPSON HR, TYSON RW, SOKOL RJ, RUSSELL DW : Identification of a new inborn error in bile acid synthesis: mutation of the oxysterol 7α-hydroxylase gene causes severe neonatal liver disease. *J Clin Invest* 1998; 102: 1690-1703

1101. OKUDA KI : Liver mitochondrial P450 involved in cholesterol catabolism and vitamin D. *J Lipid Res* 1994; 35: 361-372

1102. BABIKER A, ADERSSON O, LUND E, XIU RJ, DDEB S, RESHEF A, LEITERSDORF E, DICZFALUSY U, BJORKHEM I : Elimination of cholesterol in macrophages and endothelial cells by sterol 27-hydroxylase mechanism. *J Biol Chem* 1997; 272: 26253-26261

1103. CALI JJ and RUSSELL DW : Characterization of human sterol 27-hydroxylase A mitochondrial cytochrome P-450 that catalyzes multiple oxidation reactions in bile acid biosynthesis. *J Biol Chem* 1991; 266: 7774-7778

1104. CALI JJ, HSIEH C-L, FRANCKE U, RUSSELL DW : Mutations in the bile acid biosynthetic enzyme sterol 27-hydroxylase underlie cerebrotendinous xanthomatosis. *J Biol Chem* 1991; 266: 7779-7783

1105. LEITERSDORF E, MEINER V : Cerebrotendinous xanthomatosis. *Curr Opin Lipidol* 1994; 5: 138-142

1106. BJÖRKHEM I, BOBERG KM : Inborn errors in bile acid biosynthesis and storage of sterols other than cholesterol. In : SCRIVER CR, BEAUDET AL, SLY WS, VALLE D eds. *The metabolic and molecular bases of inherited disease* 7th edn. New York: Mc Graw Hill 1995: 2073-2099

1107. INOUE K, KUBOTA S, SEYAMA Y : Cholestanol induces apoptosis of cerebellar neuronal cells. *Biochem Biophys Res Commun* 1999; 256: 198-203

1108. VERRIPS A, HOEFSLOOT LH, STEENBERGEN GCH, THEELEN JP, WEVERS RA, GABREELS FJM, VAN ENGELEN BGM, VAN DEN HEUVEL LPWJ : Clinical and molecular genetic characteristics of patients with cerebrotendinous xanthomatosis. *Brain* 2000; 123: 908-919

1109. LEITERSDORF E, SAFADI R, MEINER V, RESHEF A, BJORKHEM I, FRIEDLANDER Y, MORKOS S, BERGINER VM : Cerebrotendinous xanthomatosis in the israeli Druze: molecular genetics and phenotypic characteristics. *Am J Hum Genet* 1994; 55: 907-9015

1110. SCHMITZ G, WILLIAMSON E : High-density lipoprotein metabolism, reverse cholesterol transport and membrane protection. *Curr Opin Lipidol* 1991; 2: 177-189

1111. DRAYNA DT, Mac LEAN J, WION KL, TRENT JM, DRABKIN HA, LAWN R : Human apolipoprotein D gene: gene sequence, chromosome localization and homology to the α2μ-Globulin superfamily. *DNA* 1987; 6: 199-204

1112. LAMBERT J, PROVOST PR, MARCEL YL, RASSART E : Structure of human apolipoprotein D gene promoter region. *Biochim Biophys Acta* 1993; 1172: 190-192

1113. MILNE RW, RASSART E, MARCEL YL : Molecular biology of Apolipoprotein D. *Curr Opin Lipidol* 1993; 4: 100-105

1114. ZENG C, SPIELMAN AI, VOWELS BR, LEYDEN JJ, BIEMANN K, PRETI G : A human axillary odorant is carried by apolipoprotein D. *Proc Natl Acad Sci USA* 1996; 93: 6626-6630

1115. STEINKASSERER A, COCKBURN DJ, BLACK DM, BOYD Y, SOLOMON E, SIM RB : Assignment of apolipoprotein H (APOH:beta-2-glycoprotein I) to human chromosome 17q23→qter; determination of the major expression site. *Cytogenet Cell Genet* 1992; 60: 31-33

1116. MEHDI H, NUNN M, STEEL D, WHITEHEAD AS, PEREZ M, WALKER L, PEEPLES ME : Nucleotide sequence and expression of the human gene encoding apolipoprotein H (ß2-glycoprotein I). *Gene* 1991; 108: 293-298

1117. SANGHERA DK, WAGENKNECHT DR, McINTYRE JA, KAMBOH MI : Identification of structural mutations in the fifth domain of apolipoprotein H (beta-2 glycoprotein) which affect phospholipid binding. *Hum Mol Genet* 1997; 6: 311-316

1118. FINK TM, ZIMMER M, TSCHOPP J, ETIENNE J, JENNE DE, LICHTER P : Human clusterin (CLI) maps to 8p21 in proximity to the lipoprotein lipase (LPL) gene. *Genomics* 1993; 16: 526-528

1119. HUMPHREYS DT, CARVER JA, EASTERBROOK-SMITH SB, WILSON MR : Clusterin has chaperone-like activity similar to that of small heat shock proteins. *J Biol Chem* 1999; 274: 6875-6881

1120. FRENCH LE, WOHLWEND A, SAPPINO AP, TSCHOPP J, SCHIFFERLI JA : Human clusterin gene expression is confined to surviving cells during in vitro progammed cell death. *J Clin Invest* 1994; 93: 877-884

1121. DUCHATEAU PN, PULLINGER CR, ORELLANA RE, KUNITAKE ST, NAYA-VIGNE J, O'CONNOR PM, MALLOY MJ, KANE JP : Apolipoprotein L, a new human high density lipoprotein apolipoprotein expressed by the pancreas. *J Biol Chem* 1997; 272: 25576-25582

1122. SELLAR GC, JORDAN SA, BICKMORE WA, FANTES JA, VAN HEYNINGEN V, WHITEHEAD AS : The human serum amyloid A protein (SAA) superfamily gene cluster: mapping to chromosome 11p15.1 by physical and genetic linkage analysis. *Genomics* 1994; 19: 221-227

1123. UHLAR CM, BURGESS CJ, SHARP PM, WHITEHEAD AS : Evolution of the serum amyloid A (SAA) protein superfamily. *Genomics* 1994; 19: 228-235

1124. DE BEER MC, YUAN T, KINDY MS, ASZTALOS BF, ROHEIM PS, DE BEER FC : Characterization of constitutive human serum amyloid A protein (SAA4) as an apolipoprotein. *J Lipid Res* 1995; 36: 526-534

1125. STEINBERG D, PARTHASARATHY S, CAREW TE, KHOO JC, WITZUM JL : Beyond cholesterol Modification of low-density lipoprotein that increase its atherogenicity. *N Engl J Med* 1989; 320: 915-924

1126. HEINECKE JW, LUSIS AJ : Paraoxonase gene polymorphisms associated with coronary heart disease: support for the oxidative damage hypothesis? *Am J Hum Genet* 1998; 62: 20-24

1127. HUMBERT R, ADLER DA, DISTECHE CM, HASSETT C, OMIECINSKI CJ, FURLONG CE : The molecular basis of the human paraoxonase activity polymorphism. *Nature Genet* 1993; 3: 73-76

1128. HEGELE RA, BRUNT HJ, CONNELLY PW : A polymorphism of the paraoxonase gene associated with variation in plasma lipoproteins in a genetic isolate. *Arterioscler Thromb* 1995; 15: 89-95

1129. SCHMIDT H, SCHMIDT R, NIEDERKORN K, GRADERT A, SCHUMACHER M, WATZINGER N, HARTUNG HP, KOSTNER GM : Paraoxonase PON1 polymorphism Leu-Met54 is associated with carotid atherosclerosis. *Stroke* 1998; 29: 2043-2048

1130. NEVIN DN, ZAMBON A, FURLONG CE, RICHTER RJ, HUMBERT R, HOKANSON JE, BRUNZELL JD : Paraoxonase genotypes, lipoprotein lipase activity and HDL. *Arterioscler Thromb Vasc Biol* 1996; 16: 1243-1249

1131. SHIH DM, GU L, HAMA S, XIA YR, NAVAB M, FOGELMAN AM, LUSIS AJ : Genetic-dietary regulation of serum paraoxonase expression and its role in atherogenesis in a mouse model. *J Clin Invest* 1996; 97: 1630-1639

1132. PHELAN SA, JOHNSON KA, BEIER DR, PAIGEN B : Characterization of the murine gene encoding Aop2 (antioxidant protein 2) and identification of two highly related genes. *Genomics* 1998; 54: 132-139

1133. KATZNELSON S : Immunosuppressive and antiproliferative effects of HMGCoA reductase inhibitors. *Transplant Proc* 1999; 31 supp 3B: 22S-24S

1134. MUNDY G, GARRETT R, HARRIS S, CHAN J, CHEN D, ROSSINI G, BOYCE B, ZHAO M, GUTTIERREZ G : Stimulation of bone formation in vitro and in rodents by statins. *Science* 1999; 286: 1946-1949

1135. CHAN KA, ANDRADE SE, BOLES M, BUIST DS, CHASE GA, DONAHUE JG, GURWITZ JH, LACROIX AZ, PLATT R : Inhibitors of hydroxymethylglutaryl-coenzyme A reductase and risk of fracture among older women. *Lancet* 2000; 355: 2185-2188

1136. ZAMBON A, TORRES A, BIJVOET S, GAGNE C, MOORJANI S, LUPIEN PJ, HAYDEN MR, BRUNZELL JD : Prevention of raised low-density lipoprotein cholesterol in a patient with familial hypercholesterolemia and lipoprotein lipase deficiency. *Lancet* 1993; 341: 1119-1121

1137. GIBSON TJ, SPRING J : Genetic redundancy in vertebrates: polyploidy and persistence of genes encoding multidomain proteins. *TIG* 1998; 14: 46-49

1138. SCHMITZ G, KAMINSKI WE, ORSÓ E : ABC transporters in cellular lipid trafficking. *Curr Opin Lipidol* 2000; 11: 493-501

1139. PATTHY L : Genome evolution and the evolution of exon-shuffling – a review. *Gene* 1999; 238: 103-114

1140. HARRIS HW, GRUNFELD C, FEINGOLD KR, READ TE, KANE JP, JONES AL, EICHBAUM EB, BLAND GF, RAPP JH : Chylomicrons alter the fate of endotoxin, decreasing tumor necrosis factor release and preventing death. *J Clin Invest* 1993; 91: 1028-1034

1141. VON SCHACKY C, FISHER S, WEBER PC : Long-term effects of marine ω-3 fatty acids upon plasma and cellular lipids, platelet function, and eicosanoid formation in humans. *J Clin Invest* 1985; 76: 1626-1631

1142. CAMPISI J : Aging, chromatin, and food restriction – connecting the dots. *Science* 2000; 289: 2062-2063

1143. SCHROEPFER GJ : Oxysterols: modulators of cholesterol metabolism and other processes. *Physiol Rev* 2000; 80: 361-554

1144. ARCHER SY, HODIN RA : Histone acetylation and cancer. *Curr Opin Genet Dev* 1999; 9: 171-174

1145. SALIMEN A, TAPIOLA T, KORHONEN P, SUURONEN T : Neuronal apoptosis induced by histone deacetylase inhibitors. *Brain Res Mol Brain Res* 1998; 61: 203-206

1146. HAMOSH A, SCOTT AF, AMBERGER J, VALLE D, McKUSICK VA : Online Mendelian Inheritance in Man (OMIM). *Hum Mutat* 2000; 15: 57-61 *http://www.ncbi.nlm.nih.gov/omim/*

1147. INTERNATIONAL HUMAN GENOME SEQUENCING CONSORTIUM : Initial sequencing and analysis of the human genome. *Nature* 2001; 409: 860-921

1148. VENTER JC, ADAMS MD, MYERS EW, LI PW, MURAL RJ, SUTTON GG, SMITH HO, YANDELL M, EVANS CA, HOLT RA, GOCAYNE JD, AMANATIDES P, BALLEW RM, HUSON DH, WORTMAN JR, ZHANG Q, KODIRA CD, ZHENG XH, CHEN L, SKUPSKI M, SUBRAMANIAN G, THOMAS PD, ZHANG J, GABOR MIKLOS GL, NELSON C, BRODER S, CLARK AG, NADEAU J, MCKUSICK VA, ZINDER N, LEVINE AJ, ROBERTS RJ, SIMON M, SLAYMAN C, HUNKAPILLER M, BOLANOS R, DELCHER A, DEW I, FASULO D, FLANIGAN M, FLOREA L, HALPERN A, HANNENHALLI S, KRAVITZ S, LEVY S, MOBARRY C, REINERT K, REMINGTON K, ABU-THREIDEH J, *et al.* : The sequence of the human genome. *Science* 2001; 291: 1304-1351

1149. OLIVIER M, AGGARWAL A, ALLEN J, ALMENDRAS AA, BAJOREK ES, BEASLEY EM, BRADY SD, BUSHARD JM, BUSTOS VI, CHU A, CHUNG TR, DE WITTE A, DENYS ME, DOMINGUEZ R FANG NY, FOSTER BD, FREUDENBERG RW, HADLEY D, HAMILTON LR, JEFFREY TJ, KELLY L, LAZZERONI L, LEVY MR, LEWIS SC, LIU X, LOPEZ FJ, LOUIE B, MARQUIS JP, MARTINEZ RA, MATSUURA MK, MISHERGHI NS, NORTON JA, OLSHEN A, PERKINS SM, PEROU AJ, PIERCY C, PIERCY M, QIN F, REIF T, SHEPPARD K, SHOKOOHI V, SMICK GA, SUN WL, STEWART EA, FERNANDO J, TEJEDA, TRAN NM, TREJO T, VO NT, YAN SC, ZIERTEN DL, ZHAO S, SACHIDANANDAM R, TRASK BJ, MYERS RM, COX DR : A high-resolution radiation hybrid map of the human genome draft sequence. *Science* 2001; 291: 1298-1302

1150. VOGEL F, MOTULSKY AG : *Human Genetics. Problems and Approaches*. Springer Berlin. 1980

1151. DIB C, FAURE S, FIZANNES C, SAMSON D, DROUOT N, VIGNAL A, MILLASSEAU P, MARC S, HAZAN J, SEBOUN E, LATHROP M, GYAPAY G, MORISETTE J, WEISSENBACH J : A comprehensive genetic map of the human genome based on 5,264 microsatellites. *Nature* 1996; 14: 152-154

1152. MORTON NE : Sequential tests for the detection of linkage. *Am J Hum Genet* 1955; 7: 277-318

1153. LATHROP GM, LALOUEL JM, JULIER C, OTT J : Strategies for multilocus linkage analysis in humans. *Proc Natl Acad Sci USA* 1984; 81: 3443-3446

1154. OTT J, SCHROTT HG, GOLDSTEIN JL, HAZZARD WR, ALLEN FH, FALK CT, MOTULSKY AG: Linkage studies in a large kindred with familial hypercholesterolemia. *Am J Hum Genet* 1974; 26: 598-603

1155. LEWIS M, KAITA H, GOGHLAN G, PHILIPPS S, BELCHER E, Mc ALPINE PJ, COOPLAND GR, WOODS RA : The chromosome 19 linkage group LDLR, C3, LW, APOC2, LU, SE in man. *Ann Hum Genet* 1988; 52: 137-144

1156. VARRET M, RABES JP, SAINT-JORE B, CENARRO A, MARIONI JC, CIVEIRA F, DEVILLERS M, KREMPF M, COULON M, THIART R, KOTZE MJ, SCHMIDT H, BUZZI JC, KOSTNER GM, BERTOLINI S, POCOVI M, ROSA A, FARNIER M, MARTINEZ M, JUNIEN C, BOILEAU C : A third major locus for autosomal dominant hypercholesterolemia maps to 1p34.1-p32. *Am J Hum Genet* 1999; 64: 1378-1387

1157. YUAN B, NEUMAN R DUAN SH, WEBER JL, KWOK PY, SACCONE NL, WU JS, LIU KY : Linkage of a gene for familial hypobetalipoproteinemia to chromosome 3p21.1-22. *Am J Hum Genet* 2000; 66: 1699-1704

1158. KORT EN, BALLINGER DG, DING W, HUNT SC, BOWEN BR, BKEVICH V, BULKA K, CAMPBELL B, CAPENER C, GUTIN A, HARSHMAN K, McDERMOTT M, THORNE T, WANG H, WARDELL B, WONG J, HOPKINS PN, SKOLNICK M, SAMUELS M : Evidence of linkage of familial hypoalphalipoproteinemia to a novel locus on chromosome 11q23. *Am J Hum Genet* 2000; 66: 1845-1856

1159. COTTON RGH, SCRIVER CR : Proof of disease causing mutation. *Hum Mut* 1998; 12: 1-3

1160. FRANCESCHINI G, CALABRESI L, CHIESA G, PAROLINI C, SIRTORI CR, CANAVESI M, BERNINI F : Increased cholesterol efflux potential of sera from ApoA-I Milano carriers and transgenic mice. *Arterioscler Thromb Vasc Biol* 1999; 19: 1257-1262

1161. GERDES LU, GERDES C, KERVINEN K, SAVOLAINEN M, KLAUSEN IC, HANSEN PS, KESANIEMI YA, FAERGEMAN O : The apolipoprotein epsilon 4 allele determines prognosis and the effect on prognosis of simvastatin in survivors of myocardial infarction : a substudy of the scandinavian simvastatin survival study. *Circulation* 2000; 101: 1366-1371

1162. ZERBA KE, SING CF : The role of genome type-environment interaction and time in understanding the impact of genetic polymorphisms on lipid metabolism. *Curr Opin Lipidol* 1993; 4: 152-162

1163. RISCH NJ : Searching for genetic determinants in the new millenium. *Nature* 2000; 405: 847-856

1164.COOPER DN, CLAYTON JF : DNA polymorphism and the study of disease associations. *Hum Genet* 1988; 78: 299-312

1165.EUROPEAN WORKING GROUP ON CYSTIC FIBROSIS GENETICS : Gradient of distribution in Europe of the major CF mutation and of its associated haplotype. *Hum Genet* 1990; 85: 436-445

1166.TROWSDALE J, POWIS SH : The MHC : relationship between linkage and function. *Curr Opin Genet Develop* 1992; 2: 492-497

1167.POCOVI M, CENARRO A, CIVEIRA F, TORRALBA MA, PEREZ-CALVO JI, MOZAS P, GIRALDO P, GIRALT M, MYERS RH, CUPPLES AL, ORDOVAS JM : β-glucocerebrosidase gene locus as a link for Gaucher's disease and familial hypo-α-lipoproteinemia. *Lancet* 1998; 351: 1919-1923

1168.BARBUJANI G, JACQUEZ GM, LIGI L : Diversity of some gene frequencies in European and Asian populations V. Steep multilocus clines. *Am J Hum Genet* 1990; 47: 867-875

1169.FUMERON F, GRANDCHAMP B, FRICKER J, KREMPF M, WOLF LM, KHAYAT MC, BOIFFARD O, APFELBAUM M : Presence of the French Canadian deletion in a French patient with familial hypercholesterolemia. *New Engl J Med* 1992; 326: 69

1170.THOMPSON MW, Mc INNES RR, WILLARD HF : *Thompson & Thompson* : *Genetics in medicine*, 5th edition. Philadelphia. Saunders 1991

1171.MA Y, LIU MS, CHITAYAT D, BRUIN T, BEISIEGEL U, BENLIAN P, FOUBERT L, DE GENNES JL, FUNKE H, FORSYTHE I, BLAICHMAN S, PAPANICOLAOU D, ERKELENS W, KASTELEIN J, BRUNZELL JD, HAYDEN MR: Recurrent missense mutations at the first and second base of codon ARG243 in human lipoprotein lipase causing chylomicronemia in patients of different ancestries. *Hum Mutat* 1994; 3: 52-58

1172.CASSANELLI S, BERTOLINI S, ROLLERI M, DE STEFANO F, CASANIRO L, ELICIO N, NASELLI A, CALANDRA S : A «de novo» point mutation of the low-density receptor gene in an Italian subject with primary hypercholesterolemia. *Clin Genet* 1998; 53: 391-395

1173.AQUADRO CF: Why is the genome variable? Insights from Drosophila. *Trends Genet* 1992; 8: 355-362

1174.ROSS RS, CHOY L, GRAVES RA, FOX N, SOLEVJEVA V, KLAUS S, RICQUIER D, SPIEGELMAN BM : Hibernoma formation in transgenic mice and isolation of a brown adipocyte cell line expressing the uncoupling protein gene. *Proc Natl Acad Sci USA* 1992; 89: 7561-7565

1175.LOCKHART DJ, WINZELER EA : Genomics, gene expression and DNA arrays. *Nature* 2000; 405: 827-836

1176.ROY CHOWDHURY J, GROSSMAN M, GUPTA S, ROY CHOWDHURY N, BAKER JR, WILSON JM : Long-term improvement of hypercholesterolemia after ex vivo gene therapy in LDLR-deficient rabbits. *Science* 1991; 254: 1802-1805

1177.GROSSMAN M, RAPER SE, WILSON JW : Transplantation of genetically modified autologous hepatocyte into nonhuman primates : feasibility and short-term toxicity. *Hum Gene Therapy* 1992; 3: 501-510

1178. GROSSMAN M, RAPER S, KOZARSKY K, STEIN E, ENGELHARDT JF, MULLER D, LUPIEN PJ, WILSON JE : Successful ex vivo gene therapy directed to liver in a patient with familial hypercholesterolemia. *Nature Genet* 1994; 6: 335-341

1179. PARISE F, SIMONE L, CROCE MA, GHISELLINI M, BATTINI R, BORGHI S, TIOZZI R, FERRARI S, CALANDRA S, FERRARI S : Construction and *in vivo* functional evaluation of a low-density lipoprotein receptor/transferrin fusion protein as a therapeutic tool for familial hypercholesterolemia. *Hum Gene Therapy* 1999; 10: 1219-1228

1180. MARSHALL E : Gene therapy on trial. *Science* 2000; 288: 951-957

1181. VEGA MA : Prospects for homologous recombination in human gene therapy. *Hum Genet* 1991; 87: 245-253

1182. KREN BT, METZ R, KUMAR R, STEER CJ : Gene repair using chimeric RNA/DNA oligonucleotide. *Seminars in liver disease* 1999; 19: 93-104

1183. LEE MH, LU K, HAZARD S, YU H, SHULENIN S, HIDAKA H, KOJIMA H, ALLIKMETS R, SAKUMA N, PEGORARO R, SRIVASTAVA AK, SALEN G, DEAN M, PATEL SB : Identification of a gene, ABCG5, important in the regulation of dietary cholesterol absorption. *Nat Genet* 2001; 27: 79-83

1184. GARCIA CK, WILUND K, ARCA M, ZULIANI G, FELLIN R, MAIOLI M, CALANDRA S, BERTOLINI S, COSSU F, GRISHIN N, BARNES R, COHEN JC, HOBBS HH : Autosomal recessive hypercholesterolemia caused by mutations in a putative LDL receptor adaptor protein. *Science* 2001; 292: 1394-1398

1185. AOUIZERAT BE, ALLAYEE H, CANTOR RM, DAVIS RC, LANNING CD, WEN P-Z, DALLINGA-THIE G, DE BRUIN TWA, ROTTER JI, LUSIS AJ : A genome scan for familial combined hyperlipidemia reveals evidence of linkage with a locus on chromosome 11. *Am J Hum Genet* 1999; 65: 397-412

1186. PAJUKANTA P, TERWILLIGER JD, PEROLA M, HIEKKALINNA T, NUOTIO I, ELLONEN P, PARKKONEN M, HARTIALA J, YLITALO K, PIHLAJAMÄKI J, PORKKA K, LAAKSO M, VIIKARI J, EHNHOLM C, TASKINEN M-R, PELTONEN L : Genomewide scan for familial combined hyperlipidemia genes in Finnish families, suggesting multiple susceptibility loci influencing triglyceride, cholesterol, and apolipoprotein B levels. *Am J Hum Genet* 1999; 64: 1453-1463

1187. AOUIZERAT BE, ALLAYEE H, BODNAR J, KRASS KL, PELTONEN L, DE BRUIN TWA, ROTTER JI, LUSIS AJ : Novel genes for familial combined hyperlipidemia. *Curr Opin Lipidol* 1999; 10: 113-122

1188. COPELAND NG, JENKINS NA, GILBERT DJ, EPPIG JT, MALTAIS LJ, MILLER JC, DIETRICH WF, WEAVER A, LINCOLN SE, STEEN RG, STEIN LD, NADEAU JH, LANDER ES : A genetic linkage map of the mouse : current applications and prospects. *Science* 1993; 262: 57-66

1189. WELSH CL, XIA Y-R, SCHECHTER I, FARESE R, MEHRABIAN M, MEHDIZADEH S, WARDEN CH, LUSIS AJ : Genetic regulation of cholesterol homeostasis: chromosomal organization of candidate genes. *J Lipid Res* 1996; 37: 1406-1421

1190. PETERFY M, PHAN J, XU P, REUE K : Lipodystrophy in the *fld* mouse results from mutation of a new gene encoding a nuclear protein, lipin. *Nat Genet* 2001; 27: 121-124

1191. REUE K, DOOLITTLE MH : Naturally occurring mutations in mice affecting lipid transport and metabolism. *J Lipid Res* 1996; 37: 1387-1405

1192. THE HUMAN GENOME PROJECT INFORMATION. *http://www.ornl.gov*

1193. GOFFEAU A, BARRELL BG, BUSSEY H, DAVIS RW, DUJON B, FELDMANN H, GALIBERT F, HOHEISEL JD, JACQ C, JOHNSTON M, LOUIS EJ, MEWES HW, MURAKAMI Y, PHILIPPSEN P, TETTELIN H, OLIVER SG : Life with 6000 genes. *Science* 1996; 274: 563-567

1194. THE C. ELEGANS SEQUENCING CONSORTIUM : Genome sequence of the nematode *C. elegans*: a platform for investigating biology. *Science* 1998; 282: 2012-2018

1195. LOOTS GG, LOCKSLEY RM, BLANKESPOOR CM, WANG ZE, MILLER W, RUBIN EM, FRAZER KA : Identification of a coordinate regulator of interleukins 4, 13, and 5 by cross-species sequence comparisons. *Science* 2000; 288; 136-140

1196. PICKAARD CS : The epigenetics of nucleolar dominance. *Trends Genet* 2000; 16: 495-500

INDEX